Introduction to Algebra

Introduction to Algebra
Second Edition

Peter J. Cameron
Queen Mary, University of London

OXFORD
UNIVERSITY PRESS

OXFORD
UNIVERSITY PRESS
Great Clarendon Street, Oxford OX2 6DP

Oxford University Press is a department of the University of Oxford.
It furthers the University's objective of excellence in research, scholarship,
and education by publishing worldwide in

Oxford New York

Auckland Cape Town Dar es Salaam Hong Kong Karachi
Kuala Lumpur Madrid Melbourne Mexico City Nairobi
New Delhi Shanghai Taipei Toronto

With offices in

Argentina Austria Brazil Chile Czech Republic France Greece
Guatemala Hungary Italy Japan Poland Portugal Singapore
South Korea Switzerland Thailand Turkey Ukraine Vietnam

Oxford is a registered trade mark of Oxford University Press
in the UK and in certain other countries

Published in the United States
by Oxford University Press Inc., New York

© Peter J. Cameron, 2008

The moral rights of the author have been asserted
Database right Oxford University Press (maker)

All rights reserved. No part of this publication may be reproduced,
stored in a retrieval system, or transmitted, in any form or by any means,
without the prior permission in writing of Oxford University Press,
or as expressly permitted by law, or under terms agreed with the appropriate
reprographics rights organization. Enquiries concerning reproduction
outside the scope of the above should be sent to the Rights Department,
Oxford University Press, at the address above

You must not circulate this book in any other binding or cover
and you must impose the same condition on any acquirer

Typeset by Newgen Imaging Systems (P) Ltd., Chennai, India
Printed in the U.S.A.

ISBN 978–0–19–856913–8

Preface

This new edition of my algebra textbook has a number of changes.

The most significant is that the book now tries to live up to its title better than it did in the previous edition: the introductory chapter has more than doubled in length, including basic material on proofs, numbers, algebraic manipulations, sets, functions, relations, matrices, and permutations. I hope that it is now accessible to a first-year mathematics undergraduate, and suitable for use in a first-year mathematics course. Indeed, much of this material comes from a course (also with the title 'Introduction to Algebra') which I gave at Queen Mary, University of London, in spring 2007.

I have also revised and corrected the rest of the book, while keeping the structure intact. In particular, the pace of the first chapter is quite gentle; in Chapters 2 and 3 it picks up a bit, and in the later chapters it is a bit faster again. Once you are used to the way I write mathematics, you should be able to take this in your stride. Since the book is intended to be used in a variety of courses, there is a certain amount of repetition. For example, concepts or results introduced in exercises may be dealt with later in the main text. New material on the Axiom of Choice, p-groups, and local rings has been added, and there are many new exercises.

I am grateful to many people who have helped me. First and foremost, Robin Chapman, for spotting many misprints and making many suggestions; and Csaba Szabó, who encouraged his students (named below) to proofread the book very thoroughly! Also, Gary McGuire spotted a gap in the proof of the Fundamental Theorem of Galois Theory, and R. A. Bailey suggested a different proof of Sylow's Theorem. The people who notified me of errors in the book, or who suggested improvements (as well as the above) are Laura Alexander, Richard Anderson, M. Q. Baig, Steve DiMauro, Karl Fedje, Emily Ford, Roderick Foreman, Will Funk, Rippon Gupta, Matt Harvey, Jessica Hubbs, Young-Han Kim, Bill Martin, William H. Millerd, Ioannis Pantelidakis, Brandon Peden, Nayim Rashid, Elizabeth Rothwell, Ben Rubin, and Amjad Tuffaha; my thanks to all of you, and to anyone else whose name I have inadvertently omitted!

P.J.C.
London
April 2007

Preface to the first edition

The axiomatic method is characteristic of modern mathematics. By making our assumptions explicit, we reduce the risk of making an error in our reasoning based on false analogy; and our results have a clearly defined area of applicability which is as wide as possible (any situation in which the axioms hold).

However, switching quickly from the concrete to the abstract makes a heavy demand on students. The axiomatic style of mathematics is usually met first in a course with a title such as 'Abstract Algebra', 'Algebraic Structures', or 'Groups, Rings and Fields'. Students who are used to factorising a particular integer or finding the stationary points of a particular curve find it hard to verify that a set whose elements are subsets of another set satisfies the axioms for a group, and even harder to get a feel for what such a group really looks like.

For this reason, among others, I have chosen to treat rings before groups, although they are logically more complicated. Everyone is familiar with the set of integers, and can see that it satisfies the axioms for a ring. In the early stages, when one depends on precedent, the integers form a fairly reliable guide. Also, the abstract factorisation theorems of ring theory lead to proofs of important and subtle properties of the integers, such as the Fundamental Theorem of Arithmetic. Finally, the path to non-trivial applications is shorter from ring theory than from group theory.

I have been teaching algebra for the whole of my professional career, and this book reflects that experience. Most immediately, it grew out of the Abstract Algebra course at Queen Mary and Westfield College. Chapters 2 and 3 are based fairly directly on the course content, and provide an introduction to rings (and fields) and to groups. The first chapter contains essential background material that every student of mathematics should know, and which can certainly stand repetition. (A great deal of algebra depends on the concept of an equivalence relation.)

Chapter 4, on vector spaces, does not try to be a complete account, since most students would have met vector spaces before they reach this point. The purpose is twofold: to give an axiomatic approach; and to provide material in a form which generalises to modules over Euclidean rings, from where two very important applications (finitely generated abelian groups and canonical forms of matrices) come.

Chapter 7 carries further the material of Chapters 2 and 3, and also introduces some other types of algebra, chosen for their unifying features: universal algebra, lattices, and categories. This follows a chapter in which the number systems are defined (so that our earlier trust that the integers form a ring can be firmly founded), the distinction between algebraic and transcendental numbers is established, and certain ruler-and-compass construction problems are shown to be impossible. The final chapter treats two important applications, drawing on much of what has gone before: coding theory and Galois Theory.

As mentioned earlier, Chapters 2 and 3 can form the basis of a first course on algebra, followed by a course based on Chapters 5 and 7. Alternatively, Chapter 3 and Sections 7.1–7.8 could form a group theory course, and Chapters 2 and 5 and Sections 7.9–7.14 a ring theory course. Sections 2.14–2.16, 6.6–6.8, 7.15–7.18, and 8.6–8.11 make up a Galois Theory course. Sections 6.1–6.5, and 6.9–6.10 could supplement a course on set theory, and Sections 2.14–2.16, 7.15–7.18, and 8.1–8.5 could be used in conjunction with some material on information theory for a coding theory course.

Some parts of the book (Sections 7.8, 7.13, and probably the last part of Chapter 7) are really too sketchy to be used for teaching a course; they are designed to tempt students into further exploration.

At the end, there is a list of books for further reading, and solutions to selected exercises from the first three chapters.

Asterisks denote harder exercises.

There is a World Wide Web site associated with this book. It contains solutions to the remaining exercises, further topics, problems, and links to other sites of interest to algebraists. The address is

$$\texttt{http://www.maths.qmul.ac.uk/~pjc/algebra/}$$

Thanks are due to many generations of students, whose questions and perplexities have helped me clarify my ideas and so resulted in a better book than I might otherwise have written.

P.J.C.
London
December 1997

Contents

1 Introduction

The purpose of this chapter is to introduce you to some of the notation and ideas that make up mathematics. Much of this may be familiar to you when you begin the study of abstract algebra. But, if it is not, I have tried to provide a friendly introduction. Your job is to practice unfamiliar skills until you are fluent. If you do not feel confident, please read this chapter carefully.

Much more than most scholarly disciplines, mathematics is structured; each subject assumes knowledge of its prerequisites and builds on them. But nobody studies mathematics starting with the logical foundations and working upwards. My view of the subject is more like a building which has basements and attics, but where you enter at the ground floor, with the knowledge you already have; then you can go upstairs to the applications or down to the foundations as you please.

This chapter, after a brief discussion of the structure and symbolism of mathematics, proceeds to give accounts of the topics which make up the common language of mathematicians: numbers, sets, functions, relations, formulae, equations, matrices, and logic. Much of the material comes back later in more serious and rigorous form. For example, in the first section, I will prove two famous theorems from Greek mathematics, about the infinitude of the primes and the irrationality of the square root of 2, even though numbers are not discussed until the second section.

What is mathematics?

> Mathematics is not best learned passively; you don't sop it up like
> a romance novel. You've got to go out to it, aggressive, and alert,
> like a chess master pursuing checkmate.
>
> Robert Kanigel (1991).

No one would doubt that a mathematics book is not like a novel. It is full of formulae using strange symbols and Greek letters, and contains words like 'theorem', 'proposition', 'lemma', 'corollary', 'proof', and 'conjecture'. Many of these words are themselves Greek in origin.

This is the legacy of Pythagoras, who was probably the first mathematician in anything like the modern sense (as opposed so somebody who used mathematics, such as a surveyor or an accountant). We know little about Pythagoras, and what we do know is unreliable, but it is clear that he cared very deeply about the subject:

the word 'theory' ... was originally an Orphic word, which
Cornford interprets as 'passionate sympathetic contemplation' ...
For Pythagoras, the 'passionate sympathetic contemplation' was
intellectual, and issued in mathematical knowledge ... To those
who have reluctantly learnt a little mathematics in school this
may seem strange; but to those who have experienced the intox-
icating delight of sudden understanding that mathematics gives,
from time to time, to those who love it, the Pythagorean view will
seem completely natural ...

Bertrand Russell (1961).

1.1   Notations.   The most important thing about mathematics is that the
assertions we make have to have proofs; in other words, we must be able to
produce a logical argument which cannot be attacked or refuted. We will see
many proofs; the next section contains two classics from the ancient Greeks.

The words 'Theorem', 'Proposition', 'Lemma', and 'Corollary' all have
the same meaning: a statement which has been proved, and has thereby
become part of the body of mathematics. There are shades of difference: a
theorem is an important statement; a proposition is one which is less important; a lemma has no importance of its own but is a stepping stone on the
way to a theorem; and a corollary is something which follows easily from a
theorem.

The word 'Proof' indicates that the argument establishing a theorem (or other
statement) will follow. The end of the argument is marked by the special symbol
□. If an exercise asks you to 'prove', 'show', or 'demonstrate' some statement,
you are being asked to construct a proof yourself.

A 'Conjecture' is a statement which is believed to be true but for which
we do not yet have a proof. Much of what mathematicians do is working to
establish a conjecture (or, since not all conjectures turn out to be true, to refute
one). Another important part of our work is to make conjectures based on our
experience and intuition, for others to prove or disprove. (The great twentieth-
century Hungarian mathematician Paul Erdős said, 'The aim of life is to prove
and to conjecture.')

Mathematicians have not always been consistent about applying these terms.
Sometimes it happens that a result which first appeared as a lemma came to be
regarded as more important than the theorem it was originally used to prove.
(See Gauss' Lemma in Chapter 2 for an example. One result in Chapter 6, Zorn's
Lemma, is really an axiom!) Also, one of the most famous conjectures (until
recently) was 'Fermat's Last Theorem', which asserted that there cannot exist
natural numbers x, y, z, n with $x, y, z > 0$ and $n > 2$ such that $x^n + y^n = z^n$.
Fermat asserted this theorem and claimed to have a proof, but no proof was
found among his papers and it is now believed that he was mistaken in thinking
he had one; but the name stuck. The conjecture was proved by Wiles in the
1990s, but we still call it 'Fermat's Last Theorem' rather than 'Wiles' Theorem'.

A 'Definition' is a precise way of saying what a word means in the mathematical context. Here is Humpty Dumpty's view (in the words of Lewis Carroll):

> When I use a word, it means exactly what I want it to mean, neither more nor less.

In mathematics, we use a lot of words with very precise meanings, often quite different from their usual meanings. When we introduce a word which is to have a special meaning, we have to say precisely what that meaning is to be. Usually, the word being defined is written in italics. For example, you may meet the definition:

> An $m \times n$ **matrix** is an array of numbers set out in m rows and n columns.

From that point, whenever you come upon the word 'matrix', it has this meaning, and has no relation to the meanings of the word in geology, in medicine, and in science fiction.

Most of the specialised notation in mathematics will be introduced as we go along. Because we use so many symbols in our arguments, one alphabet is not enough, and letters from the Greek alphabet are often called on. Table 1.1 shows the Greek alphabet.

Other alphabets including Hebrew and Chinese have been used on occasion too.

Another specialised alphabet is 'blackboard bold':

$$\mathbb{ABCDEFGHIJKLMNOPQRSTUVWXYZ}.$$

This alphabet originated because, in print, mathematicians can use bold type for special purposes, but bold type is difficult to reproduce on the blackboard with a piece of chalk. These letters are typically used for number systems:

- $\mathbb{N}$ for the **natural numbers** $1, 2, 3, \ldots$
- $\mathbb{Z}$ for the **integers** $\ldots, -2, -1, 0, 1, 2, \ldots$
- $\mathbb{Q}$ for the **rational numbers** or fractions such as $3/2$
- $\mathbb{R}$ for the **real numbers**, including $\sqrt{2}$ and π
- $\mathbb{C}$ for the **complex numbers**, including i (the square root of -1).

Most of these letters are self-explanatory, but why $\mathbb{Z}$ and $\mathbb{Q}$? The German word for numbers is *Zahlen*, which gives us the $\mathbb{Z}$. The rational numbers cannot be $\mathbb{R}$, so remember $\mathbb{Q}$ for quotients.

1.2 Proofs. The real answer to our earlier question 'What is mathematics?' is: Mathematics is about *proofs*. A proof is nothing but an argument to convince you of the truth of some assertion. Mathematical statements require proofs, which should be completely convincing, though you might have to work to understand the details. If, after a lot of effort, you are not convinced by an

Table 1.1 The Greek alphabet

Name	Capital	Lowercase
alpha	A	α
beta	B	β
gamma	Γ	γ
delta	Δ	δ
epsilon	E	ϵ
zeta	Z	ζ
eta	H	η
theta	Θ	θ
iota	I	ι
kappa	K	κ
lambda	Λ	λ
mu	M	μ
nu	N	ν
xi	Ξ	ξ
omicron	O	o
pi	Π	π
rho	P	ρ
sigma	Σ	σ
tau	T	τ
upsilon	Υ	υ
phi	Φ	ϕ
chi	X	χ
psi	Ψ	ψ
omega	Ω	ω

argument, then either the author has not made it clear, or the argument is not correct.

The proofs should ultimately be founded on logic; but we will not be too precise now about what constitutes a logically valid argument.

Here are two fine examples of proofs, from the time of ancient Greek mathematics. In each case, the statement is not at all obvious, but the proof persuades you that it must be true. In each case, the strategy is what we call 'proof by contradiction': that is, we show that assuming the opposite of what we are trying to prove leads to an absurdity or contradiction. Also, in each case, the proof has an ingenious twist.

The first theorem, probably due to Euclid, states that the series of prime numbers goes on for ever; *there is no largest prime number*. (A **prime number** is a natural number p greater than 1 which is not divisible by any natural numbers except for itself and 1. Notice that this definition says that the number 1 is not a prime number, even though it has no divisors except itself and 1. This makes sense; we will see the reason later.)

Theorem 1.1 *There are infinitely many prime numbers.*

Proof Our strategy is to show that the statement must be true because, if we assume that it is false, then we are led to an impossibility.

So we suppose that there are only finitely many primes. Let there be n primes, and let them be $p_1, p_2, \ldots, p_n$. Now consider the number $N = p_1 p_2 \cdots p_n + 1$. That is, N is obtained by multiplying together all the prime numbers and adding 1.

Now N must have a prime factor (this is a property of natural numbers which we will examine further later on). This prime factor must be one of $p_1, \ldots, p_n$ (since by assumption, these are all the prime numbers). But this is impossible, since N leaves a remainder of 1 when it is divided by any of $p_1, \ldots, p_n$.

Thus our assumption that there are only finitely many primes leads to a contradiction, so this assumption must be false; there must be infinitely many primes. □

The second theorem was proved by Pythagoras (or possibly one of his students). This theorem is surrounded by legend: supposedly Hipparchos, a disciple of Pythagoras, was killed (in a shipwreck) by the gods for revealing the disturbing truth that there are 'irrational' numbers.

Theorem 1.2 $\sqrt{2}$ *is irrational; that is, there is no number $x = p/q$ (where p and q are whole numbers) such that $x^2 = 2$.*

Proof Again the proof is by contradiction. Thus, we assume that there is a rational number p/q such that $(p/q)^2 = 2$, where p and q are integers. We can suppose that the fraction p/q is in its lowest terms; that is, p and q have no common factor.

Now $p^2 = 2q^2$. Thus, the number p^2 is even, from which it follows that p must be even. (The square of any odd number is odd: for any odd number has the form $2m + 1$, and its square is $(2m + 1)^2 = 4m(m + 1) + 1$, which is odd.) Let us write $p = 2r$. Now our equation becomes $4r^2 = 2q^2$, or $2r^2 = q^2$. Thus, just as before, q^2 is even, and so q is even.

But if p and q are both even, then they have the common factor 2, which contradicts our assumption that the fraction p/q is in its lowest terms. □

Now we look at a few proof techniques, and introduce some new terms.

Proof by contradiction We have just seen two examples of this. In order to prove a statement $\mathcal{P}$, we assume that $\mathcal{P}$ is false, and derive a contradiction from this assumption.

Proof by contrapositive The **contrapositive** of the statement 'if $\mathcal{P}$, then $\mathcal{Q}$' is the statement 'if not $\mathcal{Q}$, then not $\mathcal{P}$'. This is logically equivalent to the original statement; so we can prove this instead if it is more convenient.

Converse Do not confuse the contrapositive of a statement with its **converse**. The converse of 'if $\mathcal{P}$, then $\mathcal{Q}$' is 'if $\mathcal{Q}$, then $\mathcal{P}$'. This is *not* logically equivalent to the original statement. For example, it can be shown that the statement 'if $2^n - 1$ is prime, then n is prime' is true; but its converse, 'if n is prime, then $2^n - 1$ is prime' is false: the number $n = 11$ is prime, but $2^{11} - 1 = 2047 = 23 \times 89$.

This example by Lewis Carroll might help you remember the difference between a statement and its converse.

> 'Come, we shall have some fun now!' thought Alice. 'I'm glad they've begun asking riddles.–I believe I can guess that,' she added aloud.
>
> 'Do you mean that you think you can find out the answer to it?' said the March Hare.
>
> 'Exactly so,' said Alice.
>
> 'Then you should say what you mean,' the March Hare went on.
>
> 'I do,' Alice hastily replied; 'at least–at least I mean what I say– that's the same thing, you know.'
>
> 'Not the same thing a bit!' said the Hatter. 'You might just as well say that "I see what I eat" is the same thing as "I eat what I see"!' 'You might just as well say,' added the March Hare, 'that "I like what I get" is the same thing as "I get what I like"!'
>
> 'You might just as well say,' added the Dormouse, who seemed to be talking in his sleep, 'that "I breathe when I sleep" is the same thing as "I sleep when I breathe"!'
>
> 'It is the same thing with you,' said the Hatter, and here the conversation dropped, and the party sat silent for a minute, while Alice thought over all she could remember about ravens and writing-desks, which wasn't much.

Counterexample   Given a general statement $\mathcal{P}$, to show that $\mathcal{P}$ is true it is necessary to give a general proof; but to show that $\mathcal{P}$ is false, we have to give one specific instance in which it fails. Such an instance is called a **counterexample**. In the preceding paragraph, the number $n = 11$ is a counterexample to the general statement 'if n is prime, then $2^n - 1$ is prime'.

Sufficient condition, 'if'   We say that $\mathcal{P}$ is a **sufficient condition** for $\mathcal{Q}$ if the truth of $\mathcal{P}$ implies the truth of $\mathcal{Q}$; that is, $\mathcal{P}$ implies $\mathcal{Q}$. Another way of saying the same thing is 'if $\mathcal{P}$, then $\mathcal{Q}$', or '$\mathcal{Q}$ if $\mathcal{P}$'. In symbols, we write $\mathcal{P} \Rightarrow \mathcal{Q}$.

Necessary condition, 'only if'   We say that $\mathcal{P}$ is a **necessary condition** for $\mathcal{Q}$ if the truth of $\mathcal{P}$ is implied by the truth of $\mathcal{Q}$, that is, $\mathcal{Q}$ implies $\mathcal{P}$. (This is the converse of the statement that $\mathcal{P}$ implies $\mathcal{Q}$.) We also say '$\mathcal{Q}$ only if $\mathcal{P}$'.

Necessary and sufficient condition, 'if and only if'   We say that $\mathcal{P}$ is a **necessary and sufficient condition** for $\mathcal{Q}$ if both of the above hold, that is, each of $\mathcal{P}$ and $\mathcal{Q}$ implies the other. We also say '$\mathcal{P}$ if and only if $\mathcal{Q}$'. Note that there are two things to prove: that $\mathcal{P}$ implies $\mathcal{Q}$, and that $\mathcal{Q}$ implies $\mathcal{P}$. In symbols, we write $\mathcal{P} \Leftrightarrow \mathcal{Q}$.

Proof by induction This is a very important technique for proving things about natural numbers. We discuss it later in this chapter.

1.3 Axioms. In the proofs in the last section, we used various properties of numbers: every integer greater than 1 has a prime factor; any number is either odd or even; and any fraction can be put into its lowest terms by cancelling common factors. Later on in the book we will examine these assumptions.

The process of examining our hidden assumptions is very important in mathematics. Each assumption should be proved, but the proof will probably involve more basic assumptions. There is a sense in which everything can be built from nothing using only the processes of logic. Usually this is much too long-winded; so instead we start by making our basic assumptions explicit.

It used to be thought that the basic assumptions of mathematics were true statements about the real world. Euclid's geometry was the model for many centuries. Euclid begins with **axioms**, which he regarded as 'self-evident truths', and deduced a huge body of theorems from them. But one of his axioms, the 'axiom of parallels', is far from self-evident. Mathematicians tried hard to prove it, but eventually were forced to admit that it was possible to construct a kind of geometry in which this axiom is false. (This is now referred to as **non-Euclidean geometry**.)

Now we regard the axioms as starting points which we choose, depending on the branch of mathematics we are studying. The theorems we prove will be true in any system (including any real-world system) which happens to satisfy the axioms.

One of the advantages of this approach is that, instead of proving theorems about, say, the integers, we can prove theorems about 'principal ideal domains'; as long as the integers satisfy the axioms for principal ideal domains, our theorems will be true in the integers. This is how we shall justify the assumptions of the last section about primes and common factors.

It is very important, however, not to bring in any hidden assumptions. For example, if we are doing geometry, the axioms will probably refer to points and lines; we must only use properties of points and lines specified in the axioms, rather than our commonsense view of how points and lines behave.

The German mathematician David Hilbert put it like this:

> One must be able at any time to replace 'points, lines, and planes'
> with 'tables, chairs, and beer mugs'.

Here is a small example. Suppose that we are doing geometry with just the following three of Euclid's axioms:

(1) Any two points lie on a unique line.
(2) If the point P does not lie on the line L, then there is exactly one line L' passing through P and parallel to L.
(3) There exist three non-collinear points.

We understand that 'collinear' means 'lying on a common line', and that two lines are 'parallel' if no point lies on both. Notice that if two lines are not parallel then they have exactly one common point (for more than one common point would violate Axiom (1)).

According to Hilbert's dictum, it would be equally valid to begin

(1) Any two tables lie on a unique chair.
(2) . . .

From these axioms, we can prove the following theorem:

Theorem 1.3   *Two lines parallel to the same line are parallel to one another.*

Proof   Let L' and L'' be two lines both parallel to L. Arguing by contradiction, suppose that L' and L'' are not parallel. Then they have a point P in common. But now there are two lines L' and L'' containing P and parallel to L, contradicting Axiom (3). $\square$

This is 'obviously' true in the ordinary Euclidean plane, but we have proved it in any geometry satisfying the axioms. Here is a less obvious example:

Points: $A, B, C, D, E, F, G, H, I$
Lines: $ABC, DEF, GHI, ADG, BEH, CFI, AEI, BFG, CDH, AFH,$
    $BDI, CEG.$

It is some labour to verify the axioms, but once this is done then the conclusion of the theorem must hold. Indeed, the lines DEF and GHI are both parallel to ABC, and they are parallel to one another. Here we seem to be a long way from traditional geometry, and it does not seem so stupid to say that $A, B, C, \ldots$ are tables and $ABC, DEF, \ldots$ are chairs, and that any two tables lie on a unique chair!

An even simpler example is the following:

Points: A, B, C, D
Lines: $AB, CD, AC, BD, AD, BC.$

In this case, there is only one line parallel to a given one, so the theorem holds 'vacuously': we cannot choose two lines L' and L'' parallel to L. This is a bit puzzling at first: what is going on here?

A statement of the form 'If $\mathcal{P}$, then $\mathcal{Q}$' is true, according to the rules of logic, if $\mathcal{P}$ is false. We discuss this further on page 60. If $\mathcal{P}$ can never be true, we sometimes say that the statement is 'vacuously' true.

Non-Euclidean geometry was discovered in the nineteenth century. By the early twentieth century, the 'axiomatic method' had become the paradigm for mathematics.

Exercise 1.1   Prove from Axioms (1)–(3) the following assertions:

(a) Any line passes through at least two points.
(b) Any two lines pass through the same number of points.

Exercise 1.2 Give an example of a system of points and lines satisfying Axioms (1) and (3) but not (2) (a 'non-Euclidean geometry').

Exercise 1.3 Let n be a natural number. Show that n^2 is even if and only if n is even. (We say that n is **even** if $n = 2m$ for some natural number m, and is **odd** if $n = 2m + 1$ for some natural number n. The exercise asks you to show two things: if n is even then n^2 is even; and if n^2 is even, then n is even. In this question you are permitted to use the fact that every natural number is either even or odd: the proof of this obvious-looking assertion is the subject of Exercise 1.12 later on.)

Exercise 1.4 Let the prime numbers, in order of magnitude, be $p_1, p_2, \dots$. Prove that $p_{n+1} \leq p_1 p_2 \cdots p_n + 1$.

Exercise 1.5 (a) Prove that, for any prime number p, $\sqrt{p}$ is irrational.
  (b) Prove that the cube root of 2 is irrational.

Exercise 1.6 Fill in the details in the following argument.

Proposition 1.4 *If n is a positive integer which is not a perfect square, then $\sqrt{n}$ is irrational.*

Proof Suppose that $\sqrt{n} = a/b$, where a/b is a fraction in its lowest terms. Then $a/b = nb/a$, so the fractional parts of these two numbers are equal, say $d/b = c/a$, where $0 < c < a$ and $0 < d < b$. Then $a/b = c/d$, contradicting the assumption that a/b is in its lowest terms. $\qquad\qquad\square$

(This argument is taken from *The Book of Numbers*, by J. H. Conway and R. K. Guy.)

Exercise 1.7 Can you prove that, if $2^n - 1$ is prime, then n is prime? (We will see the proof later in this chapter.)

Exercise 1.8 (a) Write down the converse of the statement

  If n is an even integer greater than 2, then n is the sum of two prime numbers.
  (b) Is the converse true or false? Why?

Remark The statement given in (a) is a famous conjecture due to Goldbach. It is believed to be true, but this is not yet known.

Exercise 1.9 Is the following argument valid? If not, why not?
We are going to prove that a triangle whose sides have lengths 3, 4, and 5 is right-angled.
  By Pythagoras' Theorem, if a triangle with sides a, b, c is right-angled, with hypotenuse c, then $a^2 + b^2 = c^2$.
  Now $3^2 + 4^2 = 9 + 16 = 25 = 5^2$.
  So the triangle is right-angled.

Numbers

Algebraic formulae often have symbols in them: x, a, and so on. In elementary algebra we think of these as numbers. But the domain we consider has an effect on whether the equations have solutions or not.

1.4   The number systems.   We consider briefly the different kinds of number systems used in elementary algebra. You should be familiar with most of these. In Chapter 6, we will go into more detail on exactly how the different kinds of number are constructed.

The natural numbers   The natural numbers are the ones we use to count: 1, 2, 3, and so on. They are sometimes called **counting numbers**. Actually, there is no agreement among mathematicians about whether 0 should be included as a natural number or not. Historically, the positive numbers arose (for use in counting) before the dawn of history, whereas zero is a much more recent and problematic invention. It is also more difficult for children to grasp. Brian Butterworth, an expert on the development of number sense in childhood, says, in his book *The Mathematical Brain*:

> Although the idea that we have no bananas is unlikely to be a
> new one, or one that is hard to grasp, the idea that no bananas,
> no sheep, no children, no prospects are really all the same, in that
> they have the same numerosity, is a very abstract one.

Logically, however, it makes sense to count zero as the smallest natural number, as we will see.

Fortunately, it does not very much matter what view we take about this.

The set of natural numbers is denoted by $\mathbb{N}$.

The important property of natural numbers to an algebraist is that they can be added and multiplied. If one heap contains m beans and another has n beans, then together the two heaps contain $m + n$ beans. Moreover, if we arrange some beans in a rectangular array with m rows and n columns, then mn beans are required.

These operations satisfy some simple laws, sometimes called the **laws of arithmetic**:

- $m + n = n + m$ and $mn = nm$ (the **commutative laws**);
- $m + (n + p) = (m + n) + p$ and $m(np) = (mn)p$ (the **associative laws**);
- $(m + n)p = mp + np$ (the **distributive law**).

In addition, adding zero, or multiplying by one, leaves any natural number unchanged.

The bean-counting interpretation allows us to picture these laws; some people find that the pictures provide convincing explanations. For example, Figure 1.1 shows the distributive law.

The reverse operations are not always possible. Subtraction, defined by requiring that $m - n$ is a number x such that $n + x = m$, is only possible if

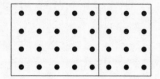

Fig. 1.1 $(5+3) \cdot 4 = 5 \cdot 4 + 3 \cdot 4$

m is at least as large as n (in symbols, $m \geq n$). Division, defined by requiring that m/n is a number y such that $ny = m$, is only possible if m is a multiple of n (in symbols, $n \mid m$). *Warning*: Be sure to distinguish betweem m/n (a number), and $n \mid m$ (a statement, which is either true or false). If n does not divide m, we write $n \nmid m$.

We already saw Euclid's proof that there are infinitely many prime numbers. Of course there are infinitely many composite numbers too: for example, every even number greater than 2 is composite. (A number $n > 1$ is **composite** if it is not prime.)

The natural numbers have a very important property, sometimes called the **induction property**.

Theorem 1.5 *Let S be any set of natural numbers. Suppose that*

(a) 0 belongs to S;
(b) for any natural number n, if n belongs to S, then $n + 1$ belongs to S.

Then $S = \mathbb{N}$, that is, S is the set of all natural numbers.

This theorem is true because the natural numbers are the 'counting numbers'; that is, given any natural number n, it is possible (at least in principle) to start at zero and count up to n: 'zero, one, two, three, $\ldots$, n'. Now the first number in the chain is in S; and as soon as we know that a number is in S then the next number is in S too. After n steps we find that n is in S.

Sometimes this is called the 'domino property'. Imagine we have an infinite number of dominoes standing in a line, labelled $0, 1, 2, \ldots$. The dominoes are arranged in such a way that, if number n falls, it will knock over number $n + 1$. Then, if we knock over domino number 0, we can be sure that all the dominoes will fall. This is exactly what the induction property says, with S as the set of labels of dominoes that fall over. See Figure 1.2.

Even if m is not a multiple of n, all is not lost. At school we learn the **division algorithm**:

Theorem 1.6 (Division algorithm for natural numbers) *Let m and n be any natural numbers with $n > 0$. Then there exist natural numbers q and r such that*

(a) $m = nq + r$;
(b) $r < n$.

Fig. 1.2 Which dominoes will fall?

Moreover, q and r are unique; that is, if also $m = nq' + r'$, where $r' < n$, then $q = q'$ and $r = r'$.

The numbers q and r are called the **quotient** and **remainder** when m is divided by n. (The numbers m and n are sometimes called the **dividend** and the **divisor**.)

Proof First we show the uniqueness. Suppose that $m = nq + r = nq' + r'$ with $r < n$ and $r' < n$. If $r = r'$, then $nq = nq'$, so $q = q'$. Suppose that $r \neq r'$. Then one of them is larger; say $r > r'$. Then

$$r - r' = n(q' - q),$$

so the same natural number is both less than n and a multiple of n, which is not possible.

Now we show the existence. Consider the multiples of n: n, $2n$, $3n$,.... Eventually we reach one which is greater than m (for certainly $(m + 1)n > m$). Let q be the last integer x for which $xn \leq m$; that is, $nq \leq m$ but $n(q + 1) > m$. (It may be that $q = 0$.) Put $r = m - nq$. Then $r \geq 0$ but $r < n$; and $m = nq + r$. $\square$

The integers   As we have seen, subtraction is not always possible for natural numbers. To get round this, we enlarge the number system to include negative numbers as well as positive numbers and zero, giving the set

$$\mathbb{Z} = \{\ldots, -2, -1, 0, 1, 2, 3, \ldots\}$$

of **integers**. Thus, we can add, subtract, and multiply integers. The laws we saw for natural numbers extend to the integers.

We enlarge the number system because we are trying to solve equations which cannot be solved in the original system. At every stage in the process, people first thought that the new numbers were just aids to calculating, and not 'proper' numbers. The names given to them reflect this: negative numbers, improper fractions, irrational numbers, imaginary numbers! Only later were they fully accepted. You may like to read the book *Imagining Numbers* by Barry Mazur, about the long process of accepting imaginary numbers.

The natural numbers $1, 2, \ldots$ are positive, while $-1, -2, \ldots$ are negative. Integers satisfy the law of signs: the product of a positive and a negative number is negative, while the product of two negative numbers is positive.

The rational numbers Similarly, division is not always possible for integers. To get round this, we enlarge the number system to the set $\mathbb{Q}$ of **rational numbers**, of the form m/n where $n \neq 0$. By cancellation, we may assume that $n > 0$ and that the fraction is in its 'lowest terms', that is, m and n have no common factor. For example, $20/(-12)$ is the same as $-5/3$.

We can write rules for adding and multiplying rational numbers:

$$\frac{a}{b} + \frac{c}{d} = \frac{ad + bc}{bd}, \quad \frac{a}{b} - \frac{c}{d} = \frac{ad - bc}{bd},$$

$$\frac{a}{b} \times \frac{c}{d} = \frac{ac}{bd}, \quad \frac{a}{b} \Big/ \frac{c}{d} = \frac{ad}{bc} \text{ if } c \neq 0.$$

The last rule says: to divide by a fraction, turn it upside down and multiply.

Thus, we can add, subtract, multiply, and divide rationals (except for division by zero). The usual laws extend to the rational numbers.

The real numbers There are still many equations we cannot solve with rational numbers. One such equation is $x^2 = 2$. (We saw Pythagoras' proof of this in Theorem 1.2.) Other equations involve functions from trigonometry (such as $\sin x = 1$, which has the irrational solution $x = \pi/2$) and calculus (such as $\log x = 1$, which has the irrational solution $x = e$).

So, we take a larger number system in which these equations can be solved, the **real numbers**. A real number is a number that can be represented as an infinite decimal. This includes all the rational numbers and many more, including the solutions of the three equations above; for example,

$$\tfrac{2}{5} = 0.4$$

$$\tfrac{1}{7} = 0.142857142857\ldots,$$

$$\sqrt{2} = 1.41421356237\ldots,$$

$$\tfrac{\pi}{2} = 1.57079632679\ldots,$$

$$e = 2.71828182846\ldots$$

In the last three cases, we cannot write out the number exactly as a decimal, but the approximation gets better as the number of digits increases.

The arithmetic operations (excluding division by zero) extend from $\mathbb{Q}$ to $\mathbb{R}$, and the laws of arithmetic continue to hold.

The completeness of $\mathbb{R}$ (the fact that there are no gaps) is shown by various results from analysis such as the **Intermediate Value Theorem**: a continuous function cannot go from negative to positive values without passing through zero.

The complex numbers Although there are no gaps in the real numbers, there are still some equations which cannot be solved. For example, the square of any real number is positive, so there is no real number x satisfying the equation

$$x^2 = -1.$$

We enlarge the real numbers to the set $\mathbb{C}$ of **complex numbers** by adjoining a special number i satisfying this equation. Thus, complex numbers are expressions of the form $x + y$i, where x and y are real numbers. The rules for addition and multiplication are exactly what you would expect, except that i^2 is replaced by -1 whenever it appears. Thus,

$$(x_1 + y_1\text{i}) + (x_2 + y_2\text{i}) = (x_1 + x_2) + (y_1 + y_2)\text{i},$$
$$(x_1 + y_1\text{i})(x_2 + y_2\text{i}) = (x_1x_2 - y_1y_2) + (x_1y_2 + x_2y_1)\text{i}.$$

The number i is sometimes called 'imaginary', since at first it seemed to mathematicians to be less 'real' than the real numbers. The term 'complex', on the other hand, is not meant to suggest that the complex numbers are more difficult to understand than the real numbers, but only that each complex number $x + y$i is made up of a kind of 'compound' of two real numbers x and y; we call x and y the **real** and **imaginary** parts of $x + y$i. The complex number $x - y$i is called the **complex conjugate** of z, and is written $\bar{z}$.

All the arithmetic operations (except, as usual, division by zero) are possible, and the laws of arithmetic hold. Here, unlike for the other forms of numbers, we do not have to take on trust that the laws hold; we can prove them for complex numbers (assuming their truth for real numbers). Here, for example, is the distributive law. Let $z_1 = x_1 + y_1\text{i}$, $z_2 = x_2 + y_2\text{i}$, and $z_3 = x_3 + y_3\text{i}$. Now

$$z_1(z_2 + z_3) = (x_1 + y_1\text{i})((x_2 + x_3) + (y_2 + y_3)\text{i})$$
$$= (x_1(x_2 + x_3) - y_1(y_2 + y_3)) + (x_1(y_2 + y_3) + y_1(x_2 + x_3))\text{i},$$

and

$$z_1z_2 + z_1z_3 = ((x_1x_2 - y_1y_2) + (x_1y_2 + x_2y_1)\text{i})$$
$$+ ((x_1x_3 - y_1y_3) + (x_1y_3 + x_3y_1)\text{i})$$
$$= (x_1x_2 - y_1y_2 + x_1x_3 - y_1y_3) + (x_1y_2 + x_2y_1 + x_1y_3 + x_3y_1)\text{i},$$

and a little bit of rearranging shows that the two expressions are the same.

Example

$$\frac{2 - 3\text{i}}{4 + \text{i}} = \frac{(2 - 3\text{i})(4 - \text{i})}{4^2 + 1^2}$$
$$= \frac{5 - 14\text{i}}{17},$$

which can be verified by multiplying the result by $4 + $i.

Moreover, quadratic, cubic, and higher-degree equations can always be solved in the complex numbers. (This is the **Fundamental Theorem of Algebra**, proved by Gauss.)

No further enlargements of the number system are possible without sacrificing some properties.

The rules for addition and subtraction can be put like this

> To add or subtract complex numbers, we add or subtract their
> real parts and their imaginary parts.

The rule for multiplication looks more complicated as we have written it out. There is another representation of complex numbers which makes it look simpler. Let $z = x + y\mathrm{i}$, and suppose that $z \neq 0$. We define the **modulus** and **argument** of z by

$$|z| = \sqrt{x^2 + y^2},$$
$$\arg(z) = \theta \text{ where } \cos\theta = x/|z| \text{ and } \sin\theta = y/|z|.$$

In other words, if $|z| = r$ and $\arg(z) = \theta$, then

$$z = r(\cos\theta + \mathrm{i}\sin\theta).$$

For example, let $z = 1 + \mathrm{i}$. Then the modulus of z is

$$|z| = \sqrt{1^2 + 1^2} = \sqrt{2},$$

and the argument θ satisfies $\cos\theta = 1/\sqrt{2}$ and $\sin\theta = 1/\sqrt{2}$, so that $\theta = \pi/4$.

Now the rules for multiplication and division are

> To multiply two complex numbers, multiply their moduli and add
> their arguments. To divide two complex numbers, divide their
> moduli and subtract their arguments.

The complex plane, or Argand diagram The complex numbers can be represented geometrically, by points in the Euclidean plane (which is usually referred to as the **Argand diagram** or the **complex plane** for this purpose). The complex number $z = x + y\mathrm{i}$ is represented as the point with coordinates (x, y). Then $|z|$ is the length of the line from the origin to the point z, and $\arg(z)$ is the angle between this line and the x-axis. See Figure 1.3.

In terms of the complex plane, we can give a geometric description of addition and multiplication of complex numbers. The addition rule is the **parallelogram rule** (see Figure 1.4).

Multiplication is a little bit more complicated. Let z be a complex number with modulus r and argument θ, so that $z = r(\cos\theta + \mathrm{i}\sin\theta)$. Then the way to multiply an arbitrary complex number by z is a combination of a stretch and a rotation: first we expand the plane so that the distance of each point from the origin is multiplied by r; then we rotate the plane through an angle θ. See Figure 1.5, where we are multiplying by $1 + \mathrm{i} = \sqrt{2}(\cos(\pi/4) + \mathrm{i}\sin(\pi/4))$; the dots represent the stretching out by a factor of $\sqrt{2}$, and the circular arc represents the rotation by $\pi/4$.

Now let us check the correctness of our rule for multiplying complex numbers. Remember that the rule is: to multiply two complex numbers, we multiply the

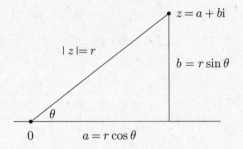

Fig. 1.3 The Argand diagram

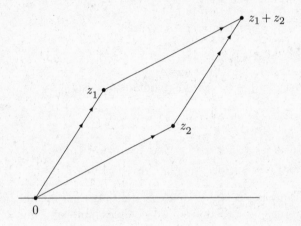

Fig. 1.4 Addition of complex numbers

moduli and add the arguments. To see that this is correct, suppose that z_1 and z_2 are two complex numbers; let their moduli be r_1 and r_2, and their arguments $\theta_1 + \theta_2$. Then

$$z_1 = r_1(\cos\theta_1 + i\sin\theta_1),$$
$$z_2 = r_2(\cos\theta_2 + i\sin\theta_2).$$

Then

$$
\begin{aligned}
z_1 z_2 &= r_1 r_2(\cos\theta_1 + i\sin\theta_1)(\cos\theta_2 + i\sin\theta_2) \\
&= r_1 r_2((\cos\theta_1\cos\theta_2 - \sin\theta_1\sin\theta_2) + (\cos\theta_1\sin\theta_2 + \sin\theta_1\cos\theta_2)i) \\
&= r_1 r_2(\cos(\theta_1 + \theta_2) + i\sin(\theta_1 + \theta_2)),
\end{aligned}
$$

which is what we wanted to show.

From this we can prove **De Moivre's Theorem**.

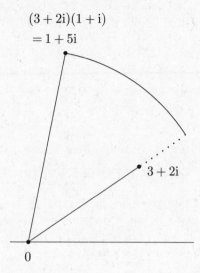

Fig. 1.5 Multiplication of complex numbers

Theorem 1.7 *For any natural number n, we have*

$$(\cos\theta + \mathrm{i}\sin\theta)^n = \cos n\theta + \mathrm{i}\sin n\theta.$$

Proof The proof is by induction. Starting the induction is easy since $(\cos\theta + \mathrm{i}\sin\theta)^0 = 1$ and $\cos 0 + \mathrm{i}\sin 0 = 1$.

For the inductive step, suppose that the result is true for n, that is,

$$(\cos\theta + \mathrm{i}\sin\theta)^n = \cos n\theta + \mathrm{i}\sin n\theta.$$

Then

$$
\begin{aligned}
(\cos\theta + \mathrm{i}\sin\theta)^{n+1} &= (\cos\theta + \mathrm{i}\sin\theta)^n \cdot (\cos\theta + \mathrm{i}\sin\theta) \\
&= (\cos n\theta + \mathrm{i}\sin n\theta)(\cos\theta + \mathrm{i}\sin\theta) \\
&= \cos(n+1)\theta + \mathrm{i}\sin(n+1)\theta,
\end{aligned}
$$

which is the result for $n+1$. So the proof by induction is complete.

Note that, in the second line of the chain of equations, we have used the inductive hypothesis, and in the third line, we have used the rule for multiplying complex numbers. □

The argument is clear if we express it geometrically. To multiply by the complex number $(\cos\theta + \mathrm{i}\sin\theta)^n$, we rotate n times through an angle θ, which is the same as rotating through an angle $n\theta$.

De Moivre's Theorem is useful in deriving trigonometrical formulae. For example,

$$\cos 3\theta + \mathrm{i}\sin 3\theta = (\cos\theta + \mathrm{i}\sin\theta)^3$$
$$= (\cos^3\theta - 3\cos\theta\sin^2\theta) + (3\cos^2\theta\sin\theta - \sin^3\theta)\mathrm{i},$$

so

$$\cos 3\theta = \cos^3\theta - 3\cos\theta\sin^2\theta,$$
$$\sin 3\theta = 3\cos^2\theta\sin\theta - \sin^3\theta.$$

These can be converted into the more familiar forms $\cos 3\theta = 4\cos^3\theta - 3\cos\theta$ and $\sin 3\theta = 3\sin\theta - 4\sin^3\theta$ by using the equation $\cos^2\theta + \sin^2\theta = 1$.

In Analysis, the definition of the exponential function is extended from the real numbers to the complex numbers so that

$$\mathrm{e}^{\mathrm{i}\theta} = \cos\theta + \mathrm{i}\sin\theta.$$

If we do this, then the modulus-argument form of a complex number is $z = r\mathrm{e}^{\mathrm{i}\theta}$, and we have

$$\mathrm{e}^{\mathrm{i}\theta_1} \cdot \mathrm{e}^{\mathrm{i}\theta_2} = \mathrm{e}^{\mathrm{i}(\theta_1 + \theta_2)}.$$

De Moivre's Theorem becomes

$$(\mathrm{e}^{\mathrm{i}\theta})^n = \mathrm{e}^{\mathrm{i}n\theta}.$$

1.5   Induction.   The induction property of the natural numbers—which says that if you start at the beginning and step through them one at a time, then you eventually reach any number—is an important proof technique.

We summarise the Principle of Induction formally in a theorem as follows. (In the domino example of Figure 1.2, $\mathcal{P}(n)$ is the proposition 'Domino number n will fall'.)

Theorem 1.8 (Principle of Induction)   *Let $\mathcal{P}(n)$ be a statement about the natural number n. Suppose that*

(a) $\mathcal{P}(0)$ is true;
(b) For any natural number n, if $\mathcal{P}(n)$ is true, then $\mathcal{P}(n+1)$ is true.

Then $\mathcal{P}(n)$ is true for every natural number n.

Proof   Let S be the set of all those natural numbers n for which $\mathcal{P}(n)$ is true. Then the hypotheses of the theorem tell us that $0 \in S$, and that, if $n \in S$, then $n + 1 \in S$. So the induction property shows us that S is the set of all natural numbers.   □

There are several variations on this principle. Perhaps, in place of knowing $P(0)$, we know $P(1)$. Then we can conclude that $P(n)$ holds for all $n \geq 1$. A similar statement would hold with 100, or any fixed number, replacing 1.

It is important to note that, in a proof by induction, we have two jobs: to prove $P(0)$ (called **starting the induction**) and to prove that the implication from $P(n)$ to $P(n+1)$ holds (called the **inductive step**). However, there is another version, called the **Principle of Strong Induction**, which appears to get by without starting the induction.

Theorem 1.9 (Principle of Strong Induction) *Let $P(n)$ be a statement about the natural number n. Suppose that, for any natural number n, if $P(m)$ is true for all $m < n$, then $P(n)$ is true. Then $P(n)$ is true for every natural number n.*

Proof This time let S be the set of natural numbers n having the property that $P(m)$ holds for all $m < n$. Now:

(a) $0 \in S$; for there are no natural numbers $m < 0$, so $P(m)$ vacuously holds for all of them!

(b) If $n \in S$, then $P(m)$ holds for all $m < n$. By hypothesis, $P(n)$ holds. Now any number $m < n+1$ either satisfies $m < n$ or $m = n$, and $P(m)$ holds in either case. So $n+1 \in S$.

By the Induction Property, S contains all natural numbers; so, given n, we have $n+1 \in S$, so $P(n)$ is true. $\qquad\square$

It is time to have an example of the use of this principle. Suppose that you are asked to find the sum of the first n squares, that is, find

$$1^2 + 2^2 + \cdots + n^2.$$

It is a daunting task without help. But suppose you are told, or guess, that the answer is $n(n+1)(2n+1)/6$. Then you can prove your guess by induction. Let $P(n)$ be the statement

$$1^2 + 2^2 + \cdots + n^2 = n(n+1)(2n+1)/6.$$

Now $P(1)$ is true: for, when $n = 1$, the left-hand side is $1^2 = 1$, and the right-hand side is $1 \cdot 2 \cdot 3/6 = 1$.

Suppose that $P(n)$ is true; that is,

$$1^2 + 2^2 + \cdots + n^2 = n(n+1)(2n+1)/6.$$

Then

$$1^2 + 2^2 + \cdots + n^2 + (n+1)^2 = n(n+1)(2n+1)/6 + (n+1)^2$$
$$= (n+1)(2n^2 + n + 6n + 6)/6$$
$$= (n+1)(n+2)(2n+3)/6.$$

But the right-hand side is what we get from our expression $n(n+1)(2n+1)/6$ by substituting $n+1$ for n. So we have verified $\mathcal{P}(n+1)$.

By the Principle of Induction, we have proved $\mathcal{P}(n)$ for all $n \geq 1$.

Study this proof carefully. It seems at first that we are assuming what we are asked to prove. If we were, the argument would not be valid. You should convince yourself that this is not the case.

Here is another example. Consider the sequence

$$\sqrt{2}, \sqrt{2+\sqrt{2}}, \sqrt{2+\sqrt{2+\sqrt{2}}}, \ldots.$$

We want to show that the terms of this sequence increase, but are never greater than 2.

Let x_n be the nth term of the sequence. The relationship between consecutive terms is

$$x_{n+1} = \sqrt{2+x_n}.$$

We prove by induction that $x_n < x_{n+1}$ and $x_n < 2$ for all n.

Both of these statements are true for $n = 0$. (Why is $\sqrt{2} < \sqrt{2+\sqrt{2}}$?)

Suppose that $x_n < x_{n+1}$. Then

$$x_{n+1} = \sqrt{2+x_n} < \sqrt{2+x_{n+1}} = x_{n+2},$$

since $\sqrt{2+x}$ is a strictly increasing function of x for positive x. (This means that, if $x < y$, then $\sqrt{2+x} < \sqrt{2+y}$.)

Now suppose that $x_n < 2$. Then

$$x_{n+1} = \sqrt{2+x_n} < \sqrt{2+2} = 2,$$

using the same fact again.

So we have proved the inductive step, and both statements follow by induction.

Incidentally, from real analysis we know that an increasing sequence which is bounded tends to a limit. What is the limit of this sequence? (If you cannot see the answer immediately, calculate a few terms of the sequence.)

Here is an example of the use of strong induction. This is a result which was used in the proof of Euclid's Theorem.

Proposition 1.10 *Any natural number greater than 1 has a prime factor.*

Proof We have to show that, if $n > 1$, then n has a prime factor. We do this by strong induction. Let n be a natural number, and assume that, if m is any natural number satisfying $1 < m < n$, then m has a prime factor.

- If $n \leq 1$ then the statement is vacuously true.
- If n is prime, then it is a prime factor of itself, and the statement is true.

- Suppose that n is composite; then $n = ab$, where $1 < a, b < n$. By the induction hypothesis, a has a prime factor p. Now p is a prime factor of n, and so again the statement is true.

We have covered all cases, and so the proof is done. $\qquad\square$

Another consequence of the induction property is the following fact about the natural numbers.

Theorem 1.11 *Let T be any non-empty subset of the natural numbers. Then T has a smallest element.*

Proof We show the contrapositive form of this statement: that is, if T is a subset of the natural numbers which has no smallest element, then T is empty.

So suppose that T has no smallest element. Let S be the complement of T, the set of all natural numbers not in T. Let n be a natural number, and suppose that every natural number m smaller than n belongs to S. Then n must belong to S also; for, if $n \in T$, then n would be the smallest element of T (since all smaller numbers are in S). By the Strong Induction principle, $S = \mathbb{N}$, and so T is empty. $\qquad\square$

This property is sometimes referred to as the **well-ordering property**.

Exercise 1.10 Show that $(x + yi)(x - yi) = x^2 + y^2$. Hence show that, if $x + yi \neq 0$, then we can divide by it:

$$\frac{u + vi}{x + yi} = \left(\frac{ux + vy}{x^2 + y^2}\right) + \left(\frac{vx - uy}{x^2 + y^2}\right)i.$$

Exercise 1.11 Show that the square root of $x + yi$ is

$$\pm \left(\left(\sqrt{\frac{1}{2}\left(x + \sqrt{x^2 + y^2}\right)} \right) + \left(\sqrt{\frac{1}{2}\left(-x + \sqrt{x^2 + y^2}\right)} \right) i \right).$$

[*Hint*: Square it and see!] Can you be sure that both the real and imaginary parts are genuine real numbers (that is, they are square roots of non-negative real numbers)?

Exercise 1.12 Prove by induction that every natural number is either even or odd. Prove also that no natural number can be both even and odd.

Exercise 1.13 Prove the following statements by induction.

(a) The sum of the first n positive integers is $n(n + 1)/2$.
(b) The sum of the cubes of the first n positive integers is equal to the square of their sum.

Exercise 1.14 When the mathematician Gauss was in primary school, his teacher asked the class to add up all the numbers from 1 to 100. Gauss saw that, if he took the sum

$$S = 1 + 2 + 3 + \cdots + 99 + 100,$$

and wrote it down reversed,

$$S = 100 + \ \ 99 + \ \ 98 + \cdots + \ \ \ 2 + \ \ \ 1,$$

then each pair of numbers in the two sums adds up to 101. So

$$2S = 100 \times 101 = 10100,$$

and $S = 5050$. Your job is to turn this into a general proof that the sum of the natural numbers from 1 to n is $n(n+1)/2$.

Exercise 1.15 Use induction to prove each of the following statements:

(a) For all $n \geq 1$,

$$\frac{1}{1 \times 3} + \frac{1}{3 \times 5} + \cdots + \frac{1}{(2n-1) \times (2n+1)} = \frac{n}{2n+1}.$$

(b) $4^n \geq 16n^2$ for all $n \geq 4$.
(c) For all $n \geq 2$,

$$\frac{1}{2^2 - 1} + \frac{1}{3^2 - 1} + \cdots + \frac{1}{n^2 - 1} = \frac{3}{4} - \frac{1}{2n} - \frac{1}{2(n+1)}.$$

Exercise 1.16 Let $a_1, a_2, a_3, \ldots$ be numbers satisfying the rules that $a_1 = 1$ and $a_n = 2a_{n-1}$ for all $n > 1$.

(a) Write down the first few numbers a_n.
(b) Guess a formula for a_n.
(c) Prove your guess by induction.

Exercise 1.17 (∗) Euclid's proof that there are infinitely many primes gives us a rule for finding a new prime, if we already have a finite number:

Suppose that we have found n primes already, say $p_1, p_2, \ldots, p_n$.

Multiply them together and add one: let N be this number, so that $N = p_1 p_2 \cdots p_n + 1$.

If N is prime, take it to be the next prime p_{n+1}. Otherwise, take p_{n+1} to be the smallest prime which divides N.
Euclid gives us a guarantee that p_{n+1} is different from all the primes $p_1, \ldots, p_n$.

Take $p_1 = 2$. Use MAPLE or a calculator to find $p_2, p_3, \ldots, p_8$.
    Experiment with taking different primes for p_1. Does the prime 2 always turn up in the list sooner or later? Does the prime 3 always turn up? What is the main difficulty in the calculation?

Exercise 1.18 Prove, using the well-ordering property, that an infinite strictly decreasing sequence of positive integers (that is, a sequence $a_1, a_2, a_3, \ldots$ satisfying $a_n > a_{n+1}$ for all n) cannot exist.

Exercise 1.19 What is wrong with the following argument?

Proposition 1.12   *All horses have the same colour.*

Proof Let $\mathcal{P}(n)$ be the proposition that, in a set of n horses, all the horses have the same colour. We start the induction with $\mathcal{P}(1)$, which is clearly true.

Now suppose that $\mathcal{P}(n)$ is true. Let $\{H_1, \ldots, H_{n+1}\}$ be a set of $n+1$ horses. Then $\{H_1, \ldots, H_n\}$ is a subset containing n horses; by $\mathcal{P}(n)$, they all have the same colour. Similarly, $\{H_2, \ldots, H_{n+1}\}$ is a set of n horses, so these also have the same colour. It follows that $\{H_1, \ldots, H_{n+1}\}$ all have the same colour; so $\mathcal{P}(n+1)$ is true.

By the Principle of Induction, $\mathcal{P}(n)$ is true for all positive integers n. $\qquad\square$

Elementary algebra

Abu Ja'far Muhammad ibn Musa al-Khwarizmi (whose name gives us the word 'algorithm') wrote an algebra textbook which included much of what is still regarded as elementary algebra today. The title of his book was *Hisab al-jabr w'al-muqabala*. The word *al-jabr* means 'restoring', referring to the process of moving a negative quantity to the other side of an equation; the word *al-muqabala* means 'comparing', and refers to subtracting equal quantities from both sides of an equation. Both processes are familiar to anyone who has to solve an equation! The word *al-jabr* has, of course, been incorporated into our language as 'algebra'.

In this section we briefly revise the techniques of elementary algebra.

1.6 Formulae and equations. A **formula**, or **expression**, is some collection of symbols like

$$x^3 \sin(\log_{10} x) + x^{x^{x^{x^x}}} + 196883.$$

This formula contains a **variable** x, and the assumption is that if we assign a numerical value to x, then we can in principle evaluate the formula and obtain a number. (We may not be able to do that in practice; if, for example, $x = 3$, then the above formula cannot be evaluated because the universe is not large enough to write down the answer!) We allow a formula to contain more than one variable. Thus, $x^2 + 2^y$ is a formula with two variables x and y.

In Algebra, for the most part, we use only formulae built up using the arithmetic operations (addition, subtraction, multiplication, and division) and sometimes others such as exponentiation and taking square roots. More complicated functions such as sines and logarithms lie in the domain of 'analysis'.

An **equation** is a mathematical statement of the form

$$F_1 = F_2,$$

where F_1 and F_2 are formulae. Now it may be that, no matter what values we substitute for the formulae F_1 and F_2, the equation is true. In this case, the equation is called an **identity**. An example of an identity is

$$x^4 + x^2 + 1 = (x^2 + x + 1)(x^2 - x + 1).$$

If an equation is not an identity, then there still may be some values of the variables for which the equation is true when these values are substituted. The procedure of finding all such values is called **solving** the equation, and the values are the **solutions**. An equation may have no solution, or one, or more than one. For example, the equation

$$x^2 = x + 2$$

has two solutions: $x = 2$ and $x = -1$.

In solving an equation, we can apply any operation to it provided that we do the same thing to both sides. For example, from the above equation, we could obtain $x^2 - x = 2$ (by subtracting x from each side), or $2x^2 = 2x + 4$ (by multiplying each side by 2). However, we cannot obtain $2x^2 = 2x + 2$, since we have failed to multiply everything on the right by 2.

Originally, the purpose of algebra was to solve equations!

1.7   Brackets.   The formulae $2x + 5$ and $2(x + 5)$ are different; when $x = 2$, the first evaluates to 9 and the second to 14.

The difference between them depends on a convention universally adopted in mathematics:

> In evaluating a formula, we perform multiplications and divisions
> before additions and subtractions.

This rule is called **precedence of operators**.

Thus, in the first formula above, we multiply 2 and x and then add 5 to the result. If we wish instead to add x and 5 and then multiply 2 by the result, we have to change the precedence of the operators. So we supplement the precedence rule by another rule asserting that, if part of a formula is enclosed in brackets, then this part is evaluated first and then treated as a single quantity in the later evaluation. The second formula above thus does exactly what we want.

Brackets can be nested, in which case they are evaluated from the inside out. For example, the formula

$$x + 2(y + 3(z + 4))$$

says: 'add z to 4, multiply the result by 3, add y to this, multiply the result by 2, and finally add x'.

Remember the distributive law we met earlier, which states that

$$a(b + c) = ab + ac.$$

Using this, if a formula contains brackets, we may replace it by a formula not containing brackets. This is called **expanding the brackets**. For example, the formula with nested brackets above can be changed into

$$x + 2y + 6z + 24.$$

Brackets may contain arbitrarily complicated expressions. If you are expanding brackets, remember to multiply everything inside the brackets. So $2(3x + 4y + 5z) = 6x + 8y + 10z$, for example.

If several brackets have to be multiplied together, the work should be done in stages:

$$(a + b)(c + d) = (a + b)c + (a + b)d = ac + bc + ad + bd.$$

Finally, note that mathematicians use several types of brackets, such as ( ), [ ], { }. In the past, these were all used in formulae; some mathematical publishers even had rules about the order in which they were to be used in a nested expression!

The rule now is only to use the ordinary 'parentheses' ( ) for this job, as the others have different meanings. We have seen that 'braces' { } are used for sets, while 'square brackets' [ ] are sometimes used to denote the integer part or 'round-down' of the expression, as in $[9/2] = 4$. (It is better to use the more specialised brackets $\lfloor \ \rfloor$ for this, so that $\lfloor 9/2 \rfloor = 4$. Then we can use $\lceil \ \rceil$ for 'round-up', as in $\lceil 9/2 \rceil = 5$.)

Still other brackets such as 'angle brackets' $\langle \ \rangle$ are used with other specialised meanings. We can also think of modulus signs $|\ |$ as a kind of brackets.

Some mathematical expressions have **implicit brackets**. In the formulae

$$\frac{a + b}{c + d} - \sqrt{e + f},$$

the three additions must be performed before the division and taking the square root, even though there are no actual brackets in the formulae.

1.8 Fractions. Formulae may contain fractions, such as $\dfrac{a + b}{c + d}$ above. This can also be written (to save space) as $(a + b)/(c + d)$.

The rules for manipulating fractions are

- $\dfrac{a}{b} + \dfrac{c}{d} = \dfrac{ad + bc}{bd}$;
- $\dfrac{a}{b} - \dfrac{c}{d} = \dfrac{ad - bc}{bd}$;
- $\dfrac{a}{b} \cdot \dfrac{c}{d} = \dfrac{ac}{bd}$;
- $\dfrac{a}{b} \bigg/ \dfrac{c}{d} = \dfrac{ad}{bc}$;
- $\dfrac{ax}{bx} = \dfrac{a}{b}$.

The addition and subtraction rules involve putting the fractions over a common denominator. They are easily proved using the last rule (the cancellation rule) in reverse. Thus $\dfrac{a}{b} = \dfrac{ad}{bd}$, and $\dfrac{c}{d} = \dfrac{bc}{bd}$; now they have the same denominator and can be added.

Do not learn these rules. Rather, you should practice until you can manipulate fractions without thinking. Also, fractions can be cancelled at any time, not just the end of the calculation. If you have to work out $\dfrac{1}{5} - \dfrac{3}{40}$, it is better to write them with a denominator of 40 to get

$$\frac{1}{5} - \frac{3}{40} = \frac{8-3}{40} = \frac{5}{40} = \frac{1}{8}$$

than to apply the rules literally:

$$\frac{1}{5} - \frac{3}{40} = \frac{1 \cdot 40 - 3 \cdot 5}{40} = \frac{25}{200} = \frac{1}{8};$$

the bigger numbers in the second calculation make mistakes more likely.

1.9   Square roots. The **square root** of a non-negative number x is the non-negative number y such that $y^2 = x$. Notice that, at least for real numbers, only non-negative numbers have square roots, and that the square root is itself non-negative. So, even though it is true that $(-4)^2 = 16$, yet the square root of 16 is 4 and not -4.

There is no simple formula for adding square roots. The multiplication rule is

$$\sqrt{a} \cdot \sqrt{b} = \sqrt{ab}.$$

This means that $x\sqrt{a} = \sqrt{x^2 a}$ if x is non-negative.

Remember that square roots are implicitly bracketed. Thus when we multiply out, everything under the square root sign must be multiplied:

$$x\sqrt{a+b} = \sqrt{x^2 a + x^2 b}.$$

But do not make the mistake of thinking that $\sqrt{a+b} = \sqrt{a} + \sqrt{b}$; this is almost always wrong!

Similar principles hold for cube (and other) roots.

1.10   Powers. If n is a positive integer, then x^n means the expression obtained by taking n factors x and multiplying them together: for example, $x^4 = x \cdot x \cdot x \cdot x$.

The rules for expressions with powers are:

- $x^{m+n} = x^m \cdot x^n$;
- $x^{mn} = (x^m)^n$;
- $(xy)^n = x^n \cdot y^n$.

Mathematicians have extended this definition: if x is positive, it is possible to give a meaning to x^r for any real number r, in such a way that the three laws just stated continue to hold. The important cases to remember are

$$x^0 = 1, \qquad x^{-1} = \frac{1}{x}, \qquad x^{1/2} = \sqrt{x}.$$

In general, $x^{-r} = 1/x^r$.

1.11   Polynomials. A **polynomial** in the variable x is a formula which is a sum of a number of terms each of the form ax^n, where a is a number and n a non-negative integer. (Remember that $x^0 = 1$, so that ax^0 is just a.) In this section we take the word 'number' to mean 'real number'. An example of a polynomial is

$$27x^5 + 203x^2 - 31x + 5.$$

Often we use the function notation $f(x)$ to denote a polynomial in the variable x. Then, if c is any number, $f(c)$ denotes the evaluation of $f(x)$ when x is given the value c.

The expressions ax^n making up a polynomial are its **terms**, and the **degree** of the term ax^n is the number n. A **constant term** is one whose degree is zero.

We assume that the coefficient a of any term is non-zero (we omit any terms with zero coefficients), and that different terms have different degrees (as several terms with the same degree can be combined into a single term). The only problem here is that there is a polynomial with no terms at all, which we write as 0 (the alternative would be not to write anything, which may be confusing!).

Polynomials are added and multiplied as formulae. This means that, if the same power of x occurs in two polynomials, then when we add them we can combine the corresponding terms. For example,

$$(27x^5 + 203x^2 - 31x + 5) + (x^3 - 200x^2 + 31x + 7) = 27x^5 + x^3 + 3x^2 + 12.$$

A polynomial can also be thought of as a function, whose value for a given value of x is obtained by substituting the value of x and then evaluating the result.

In fact, the question 'what exactly is a polynomial?' is much more difficult to answer than indicated, here. To mention just two problems:

- If two polynomials are identical apart from the fact that the variables have been given different names, are they the same polynomial or not?
- If two polynomials give rise to identical functions, are they the same polynomial or not?

In the next chapter, we will see how mathematicians currently view these questions.

Addition and multiplication of polynomials satisfy many of the same laws as the same operations for numbers: the commutative, associative, and distributive

laws all hold. These statements do not strictly require that the coefficients of
the polynomials are real numbers: it is enough that the coefficients themselves
should satisfy these three laws, so any of the standard number systems will
do (See Exercise 1.25.) Also, adding the polynomial 0, or multiplying by the
polynomial 1, has no effect.

The **degree** of a polynomial is the largest degree of any of its terms. (According
to this definition, the zero polynomial 0 does not have a degree. Some people
arbitrarily set its degree to be -1, or $-\infty$, but my convention seems simpler.) A
polynomial with degree 0, 1, 2, 3, 4, or 5 is called **constant**, **linear**, **quadratic**,
cubic, **quartic**, or **quintic**, respectively.

As for integers, there is a **division algorithm** for polynomials: if $f(x)$ and
$g(x)$ are polynomials, then there exists a **quotient** $q(x)$ and a **remainder** $r(x)$
such that

- $f(x) = g(x)q(x) + r(x)$;
- either $r(x) = 0$, or $r(x)$ has degree smaller than the degree of $g(x)$.

The way of finding the quotient and remainder is very similar to the division
algorithm for integers. If the remainder $r(x)$ is the zero polynomial, we say that
$g(x)$ **divides** $f(x)$.

Here is an example: Divide $x^4 + 4x^3 - x - 5$ by $x^2 + 2x - 1$.

$$
\begin{array}{r}
x^2 + 2x - 3 \\
x^2 + 2x - 1 \;) \; \overline{x^4 + 4x^3 - x \; -5} \\
\underline{x^4 + 2x^3 - x^2 } \\
2x^3 + x^2 - x \\
\underline{2x^3 + 4x^2 - 2x } \\
-3x^2 + x \; - 5 \\
\underline{-3x^2 - 6x \; + 3} \\
7x \; - 8
\end{array}
$$

This calculation shows that when we divide $x^4 + 4x^3 - x + 5$ by $x^2 + 2x - 1$, the
quotient is $x^2 + 2x - 3$ and the remainder is $7x - 8$.

In one particular case, it is easy to calculate the remainder:

Theorem 1.13 (Remainder Theorem) *If $f(x)$ is divided by $x - c$, the
remainder is $f(c)$.*

Proof Suppose that $f(x) = (x - c)q(x) + r(x)$. Since $r(x)$ has degree less
than 1, it is a constant polynomial. Then substituting $x = c$ we find that $f(c) =
r(c)$, so that $r(x)$ is the constant polynomial $f(c)$ (or the zero polynomial, if
$f(c) = 0$).   □

From this we immediately obtain Theorem 1.14.

Theorem 1.14 (Factor Theorem) *Let $f(x)$ be a polynomial and c a number.
Then $x - c$ divides $f(x)$ if and only if $f(c) = 0$.*

A non-constant polynomial is called **irreducible** if it cannot be written as the product of two polynomials of smaller degree. Any linear polynomial is obviously irreducible, since the only polynomials of smaller degree are constants.

Over the real numbers, the polynomial $x^2 + 1$ is irreducible. For, if it is not irreducible, it must have a factor of degree 1, which we can take to be $x - c$ for some real number c; then the Factor Theorem shows that $c^2 + 1 = 0$, which is impossible.

This argument would fail if our numbers were complex numbers. Indeed, we would have

$$x^2 + 1 = (x + \mathrm{i})(x - \mathrm{i}).$$

Irreducible polynomials play a similar role to prime numbers, with one main difference: any non-constant polynomial can be factorised into irreducible polynomials, but the factors are not unique. For example,

$$6x^2 - 6 = (2x + 2)(3x - 3) = (3x + 3)(2x - 2).$$

We will see in the next chapter how, in a much more general situation, we can prove a 'unique factorisation theorem' for polynomials.

To conclude this section, I mention that it is possible to have polynomials in more than one variable. For example, a polynomial in x and y is a sum of terms of the form $ax^m y^n$, where a is a number and m and n are non-negative integers.

1.12   Quadratic and cubic equations.   A **polynomial equation** is an equation of the form $f(x) = g(x)$, where $f(x)$ and $g(x)$ are polynomials. By subtracting $g(x)$ from both sides, we can write this as $h(x) = 0$, where $h(x)$ is the polynomial $f(x) - g(x)$. In this form, we say that the equation is **quadratic**, **cubic**, ..., if $h(x)$ is quadratic, cubic, ...

Throughout the history of mathematics, one of the most important topics has been the problem of solving polynomial equations.

Here is Al-Khwarizmi's solution of the quadratic equation $x^2 + 10x = 39$. In the quotation, the 'root' is x and the 'square' is x^2; according to the conventions of his day, Al-Khwarizmi did not consider the possibility of negative solutions. In modern terminology his solution is

$$x = \sqrt{5^2 + 39} - 5 = 3.$$

What is the square which combined with ten of its roots will give a sum total of 39? The manner of solving this type of equation is to take one-half of the roots just mentioned. Now the roots in the problem before us are 10. Therefore take 5, which multiplied by itself gives 25, an amount which you add to 39 giving 64. Having taken then the square root of this which is 8, subtract from it half the roots, 5 leaving 3. The number three therefore represents one root of this square, which itself, of course is 9. Nine therefore gives the square.

Abu Ja'far Muhammad ibn Musa al-Khwarizmi (about 810)

Notice how Al-Khwarizmi does not use symbols (which were a long way in the future in his time), but explains the method clearly with an example. In essence, he gives us an algorithm to solve any equation of the form $x^2 + ax = b$: halve the coefficient of x, square it, add the constant term, take the square root, subtract half the coefficient of x.

The only differences between this and the modern formula is that now we do not mind negative numbers in our equations, so we write any quadratic as

$$x^2 + ax + b = 0,$$

and similarly we allow both positive and negative solutions. Thus, our b is the negative of Al-Khwarizmi's, and we have to subtract it instead of adding, giving the well-known formula

$$x = \pm\sqrt{(a/2)^2 - b} - a/2 = \frac{-a \pm \sqrt{a^2 - 4b}}{2}.$$

The key idea in obtaining this solution is **completing the square**. We recognise that

$$x^2 + ax + (a/2)^2 = (x + a/2)^2,$$

and so by adding $(a/2)^2 - b$ to both sides of the equation $x^2 + ax + b = 0$ we obtain

$$(x + a/2)^2 = (a/2)^2 - b,$$

so that $x + a/2 = \pm\sqrt{(a/2)^2 - b}$, and subtracting $a/2$ from both sides we obtain the solution.

Can we do anything similar for more complicated equations? Consider the cubic equation

$$x^3 + ax^2 + bx + c = 0.$$

We try **completing the cube**, using the fact that

$$x^3 + ax^2 + (a^2/3)x + a^3/27 = (x + a/3)^3,$$

so that the given equation becomes

$$(x + a/3)^3 = (a^2/3 - b)x + (a^3/27 - c).$$

Now it is not so simple, since there is still an x on the right. It took mathematicians hundreds of years to figure out what to do next!

We continue this story later in the book.

Exercise 1.20 This exercise contains a few 'drill' questions on manipulation of formulae. If you cannot do them quickly and accurately, it is very important that you should practice similar examples until you can. These examples are taken from a course at Queen Mary, University of London, entitled 'Essential Mathematics'; the name of the course is chosen for a good reason!

(a) Evaluate

$$\frac{5}{30} - \frac{1}{7} \times \left(\left(4 - \frac{18}{27} \right) \Big/ \frac{8}{3} + \left(\frac{3}{8} \right)^2 \times \left(\frac{7}{4} \times \frac{7}{9} + \frac{5}{12} \right) \right).$$

(b) Simplify

$$\left(\frac{-ab^2c^3}{cb^3} \right)^3 \left(\frac{a^5}{-b^7c} \right)^{-2}.$$

(c) Compute the remainder when $-x^4 + 3x^2 + 2x - 1$ is divided by $x^2 + 2$.

(d) Add and simplify

$$\frac{1}{y^2 - 2y - 15} + \frac{3}{y^2 - 10y + 25}.$$

(e) Compute $f\left(-1/a^3 \right)$, where

$$f(x) = \frac{x^2 - x - 1}{x - 1}.$$

(f) Simplify

$$\frac{1}{\sqrt{5}} \frac{\sqrt{30} - \sqrt{12}\sqrt{15}}{(\sqrt{2} - \sqrt{3})^2}$$

to the form $m + n\sqrt{d}$, where m, n, and d are integers.

(g) Simplify

$$\sqrt{x^2 - x^3} - \sqrt{4 - 4x}.$$

(h) Solve the equation

$$4x - \frac{x^2 - 9}{x + 3} = \frac{6}{3x - 1}.$$

Exercise 1.21 Show by means of an example that the distributive law fails for the 'round-down' brackets $\lfloor \ \rfloor$; that is, it is not true that $a\lfloor b + c \rfloor = ab + ac$. Indeed it is not even equal to $\lfloor ab \rfloor + \lfloor ac \rfloor$.

Exercise 1.22 Show directly that $x^n - c^n$ is divisible by $x - c$ for any natural number n. Deduce the Remainder Theorem from this.

Exercise 1.23 Use the preceding exercise to show that $x - 1$ divides $x^k - 1$ for any natural number k, and deduce that $m - 1$ divides $m^k - 1$ for any natural number m. By taking $m = 2^l$, show that $2^l - 1$ divides $2^{kl} - 1$. Deduce that, if $2^n - 1$ is prime, then n is prime. (Compare page 6, where we showed that the converse is false.)

Exercise 1.24 If you know the Intermediate Value Theorem, use it to show that (over the real numbers) there is no irreducible polynomial of degree 3.

Exercise 1.25 Show that the proof of commutativity of multiplication of polynomials requires the commutativity and associativity of addition for the coefficients, as well as the commutativity of their multiplication.

Exercise 1.26 Let f be a polynomial, and let $f(t)$ be the result of substituting the real number t for the indeterminate x. Are the following statements true or false? If false, what is the correct rule?

(a) $(f + g)(t) = f(t) + g(t)$;
(b) $(fg)(t) = g(f(t))$.

Sets

In this section, we take a quick look at sets, mainly to introduce the notation for unions, intersections, Cartesian products, and so forth, and to examine the concepts of relations and functions, especially equivalence relations. Although we cannot say what a set is without going round in circles, sets provide the accepted basis for mathematics.

1.13   Introduction.   It is very difficult to define a set. On the other hand, everybody knows what a set is, and the explanation 'a set is a collection of objects', though no good as a definition (What is a collection?), is quite clear. What is important is that we can tell, of any particular element, whether or not it belongs to the set in question. A set is completely determined when we know all of its members.

Often, we use capital letters for sets, and lower case for their elements. We write $x \in X$ to denote that the element x is a member of (or belongs to) the set X, and $x \notin X$ for the negation of this.

How do we specify a set?

If it is finite, we can just list its elements inside braces, or curly brackets {}. Thus, $\{1, 3, 4, 5, 9\}$ is a set with five elements.

Certain familiar infinite sets have names. Thus, $\mathbb{N}$, $\mathbb{Z}$, $\mathbb{Q}$, $\mathbb{R}$, and $\mathbb{C}$ denote the sets of numbers, which we met earlier. Other infinite sets can be described by giving a test for membership in the set. For example,

$$\{x \in \mathbb{Z} : x = 2y + 1 \text{ for some } y \in \mathbb{Z}\}$$

is the set of odd integers.

1.14   Sets and set operations.   Two sets are equal if they have the same members. For example,

$$\{2\} = \{x \in \mathbb{Z} : x > 0, x \text{ is even}, x \text{ is prime}\}.$$

So the **basic test for equality** is

$$A = B \quad \text{means that} \quad (x \in A) \Leftrightarrow (x \in B) \text{ for all elements } x.$$

Notice that, to prove two sets equal, we have to prove two things: every element of A is in B; and every element of B is in A. (Remember that $\mathcal{P} \Leftrightarrow \mathcal{Q}$ means $\mathcal{P} \Rightarrow \mathcal{Q}$ and $\mathcal{Q} \Rightarrow \mathcal{P}$.)

The **empty set** is the set which has no members. It is written as $\emptyset$, a zero with a slash; not the Greek letter ϕ (phi).

The empty set is a notorious source of trouble in mathematics. Arguing about it requires cool-headed logic rather than intuition. If you have proved a general fact about sets, it is worth checking that it holds for the empty set. Here is an example of how to argue with the empty set:

There is only one empty set. For suppose that E_1 and E_2 are empty sets. For any element x, the statements $(x \in E_1)$ and $(x \in E_2)$ are both false, since E_1 and E_2 have no members. So, according to the rule for 'if and only if', the statement $(x \in E_1) \Leftrightarrow (x \in E_2)$ is true, and by the basic test for equality, we have $E_1 = E_2$. Informally, E_1 and E_2 have the same members (viz., none at all).

A is a **subset** of B if A consists of some (perhaps all) of the elements of B. This is written as $A \subseteq B$, like a curved 'less than or equal' sign. So the **basic test for a subset** is

$$A \subseteq B \quad \text{means that} \quad (x \in A) \Rightarrow (x \in B) \text{ for all elements } x.$$

Note that this involves half the work of proving that $A = B$. Any set is a subset of itself. Also, the empty set is a subset of any set. (For consider the proposition $(x \in \emptyset) \Rightarrow (x \in A)$. For any x, the statement $x \in \emptyset$ is false; and a false proposition implies any proposition, so the implication is true.) These two subsets, which always exist, are the **trivial subsets**.

If $A \subseteq B$ and $A \neq B$, we say that A is a **proper subset** of B, and write $A \subset B$.

Now we turn to some ways of building new sets from old.

Definition The **union** of two sets A and B is the set of all elements lying in either A or B:

$$A \cup B = \{x : x \in A \text{ or } x \in B\}.$$

The **intersection** of A and B is the set of all elements lying in both:

$$A \cap B = \{x : x \in A \text{ and } x \in B\}.$$

The **difference** $A \setminus B$ consists of the elements which lie in A but not in B:

$$A \setminus B = \{x : x \in A \text{ and } x \notin B\}.$$

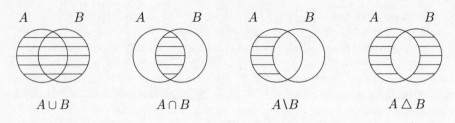

$$A \cup B \qquad\qquad A \cap B \qquad\qquad A \backslash B \qquad\qquad A \triangle B$$

Fig. 1.6 Venn diagrams for set operations

The **symmetric difference** $A \triangle B$ consists of the elements which lie in one of A and B but not in both, so that we have

$$A \triangle B = (A \setminus B) \cup (B \setminus A) = (A \cup B) \setminus (A \cap B).$$

For example, if $A = \{2,3,5\}$ and $B = \{1,2,4\}$, then $A \cup B = \{1,2,3,4,5\}$, $A \cap B = \{2\}$, $A \setminus B = \{3,5\}$, and $A \triangle B = \{1,3,4,5\}$.

These concepts are conveniently illustrated by Venn diagrams (see Figure 1.6).

An **ordered pair** (a,b) has a **first element** a and a **second element** b. This means that two ordered pairs (a,b) and (c,d) are equal if and only if both $a = c$ and $b = d$. So the ordered pair (a,b) is not the same as the set $\{a,b\}$: the elements of a set do not come in any particular order, so $\{a,b\}$ is the same as $\{b,a\}$. (It does not matter exactly how ordered pairs are defined.)

Definition   Let A and B be sets. Their **Cartesian product** is the set

$$A \times B = \{(a,b) : a \in A, b \in B\},$$

the set of all ordered pairs with first element in A and second element in B. If $A = B$, we write A^2 for $A \times A$.

The name 'Cartesian' commemorates Descartes, who unified algebra and geometry by the insight that the Euclidean plane (a geometric object) is essentially the same as the set $\mathbb{R} \times \mathbb{R} = \mathbb{R}^2$: each point of the plane can be represented by its Cartesian coordinates, which are an ordered pair (x,y) of real numbers; and every pair of real numbers represents a point.

More generally, if n is a positive integer, then an ordered n-tuple, written $(a_1, a_2, \ldots, a_n)$, has a first, second, $\ldots$, nth element; and $(a_1, \ldots, a_n) = (b_1, \ldots, b_n)$ if and only if $a_1 = b_1$, $\ldots$, $a_n = b_n$. The Cartesian product of $A_1, A_2, \ldots, A_n$ is

$$A_1 \times A_2 \times \cdots \times A_n = \{(a_1, a_2, \ldots, a_n) : a_1 \in A_1, a_2 \in A_2, \ldots, a_n \in A_n\},$$

and $A^n = A \times A \times \cdots \times A$ (n factors). Thus, according to Descartes, $\mathbb{R}^n$ is n-dimensional space.

The number of elements in a set A is called the **cardinality** of A, and written as $|A|$. Although we use the same notation as for the absolute value or modulus of a number, the meaning is quite different. For example, $|-3| = 3$, but $|\{-3\}| = 1$.

There is a theory of cardinalities of infinite sets, developed by Cantor in the nineteenth century, but we do not need this yet. If the set A is finite, then its cardinality is a non-negative integer. Here are a couple of basic results about cardinalities of finite sets.

Proposition 1.15 *Let A and B be finite sets. Then*

(a) $|A \cup B| = |A| + |B| - |A \cap B|$;
(b) $|A \times B| = |A| \cdot |B|$.

Finally, note that the elements of a set may themselves be sets. For example, $\{1, \{1, 2\}\}$ is a set with two elements: the number 1 and the set $\{1, 2\}$. This process can be continued as far as we like: there are sets like $\{\{\{\{1\}\}\}\}$; and this is not the same set as $\{\{\{1\}\}\}$. Somewhere in the Galaxy there is a race of set theorists to which this comment seems natural and obvious. But, for human beings, the concept of a set of sets seems to cause panic and confusion, especially when we have to perform operations on its members (which are sets) as if they were single objects. This is unavoidable in algebra, as you will see when we reach factor rings in the next chapter. You have been warned!

1.15   Functions.   Until fairly recently, mathematicians thought of functions as formulae: the logarithm function $\log_{10} x$, the sine function $\sin x$, or more complicated compounds such as the one given on page 23. Later they introduced functions with 'split' definitions, such as

$$f(x) = \begin{cases} 1 & \text{if } x \text{ is rational;} \\ 0 & \text{if } x \text{ is irrational.} \end{cases}$$

Eventually it seemed impossible to make a definition of a formula general enough for everything that mathematicians wanted to consider; so a completely different approach was taken.

Think of a function F as a black box, where we can feed any element of A into the box, and an element of B will emerge at the other end.

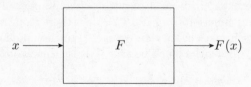

The name of the function is F; we put x into the black box and $F(x)$ comes out. Be careful not to confuse F, the name written on the black box, with $F(x)$, which is what comes out when x is put in. Sometimes the language makes it hard to keep this straight. For example, there is a function which, when you put in x, outputs x^2. We tend to call this 'the function x^2', but it is really 'the squaring function', or 'the function $x \mapsto x^2$'. (You see that we have a special symbol $\mapsto$ to denote what the black box does.)

Black boxes are not really mathematical notation, so we reformulate this definition in more mathematical terms. We have to define what we mean by a function F. Now there will be a set X of allowable inputs to the black box; X is called the domain of F. Similarly, there will be a set Y which contains all the possible outputs; this is called the codomain of F. (We do not necessarily require that every value of Y can come out of the black box. For the squaring function, the domain and the codomain are both equal to $\mathbb{R}$, even though none of the outputs can be negative.)

The important thing is that every input $x \in X$ produces exactly one output $y = F(x) \in Y$. The ordered pair (x, y) is a convenient way of saying that the input x produces the output y. Then we can take all the possible ordered pairs as a description of the function. Thus we come to the formal definition:

Definition A **function** $F : A \to B$ is a subset of $A \times B$ such that, for every element $a \in A$, there is a unique element $b \in B$ for which $(a, b) \in F$. We write $b = F(a)$ if $(a, b) \in F$. (Note that the definition says nothing at all about the way in which the function values are actually calculated.) We call $F(a)$ the **image** of a under F.

A function is also often called a **mapping** or **map**: if $F : A \to B$, we say that F **maps** elements of A to elements of B (or **maps** A to B). The sets A and B are called the **domain** and **codomain** of F, respectively. The **image** or **range** of F is the set of values that come out of the black box when all elements of the codomain are fed in:

$$\mathrm{Im}(F) = \{F(a) : a \in A\}.$$

It is a subset of the codomain.

A function $F : A \to B$ is said to be **one-to-one** or **injective** if $a \neq b$ implies $F(a) \neq F(b)$ (so that different points have different images under F). It is **onto** or **surjective** if every point of B is the image of some point of A. It is **bijective**, or a **one-to-one correspondence**, if it is both injective and surjective. If F is a bijection, then the elements of A are 'paired up' with elements of B by F, so that, if the sets are finite, they must have the same number of elements.

If F is a bijective function from X to Y, then there is an **inverse function** G from Y to X which takes every element $y \in Y$ to the unique $x \in X$ for which $F(x) = y$. In other words, the black box for G is the black box for F in reverse:

$$x = G(y) \text{ if and only if } y = F(x).$$

The inverse function G is also bijective. Thus a bijective function F and its inverse G satisfy

- $G(F(x)) = x$ for all $x \in X$;
- $F(G(y)) = y$ for all $y \in Y$.

Notice that F is the inverse function of G.

Proposition 1.16 *If $|A| = m$ and $|B| = n$, then the number of functions from A to B is n^m.*

Proof We can represent a function from A to B as a table with two columns, where we write all the elements of A in the first column, and the value of $F(a)$ opposite the entry a. Now there are $|B| = n$ choices for each entry in the second column of the table; since there are m entries, there are n^m possible tables. Each table specifies a unique function. □

Sometimes we represent a function $F : A \to B$ by a picture, where we show the two sets A and B, and draw an arrow from each element a of A to the element $b = F(a)$ of B. For such a picture to show a function, each element of A must have exactly one arrow leaving it. Now

- F is one-to-one (injective) if no point of B has two or more arrows entering it;
- F is onto (surjective) if every point of B has at least one arrow entering it;
- F is one-to-one and onto (bijective) if every point of B has exactly one arrow entering it; in this case, the arrows match up the points of A with the points of B.

Here are some illustrations. The first is not a function because some elements of A have more than one arrow leaving them while some have none.

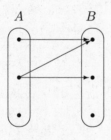

Not a function

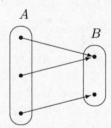

Onto, not one-to-one

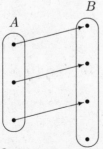

One-to-one, not onto

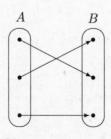

One-to-one and onto

1.16 Binary operations.

Definition A **binary operation** on a set A is a function F from $A \times A$ to A.

Often we write such an operation differently: either with **infix notation**, using a symbol such as $\circ$, where we put $a \circ b$ for $F(a,b)$; or with **juxtaposition**, where we simply write ab for $F(a,b)$. (Of course, addition in a number system is a binary operation, written in infix notation with the symbol $+$; and multiplication is usually written as juxtaposition.) Other symbols used for operations include $+$, $\cdot$, $\times$, $-$, $*$.

More generally, if n is a positive integer, an n-**ary operation** on A is a function from A^n to A. We use the terms 'unary' and 'binary' for $n = 1, 2$ respectively. But for $n \neq 2$, there is no analogue of infix notation or juxtaposition.

We can think of a binary operation as a black box with two inputs and one output: we put in a and b, and $a \circ b$ comes out.

Proposition 1.17 *If $|A| = n$, then the number of binary operations on A is n^{n^2}.*

Proof An operation is a function from $A \times A$ to A. Since $|A \times A| = n^2$ and $|A| = n$, the result follows from Proposition 1.16. □

Suppose that we have a binary operation on A, denoted by $\circ$ in infix notation. Then we have the 'closure condition'

(C) (**Closure law**): For all $a, b \in A$, $a \circ b \in A$.

This is really trivial because the operation is a function to the set A. It seems less trivial if we ask the question: When does $\circ$ give an operation on a subset X of A? This will occur if and only if X satisfies the closure law; that is, for all $a, b \in X$, $a \circ b \in X$. We will ask this question very often!

A binary operation on a finite set can be described by an **operation table**. Let $A = \{a_1, \ldots, a_n\}$. Take an $n \times n$ table, and label the rows and columns with the elements $a_1, \ldots, a_n$. In the position in row i and column j, put the element $a_i \circ a_j$. The whole table can be labelled with the name of the operation.

Example Let $A = \{\alpha, \beta, \gamma\}$, and define an operation $\circ$ by

$$x \circ y = \begin{cases} x & \text{if } x \neq y, \\ \alpha & \text{if } x = y. \end{cases}$$

The operation table is

$\circ$	α	β	γ
α	α	α	α
β	β	α	β
γ	γ	γ	α

In this case there is a simple rule describing the operation. But even when there is no obvious rule, the operation table describes an operation completely.

Example In the above example, which subsets X of A have the property that $\circ$ defines an operation on X? That is, which satisfy the closure law?

The only such subsets of X are $\emptyset$, $\{\alpha\}$, $\{\alpha,\beta\}$, $\{\alpha,\gamma\}$, and A. For the empty set certainly satisfies closure. If X is non-empty and $x \in X$, then $x \circ x = \alpha \in X$. There are four sets containing α, the other four listed above; they all satisfy closure. For example, for $\{\alpha,\gamma\}$, the operation table is

$$
\begin{array}{c|cc}
\circ & \alpha & \gamma \\
\hline
\alpha & \alpha & \alpha \\
\gamma & \gamma & \alpha
\end{array}
$$

1.17 Relations. A binary relation can be thought of as a statement which, given any two elements of a set A, is either true or false for that pair. For example, the relation 'less than' on the integers is true for the pair $(5,17)$, and false for the pair $(-1,-2)$. We know the relation completely if we know the set of pairs for which it is true. So we make a formal definition:

Definition A **binary relation** R on a set A is a subset of the Cartesian product $A \times A$.

A binary relation can be thought of as a black box with two inputs, for which the only possible outputs are 'yes' and 'no'. If we put in a and b, the output is 'yes' if the pair (a,b) satisfies the relation, and 'no' if it does not.

Proposition 1.18 *If $|A| = n$, then the number of binary relations on A is 2^{n^2}.*

Proof Proposition 1.16, since a binary operation is a function from $A \times A$ to $\{\text{yes}, \text{no}\}$. $\qquad\square$

Often we represent a binary relation by **infix notation**, in which we place a symbol between a and b to indicate that $(a,b) \in R$. Typical symbols used are $\sim$, $\equiv$, as well as the more familiar $=$, $\leq$, $<$, and so on. So, for example, we might use $a \sim b$ to denote $(a,b) \in R$.

For example, if $A = \{1,2,3\}$, then the relation 'less than' on A is the set $\{(1,2),(1,3),(2,3)\}$.

More generally, for any positive integer n, an n-**ary relation** on A is a subset of A^n.

1.18 Equivalence relations. Let R be a binary relation on A. Here are three laws which R may or may not satisfy:

(Eq1) (**Reflexive law**): $(a,a) \in R$ for all $a \in A$.
(Eq2) (**Symmetric law**): If $(a,b) \in R$ then $(b,a) \in R$.
(Eq3) (**Transitive law**): If $(a,b) \in R$ and $(b,c) \in R$ then $(a,c) \in R$.

For example, the relation 'less than' on the set $\{1, 2, 3\}$ satisfies (Eq3)—the only possible choice of a, b, c with $(a, b), (b, c) \in R$ is $a = 1$, $b = 2$, $c = 3$, and indeed $(1, 3) \in R$—but neither (Eq1) nor (Eq2). The relation 'less than or equal' on the set $\{1, 2, 3\}$ satisfies (Eq1) and (Eq3) but not (Eq2).

A very important type of relation consists of equivalence relations:

Definition An **equivalence relation** on A is a binary relation satisfying (Eq1), (Eq2), and (Eq3); that is, one which is reflexive, symmetric, and transitive.

If R is an equivalence relation, we define the **equivalence class** of the element $a \in A$ to be $\{b \in A : (a, b) \in R\}$, the set of all elements related to a by R.

Example Let R be the relation 'congruent mod 4' on the set of integers. That is,

$$R = \{(a, b) : a, b \in \mathbb{Z}, a - b = 4x \text{ for some } x \in \mathbb{Z}\}.$$

Now R is

- reflexive, since $a - a = 4 \cdot 0$;
- symmetric, since if $a - b = 4x$ then $b - a = 4(-x)$;
- transitive, since if $a - b = 4x$ and $b - c = 4y$ then $a - c = 4(x + y)$.

So R is an equivalence relation. Among its equivalence classes, we find

$$E(0) = \{\ldots, -8, -4, 0, 4, 8, \ldots\},$$
$$E(1) = \{\ldots, -7, -3, 1, 5, 9, \ldots\},$$
$$E(2) = \{\ldots, -6, -2, 2, 6, 10, \ldots\},$$
$$E(3) = \{\ldots, -5, -1, 3, 7, 11, \ldots\}.$$

Note that these four classes cover all the integers without any overlap. Moreover, $E(4) = E(0)$, and so on; no new classes are obtained. We will see that these are characteristic properties of an equivalence relation.

Definition A **partition** of a set A is a set $\{A_1, A_2, \ldots\}$ of subsets of A such that

(a) $A_i \neq \emptyset$ for all i;
(b) each element of A lies in exactly one of the sets A_i—in other words,
    (b1) $A_1 \cup A_2 \cup \cdots = A$, and
    (b2) $A_i \cap A_j = \emptyset$ for $i \neq j$.

So, in the above example, $\{E(0), E(1), E(2), E(3)\}$ is a partition of $\mathbb{Z}$.

Theorem 1.19 (Equivalence Relation Theorem) *(a) Let R be an equivalence relation on a set A. Then the set of equivalence classes is a partition of A.*

(b) Conversely, let $\{A_1, A_2, \ldots\}$ be a partition of A. Then there is an equivalence relation on A whose equivalence classes are $A_1, A_2, \ldots$.

In other words, equivalence relations and partitions are the same thing.

Proof (a) Let R be an equivalence relation on A. We have to show that the equivalence classes form a partition (satisfying conditions (a) and (b) of the definition). So take $a \in A$. Then $(a, a) \in R$ by (Eq1), so $a \in E(a)$ by definition of $E(a)$. This has two consequences. First, $E(a)$ is non-empty (since at least it contains a). Second, since a was arbitrary, every element of A is in some equivalence class. It remains to show that different classes don't overlap. So suppose that the element a lies in two classes $E(b)$ and $E(c)$. Then $(b, a), (c, a) \in R$. By (Eq2), $(a, c) \in R$; then by (Eq3), $(b, c) \in R$, and by (Eq2) again, $(c, b) \in R$. We want to show that $E(b) = E(c)$; by the test for equality of sets, we must show that every element of $E(b)$ is in $E(c)$ and vice versa. So take $x \in E(b)$. Then $(b, x) \in R$. Since $(c, b) \in R$, (Eq3) gives $(c, x) \in R$, so $x \in E(c)$. The reverse implication is similar.

(b) Now let $\{A_1, A_2, \ldots\}$ be a partition of A. Define a relation R on A by the rule

$$R = \{(a, b) : a, b \in A_i \text{ for some } i\}.$$

This relation R is

- *reflexive*: for, given $a \in A$, some set A_i contains a, and so $a, a \in A_i$, whence $(a, a) \in R$;
- *symmetric*, trivially;
- *transitive*: for suppose that $(a, b), (b, c) \in R$. Then $a, b \in A_i$ for some i, and $b, c \in A_j$ for some j. But only one set of the partition contains b, so $A_i = A_j$. Then $a, c \in A_i$, and $(a, c) \in R$.

So R is an equivalence relation.

Suppose that $a \in A_i$ (there is exactly one such set A_i for any a). We claim that $E(a) = A_i$. To show this, first take any $x \in E(a)$. Then $(a, x) \in R$, so $a, x \in A_j$ for some j. Since only one set contains a, we must have $A_j = A_i$, and $x \in A_i$. Conversely, if $x \in A_i$, then $a, x \in A_i$, and so $(a, x) \in R$, whence $x \in E(a)$. The claim is proved.

In fact, more is true. If we take an equivalence relation R, apply the construction of (a) to obtain a partition, and then apply the construction of (b), the equivalence relation that we obtain is just R. □

We will later meet the term 'canonical form'. If R is an equivalence relation on a set A, a **canonical form** is just a choice of one element from an equivalence class of R. So we can say that every element of A is equivalent to a unique element 'in canonical form'. In the example before the Equivalence Relation Theorem, the elements $\{0, 1, 2, 3\}$ can be taken as canonical forms. The term also carries a suggestion that the canonical forms are in some way 'simpler' or 'more natural' than arbitrary elements of A.

We now show how equivalence relations are important in the study of functions which are not necessarily one-to-one or onto. Let $F : A \to B$ be such a

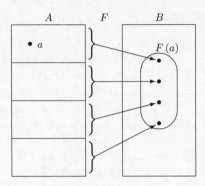

Fig. 1.7 A function

function. We defined the image of F written $\text{Im}(F)$, as the subset of B consisting of values of F:

$$\text{Im}(F) = \{b \in B : b = F(a) \text{ for some } a \in A\}.$$

Thus

- F is onto if and only if $\text{Im}(F) = B$.

The **kernel** of F is the relation R on A in which two elements of A are related if they map to the same element of B:

$$\text{KER}(F) = \{(a_1, a_2) : a_1, a_2 \in A, F(a_1) = F(a_2)\}.$$

Then it is easy to show that

- $\text{KER}(F)$ is an equivalence relation on A, which is equal to the relation of equality if and only if F is one-to-one.

By the Equivalence Relation Theorem 1.19, we can think of $\text{KER}(F)$ as either an equivalence relation or a partition. Later, we will meet a different kind of kernel (which we will write as $\text{Ker}(F)$) which is a subset (rather than a partition) of A; and we will stop to examine the relationship. Only very special functions have kernels in the second sense!

In Figure 1.7, the 'blocks' on the left-hand side are the equivalence classes of $\text{KER}(F)$, and the 'oval' on the right is $\text{Im}(F)$.

Let us look at an example. Let A and B both be the set $\{1, 2, 3, 4, 5\}$, and let F be the function given by the formula $F(x) = x^2 - 6x + 10$. (We do not insist that a function should be given by a formula; but if it is, then this is a convenient way to specify it.) Then, as a set of ordered pairs (a subset of $A \times B$), we have

$$F = \{(1, 5), (2, 2), (3, 1), (4, 2), (5, 5)\}.$$

Note that, as required, every member of A occurs as the first component in exactly one of these pairs. We have

$$\text{Im}(F) = \{1, 2, 5\},$$

and the equivalence classes of $\text{KER}(F)$ are the three sets

$$\{1, 5\}, \{2, 4\}, \{3\}.$$

1.19 Permutations and combinations. The term 'permutation' has two different meanings in mathematical language, neither of which is the same as its everyday meaning in the world of football pools.

In the nineteenth century, a permutation of a finite set $S = \{a_1, \ldots, a_n\}$ meant an arrangement of the elements of the set in order. For example, if $n = 5$ and $S = \{1, 2, 3, 4, 5\}$, then $(3, 5, 4, 1, 2)$ is a permutation.

There is a 'natural' order on this particular set, namely $(1, 2, 3, 4, 5)$. Now the permutation can be produced by a 'substitution', the function f from the set S to itself which takes 1 to 3, 2 to 5, 3 to 4, and so on. Such a substitution is one-to-one and onto (that is, bijective); and conversely, every bijective function from S to itself defines a permutation. So we can regard the function f and the n-tuple as different aspects of the same thing.

So we make a definition:

Definition A **permutation** of a set S to be a bijective function from S to S. If $S = \{1, 2, \ldots, n\}$, then the n-tuple $(f(1), f(2), \ldots, f(n))$ is called the **passive form** of the permutation.

We can write a permutation in so-called **two-line notation** by writing the elements of the domain S in the first line and their images in the second. In our example above, the two-line notation for f is

$$\begin{pmatrix} 1 & 2 & 3 & 4 & 5 \\ 3 & 5 & 4 & 1 & 2 \end{pmatrix},$$

so that the function f maps each element in the first line to the element immediately below it. In this notation, it is not compulsory to write the first line in the usual order. We could, for example, write the same permutation as

$$\begin{pmatrix} 2 & 5 & 1 & 3 & 4 \\ 5 & 2 & 3 & 4 & 1 \end{pmatrix}.$$

If the first row is in its usual order, then the second row is the passive form of the permutation.

How many permutations of $\{1, \ldots, n\}$ are there? We can answer this by counting the passive forms. The first element $f(1)$ may be any one of $\{1, \ldots, n\}$, so there are n choices. The second element cannot be equal to the first (since f is one-to-one), so there are $n - 1$ choices. Similarly there are $n - 2$ choices for the third, and so on. Finally, the last element must be the only one not yet used. So the total number of permutations is

$$n \cdot (n-1) \cdot (n-2) \cdots 2 \cdot 1,$$

the product of all natural numbers from 1 to n inclusive. This number is called n **factorial**, and written as $n!$.

For convenience in later formulae, we extend the definition of the factorial function by setting $0! = 1$.

The notion of permutation is sometimes extended. For $k \leq n$, we define a k-**permutation** of $S = \{1, \ldots, n\}$ to be a k-tuple $(a_1, \ldots, a_k)$ whose members are distinct elements from S. (This could be regarded as the passive form of a one-to-one function from $\{1, \ldots, k\}$ to $\{1, \ldots, n\}$.) The number of k-permutations is, by a similar argument,

$$n \cdot (n-1) \cdots (n-k+1)$$

(where there are k terms in the product); this number is denoted by ${}^nP_k$. note that

$$^nP_k = \frac{n!}{(n-k)!},$$

and in particular ${}^nP_n = n!$.

The essential property of permutations is that 'order is important'; a k-permutation of S is an ordered selection of k distinct elements of S. If we do not care about the order in which the objects are selected, we obtain a k-**combination** of S. Thus, a k-combination is just a subset containing k elements.

The number of k-combinations of an n-element set is denoted by ${}^nC_k$, or, more commonly, by $\binom{n}{k}$. Since each k-combination can be ordered in $k!$ ways, we have

$$\binom{n}{k} = \frac{{}^nP_k}{k!} = \frac{n!}{k!\,(n-k)!}.$$

If $k = 0$ or $k = n$, there is only one k-combination (the empty set or the whole of S, respectively). Because of our convention that $0! = 1$, the formula does give the right answer in this case.

The numbers $\binom{n}{k}$ are usually called **binomial coefficients**. This is because they occur as coefficients in the **Binomial Theorem**:

Theorem 1.20 (Binomial Theorem) *For any natural number n,*

$$(x+y)^n = \sum_{k=0}^{n} \binom{n}{k} x^k y^{n-k}.$$

Proof We have

$$(x + y)^n = (x + y)(x + y) \cdots (x + y) \ (n \text{ factors}).$$

When the brackets are all expanded, we obtain a sum of many terms, each of the form $x^k y^{n-k}$ for some k: such a term is obtained by multiplying together xs chosen from k of the brackets and ys from the remaining brackets. The number of ways we can choose k brackets to pick the xs from (and hence the coefficient of $x^k y^{n-k}$) is $\binom{n}{k}$. $\square$

1.20 More on permutations. There is another convenient way to represent a permutation, namely **cycle notation**. Let f be a permutation of $S = \{1, \dots, n\}$. Choose any point of S (let us say a_0), and follow what happens to it as we apply f repeatedly, Since S is finite, we must eventually come back to an element we have met before. Now this must be our starting point a_0. For let us suppose that this procedure generates successively $a_0, a_1, \dots, a_{r-1}$ which are all distinct, and then a_r which is equal to some a_s with $s < r$. If $s > 0$ then we have

$$f(a_{r-1}) = a_r = a_s = f(a_{s-1});$$

since f is one-to-one, we must have $a_{r-1} = a_{s-1}$, which contradicts our choice of a_r as the first repeat. So $a_r = a_0$. Now the r-tuple $(a_0, a_1, a_2, \dots, a_{r-1})$ is called a **cycle** of f. The notation tells us that f maps each point of the cycle except the last to the next one along, while the last point comes back to the first.

If $f(a_0) = a_0$, then we have a cycle with just one element, namely (a_0).

If every element of S occurs in the cycle, we are finished. Otherwise, choose a point which has not yet been used, and repeat the procedure. A similar argument shows that the cycle we generate has no elements in common with the previous one. Continue like this until we are finished. Then we simply juxtapose the cycles to obtain the cycle notation for f.

Here is an example. Let f be the function which maps $1 \mapsto 4$, $2 \mapsto 7$, $3 \mapsto 3$, $4 \mapsto 8$, $5 \mapsto 1$, $6 \mapsto 5$, $7 \mapsto 2$, and $8 \mapsto 6$. This is a permutation: in two-line notation it is

$$f = \begin{pmatrix} 1 & 2 & 3 & 4 & 5 & 6 & 7 & 8 \\ 4 & 7 & 3 & 8 & 1 & 5 & 2 & 6 \end{pmatrix}.$$

Start with the first element, 1. Follow its successive images under f until it returns to its starting point:

$$f : 1 \mapsto 4 \mapsto 8 \mapsto 6 \mapsto 5 \mapsto 1.$$

This gives us a cycle $(1, 4, 8, 6, 5)$.

If this cycle contains all the elements of the set $\{1, \dots, n\}$, then stop. Otherwise, choose the smallest unused element (in this case 2, and repeat the procedure:

$$f : 2 \mapsto 7 \mapsto 2,$$

so we have a cycle $(2,7)$ disjoint from the first.

We are still not finished, since we have not seen the element 3 yet. Now $f : 3 \mapsto 3$, so (3) is a cycle with a single element. Now we have the cycle decomposition:

$$f = (1,4,8,6,5)(2,7)(3).$$

By convention, in the cycle notation, we usually omit any cycle containing just one element. Thus the permutation $(1,4,2)$ of $\{1,\ldots,5\}$ fixes 3 and 5. If every point is fixed, this convention would tell us to write nothing: instead, in this case, we write (1).

Associated with any permutation f is a number, either $+1$ or -1, called the **sign** of f. It is defined to be $(-1)^{n-c(f)}$, where $c(f)$ is the number of cycles of f in cycle notation (including cycles with just one element). For example, the sign of $(1,3,4)(2,5)$ is $(-1)^{5-2} = -1$, while the sign of $(1,4,2)$ is $(-1)^{5-3} = +1$ (don't forget the two cycles (3) and (5)). We denote the sign of the permutation f by $\text{sign}(f)$. We sometimes use instead the **parity** of f, which is defined to be the parity (even or odd) of $n - c(f)$. Thus even parity is the same as sign $+1$, and odd parity is the same as sign -1.

Another parameter of a permutation f is its **order**, defined to be the smallest positive number m such that, if f is applied m times, then every element of S returns to its original position. Now if f has a cycle of length r, then the elements of this cycle return to their original positions whenever the number of applications of f is a multiple of r (and only then). So every point returns to its starting position if and only if the number of applications is a multiple of every cycle length. We conclude that the order of a permutation is the least common multiple of its cycle lengths.

For example, the order of $(1,3,4)(2,5)$ is 6.

We will say more about permutations in Chapter 3.

Exercise 1.27 Let A and B be sets. Prove that $A = B$ if and only if both $A \subseteq B$ and $B \subseteq A$ hold.

Exercise 1.28 (a) Prove Proposition 1.15.

(b) Prove by induction that, if A is a finite set, then $|A^n| = |A|^n$ for all positive integers n.

Exercise 1.29 Let A be a set with m elements. Then each of the following two sets contains m^n elements:

- the set A^n;
- the set of functions from $\{1,2,\ldots,n\}$ to A.

Find a bijection between these two sets.

Exercise 1.30 How many functions are there from $\{1, 2, 3\}$ to $\{1, 2, 3, 4\}$? How many of these are (a) one-to-one, or (b) onto? Repeat this question for functions from $\{1, 2, 3, 4\}$ to $\{1, 2, 3\}$.

Exercise 1.31 Each of the following attempts *fails* to define a function on the specified set. Explain why it fails, and how you could make the definition into a function.

(a) $F : \mathbb{R} \to \mathbb{R}$, $F(x) = 1/x$.
(b) $F : \mathbb{C}^2 \to \mathbb{C}^2$, $F(a, b) = (c, d)$ if c and d are the roots of the quadratic

$$x^2 + ax + b = 0.$$

Exercise 1.32 For each of the eight combinations of (Eq1), (Eq2), and (Eq3), find a relation on the set $\{1, 2, 3\}$ which satisfies precisely that combination of axioms.
    Can this be done on the set $\{1, 2\}$?

Exercise 1.33 Does the Equivalence Relation Theorem hold if A is the empty set? [How many equivalence relations are there on $\emptyset$? How many partitions of the empty set are there?]

Exercise 1.34 (a) Show that $A \times A$ is an equivalence relation on A.
    (b) Show that $\{(a, a) : a \in A\}$ is an equivalence relation on A. (This is the **relation of equality**.)

Exercise 1.35 Show that there are exactly five equivalence relations on a set of three points. How many are there on a set of four points?

Exercise 1.36 In each of the following cases, state whether the relation $\sim$ on the set X is (i) reflexive, (ii) symmetric, or (iii) transitive:

(a) X is the set of positive integers, $x \sim y$ if x divides y.
(b) X is the set of countries of Europe, $x \sim y$ if x and y have a common border.
(c) X is the set of capital cities of Europe, $x \sim y$ if it is possible to travel from x to y by train.
(d) X is the set of integers, $x \sim y$ if $x \leq y$.
(e) X is the set of integers, $x \sim y$ if $x - y$ is divisible by 4.

In those cases where the relation is an equivalence, describe its equivalence classes.

Exercise 1.37 Let $F : A \to B$ be a function. Show that F induces a bijection between the set of equivalence classes of the kernel $\mathrm{KER}(F)$ and the set $\mathrm{Im}(F)$.

Exercise 1.38 Consider the following argument:

False proposition If a relation is symmetric and transitive then it is reflexive.

Proof Let R be a symmetric and transitive relation. Take $(x, y) \in R$. Then $(y, x) \in R$ (since R is symmetric), and so $(x, x) \in R$ (since R is transitive; put $z = x$ in the transitive law). So R is reflexive. $\qquad\square$

(a) Say what is wrong with this argument.
(b) Give a counterexample to the false proposition.

Exercise 1.39 (∗) Let X be a set, and let $\sim$ be a relation on X which is reflexive and transitive. Write $x \equiv y$ to mean that both $x \sim y$ and $y \sim x$ hold.

(a) Prove that $\equiv$ is an equivalence relation.
(b) Suppose that $x \sim y$, and suppose that x_1 and y_1 belong to the equivalence classes of x and y respectively. Prove that $x_1 \sim y_1$.

Exercise 1.40 (∗) Show that ordered pairs can be defined by the rule

$$(a, b) = \{\{a\}, \{a, b\}\}.$$

[You are asked to show that $\{\{a\}, \{a, b\}\} = \{\{c\}, \{c, d\}\}$ if and only if $a = c$ and $b = d$. Be sure to cover all cases in your argument.]

Exercise 1.41 Recall that, formally, a function is a set of ordered pairs. How many functions are there from the empty set to the empty set? How many of them are one-to-one? How many are permutations?

Exercise 1.42 Write down the orders of all the permutations of the set $\{1, \ldots, 5\}$. (You should not attempt to write down all the permutations and find the order of each one!)

Exercise 1.43 Show that a permutation which has odd order must be an even permutation.
    Is the converse true?

Modular Arithmetic

In this section we define the arithmetic of 'integers mod m' for any positive integer m. First, we look at Euclid's Algorithm.

1.21   Euclid's Algorithm.

Definition   The **greatest common divisor**, or **g.c.d.**, of two positive integers m and n is the largest positive integer which divides both. We write it as $\gcd(m, n)$.

Thus, $\gcd(12, 18) = 6$.
    We can extend the notion of greatest common divisor to the case where one of the integers is equal to 0. Since any positive integer divides 0, we see that $\gcd(m, 0) = m$ if $m \neq 0$. If both m and n are zero, then $\gcd(0, 0)$ is undefined (according to our definition above), so we adopt the convention that $\gcd(0, 0) = 0$.
    Euclid gave a rule for finding the greatest common divisor of two natural numbers, based on the division algorithm.

Theorem 1.21   *Let m and n be natural numbers. Then*

$$\gcd(m, n) = \begin{cases} m & \text{if } n = 0, \\ \gcd(n, r) & \text{if } m = nq + r \text{ with } 0 \leq r < n. \end{cases}$$

Proof The first statement is true by definition. For the second, suppose that $m = nq + r$. Then also $r = m - nq$. So any integer which divides m and n also divides n and r; and *vice versa*; so the greatest common divisor of m and n is equal to the greatest common divisor of n and r. □

This is all very well, but the second line simply replaces the calculation of one g.c.d. by another; why does this help? Notice that $r < n$. This means that, if we apply Euclid's Theorem repeatedly, the second number of the pair of numbers whose g.c.d. we are finding gets smaller at each step. This cannot go on indefinitely; sooner or later, it becomes zero, and the first line applies. So Euclid's Theorem gives us a constructive method to calculate the g.c.d. of two natural numbers. We refer to it as **Euclid's Algorithm**. (An 'algorithm' is just a constructive method, like a recipe or a set of directions, for achieving some result.) Here it is more formally

> To find $\gcd(m, n)$:
> Put $a_0 = m$ and $a_1 = n$.
> As long as the last number a_k found is non-zero, put a_{k+1} equal to the remainder when a_{k-1} is divided by a_k.
> When the last number a_k is zero, then the g.c.d. is a_{k-1}.

An example should make it clear.

Example Find $\gcd(198, 78)$.
$a_0 = 198$, $a_1 = 78$.
$198 = 2 \cdot 78 + 42$, so $a_2 = 42$.
$78 = 1 \cdot 42 + 36$, so $a_3 = 36$.
$42 = 1 \cdot 36 + 6$, so $a_4 = 6$.
$36 = 6 \cdot 6 + 0$, so $a_5 = 0$.
So $\gcd(198, 78) = 6$.

Euclid's Algorithm actually does more than this. It expresses the greatest common divisor of m and n in terms of the original numbers.

Theorem 1.22 *For any two natural numbers m and n, there exist integers x and y such that $\gcd(m, n) = xm + yn$.*

I will not give a proof here: we will see this in much greater generality in the next chapter. But here is an example to show how it works. Refer to the preceding example showing that $\gcd(198, 78) = 6$.

Example

$$6 = 42 - 36$$
$$= 42 - (78 - 42) = 2 \cdot 42 - 78$$
$$= 2(198 - 2 \cdot 78) - 78 = 2 \cdot 198 - 5 \cdot 78,$$

so $x = 2$, $y = -5$.

1.22   The integers mod m.

Definition   Let m be a positive integer. We define a relation on the set $\mathbb{Z}$, called 'congruence modulo m' and written $\equiv_m$, by the rule

$$x \equiv_m y \Leftrightarrow m \mid (y - x).$$

Often, instead of $a \equiv_m b$, we write $x \equiv y \pmod{m}$. The meaning is exactly the same.

Proposition 1.23   *The relation $\equiv_m$ is an equivalence relation on $\mathbb{Z}$ with m equivalence classes. The numbers $0, 1, \ldots, m - 1$ are representatives of the equivalence classes.*

The proof is left as an exercise.

Now we denote the set of equivalence classes by $\mathbb{Z}_m$; this is a set with m elements. We denote the equivalence class containing the integer x by $[x]_m$. So we can write

$$\mathbb{Z}_m = \{[0]_m, [1]_m, \ldots, [m - 1]_m\}.$$

Sometimes we will be lazy and just write x instead of $[x]_m$.

Now we can do arithmetic with the elements of $\mathbb{Z}_m$: we add and multiply them by the rules

$$[x]_m + [y]_m = [x + y]_m, \qquad [x]_m \cdot [y]_m = [xy]_m.$$

There is a problem with these definitions. Since $[x]_m$ means 'the equivalence class containing x', you would be within your rights to use different representatives for the two equivalence classes. Then adding and multiplying them will give different representatives for the classes $[x]_m + [y]_m$ and $[x]_m \cdot [y]_m$. Is it possible that we could actually arrive at different classes? If so, our definitions would be no good! In fact, this is not possible. Here is the argument for multiplication; try addition for yourself.

Suppose that $[x]_m = [x']_m$ and $[y]_m = [y']_m$. Then, by definition, $x \equiv_m x'$ and $y \equiv_m y'$, so $x' = x + um$ and $y' = y + vm$ for some integers u and v. But then

$$x'y' = (x + um)(y + vm) = xy + (xy + vx + muv)m,$$

so $xy \equiv_m x'y'$, and $[xy]_m = [x'y']_m$, as required.

Here, for example, are the addition and multiplication tables of $\mathbb{Z}_4$. We write x instead of $[x]_4$ in these tables, and we use the representatives $0, 1, 2, 3$ for the equivalence classes.

+	0	1	2	3
0	0	1	2	3
1	1	2	3	0
2	2	3	0	1
3	3	0	1	2

·	0	1	2	3
0	0	0	0	0
1	0	1	2	3
2	0	2	0	2
3	0	3	2	1

Subtraction is always possible in $\mathbb{Z}_m$, but division is not. We say that y is the inverse of x mod m if $[x]_m[y]_m = [1]_m$. (If $[x]_m$ has an inverse, then we can divide by it simply by multiplying by its inverse.)

Theorem 1.24 *In $\mathbb{Z}_m$, the element $[x]_m$ has an inverse if and only if* $\gcd(x, m) = 1$.

Proof If $[x]_m$ has an inverse $[y]_m$, then $xy \equiv_m 1$, so $xy = 1 + um$. Let $d = \gcd(x, m)$. Then $d \mid x$ and $d \mid m$, so $d \mid xy - um = 1$; thus we must have $d = 1$.

Conversely, suppose that $\gcd(x, m) = 1$. Now Euclid's Algorithm gives us numbers y and v such that $xy + vm = 1$; hence $xy \equiv_m 1$, or $[x]_m[y]_m = [1]_m$. □

Now calculations can be done in $\mathbb{Z}_m$ as if the elements were ordinary numbers (but remembering that x and y represent the same element if they are congruent mod m).

Example Find $\dfrac{2}{3} + \dfrac{3}{5}$ in $\mathbb{Z}_{13}$.

First method Find the inverses of 3 and 5 mod 13:

$$3 \cdot 9 \equiv_{13} 1, \text{ so } 1/3 = 9;$$

$$5 \cdot 8 \equiv_{13} 1, \text{ so } 1/5 = 8.$$

(How did I find these? Either by trial and error, or by the method based on Euclid's Algorithm explained in the last subsection.)

Hence $2/3 + 3/5 = 2 \cdot 9 + 3 \cdot 8 = 42 = 3$, or $[3]_{13}$ to be more accurate.

Second method

$$\frac{2}{3} + \frac{3}{5} = \frac{2 \cdot 5 + 3 \cdot 3}{2 \cdot 3} = \frac{19}{15} = \frac{6}{2} = 3.$$

Exercise 1.44 Find all solutions of the equation $x^2 = 2$ (a) in $\mathbb{Z}_{17}$; (b) in $\mathbb{Z}_{19}$.

Exercise 1.45 (a) Let p be a prime number. Prove that the binomial coefficient $\dbinom{p}{k}$ is a multiple of p for $1 \leq k \leq p - 1$.

(b) Use the Binomial Theorem and induction on n to show that, if p is a prime number, then $n^p \equiv_p n$ for all natural numbers n. [*Hint*: Expand $(n+1)^p$ by the Binomial Theorem; all terms except the first and the last are divisible by p.]

Remark This result is known as **Fermat's Little Theorem**.

Exercise 1.46 (∗) In this question we look at the converse of Fermat's Little Theorem.

(a) Show that $2^{341} \equiv_{341} 2$, but 341 is not a prime. (The number 341 is called a **pseudoprime to base** 2.)
(b) Show that $a^{561} \equiv_{561} a$ for any integer a, but 561 is not a prime. (The number 561 is called a **Carmichael number**, after Robert D. Carmichael, who first studied numbers with this property.) *Hint*: Show that
   - if p is prime and $m = k(p-1)+1$, then $a^m \equiv_p a$;
   - if m is a product of distinct prime numbers then $a \equiv_m b$ holds if and only if $a \equiv_p b$ for each prime number dividing m.
(c) Can you find any more Carmichael numbers?

Matrices

Another familiar system of objects which make up part of the subject-matter of Algebra consists of matrices.

Anyone who has used a spreadsheet program knows the importance of 2-dimensional tables of numbers. A matrix is just such a table.

Definition A **matrix** of size $m \times n$, or an $m \times n$ matrix, is an array of numbers with m rows and n columns.

We denote the entry in row i and column j of the matrix A by $(A)_{ij}$, or often in lower-case form by a_{ij}. So

$$A = \begin{pmatrix} a_{11} & a_{12} & \cdots & a_{1n} \\ a_{21} & a_{22} & \cdots & a_{2n} \\ \cdots & \cdots & \cdots & \cdots \\ a_{m1} & a_{m2} & \cdots & a_{mn} \end{pmatrix}.$$

For small matrices, we may want to avoid writing lots of subscripts by writing, for example, a 2×2 matrix A as

$$A = \begin{pmatrix} a & b \\ c & d \end{pmatrix}.$$

1.23 Matrices and determinants. A matrix is simply a table of numbers. However, if the matrix is square, there is a single number which can be calculated from it (whose theoretical significance we will see later on). This is the **determinant** of the matrix.

Two notations are commonly used. We write the determinant of the matrix A as $\det(A)$ or $\det A$. Alternatively, if the matrix is written out as a table (in round brackets), we denote the determinant by the same table surrounded by vertical bars; thus

$$\det \begin{pmatrix} a & b \\ c & d \end{pmatrix} = \begin{vmatrix} a & b \\ c & d \end{vmatrix}.$$

How is it defined? The rule can be written down easily in small cases:

$$\det \begin{pmatrix} a & b \\ c & d \end{pmatrix} = ad - bc,$$

$$\det \begin{pmatrix} a & b & c \\ d & e & f \\ g & h & i \end{pmatrix} = aei + bfg + cdh - ceg - bdi - afh.$$

The process can be described as follows. We form all the terms which can be made by choosing one element from each row and column of the matrix. Some of these are given minus signs, and the results added up. Which terms get which signs?

For 2×2 matrices, the term on the north-west to south-east diagonal is given a $+$ sign and the term on the north-east to south-west diagonal has a $-$ sign. The same rule holds in the 3×3 case if we imagine the matrix as a tile which is repeated. One negative term is highlighted.

$$
\begin{array}{ccccc}
a & b & c & a & b & \cdots \\
d & e & f & \underline{d} & e \\
g & h & \underline{i} & g & h \\
a & \underline{b} & c & a & b & \cdots \\
d & e & f & d & e
\end{array}
$$

$$\vdots \qquad \qquad \ddots$$

This rule does not extend to larger matrices, however. The correct rule uses the notion of the sign of a permutation, which we defined in the preceding section. Notice that any term obtained by choosing one element from each row and one from each column can be specified by a function f: we choose the element from row k and column $f(k)$, for $k = 1, \ldots, n$. This function must be a permutation, and so has a sign, which we denote by $\operatorname{sign}(f)$. Now this sign is affixed to the term in the expansion.

Thus, the general formula is: if $A = (a_{ij})$ is an $n \times n$ matrix, then

$$\det(A) = \sum_{f \in S_n} \operatorname{sign}(f) a_{1\,f(1)} a_{2\,f(2)} \cdots a_{n\,f(n)},$$

where S_n is the set of all permutations of $\{1, \ldots, n\}$. So the number of terms in the sum is $n!$.

In the case $n = 3$, the term bdi is selected by the permutation mapping 1 to 2, 2 to 1, and 3 to 3, that is, $(1, 2)(3)$ in cycle notation; its sign is $(-1)^{3-2} = -1$, as indeed we saw.

1.24   Addition and multiplication.   As mentioned earlier, matrices can be added and multiplied. Here are the rules for 2×2 matrices. Let A and B be 2×2 matrices: say

$$A = \begin{pmatrix} a & b \\ c & d \end{pmatrix},$$

$$B = \begin{pmatrix} a' & b' \\ c' & d' \end{pmatrix}.$$

We define addition and multiplication by the rules

$$A + B = \begin{pmatrix} a + a' & b + b' \\ c + c' & d + d' \end{pmatrix},$$

$$AB = \begin{pmatrix} aa' + bc' & ab' + bd' \\ ca' + dc' & cb' + dd' \end{pmatrix}.$$

The rule for addition is straightforward: just add the entries in corresponding positions in the two matrices. The rule for multiplication seems to have little rhyme or reason to it. You may well have met this rule in geometry and seen how it arises from geometric transformations. If not, you will have to wait until Chapter 4 for an algebraic representation. But you should not try to memorise the rule; it is easy to explain how it works. To find a particular entry in AB, for example the entry in the first row and second column, we look at the first row of A (containing entries a and b) and the second column of B (with entries b' and d'). Now multiply each entry in the chosen row of A by the corresponding entry in the chosen column of B (giving ab' and bd'), and add these to obtain the entry of AB in the required position.

For example:

$$\rightarrow \begin{pmatrix} 1 & 2 \\ 3 & 4 \end{pmatrix} \begin{pmatrix} 5 & 6 \\ 7 & 8 \end{pmatrix} \overset{\downarrow}{=} \begin{pmatrix} & 22 \\ & \end{pmatrix}$$

$$1 \cdot 6 + 2 \cdot 8 = 22.$$

Now it is possible to show that most of the now-familiar properties hold for matrices: addition is commutative and associative, multiplication is associative, and the distributive law holds. But there is a surprise: multiplication is not commutative. Take the matrices in the above example: let

$$A = \begin{pmatrix} 1 & 2 \\ 3 & 4 \end{pmatrix}, \qquad B = \begin{pmatrix} 5 & 6 \\ 7 & 8 \end{pmatrix}.$$

Now check that

$$AB = \begin{pmatrix} 19 & 22 \\ 43 & 50 \end{pmatrix}, \qquad BA = \begin{pmatrix} 23 & 34 \\ 31 & 46 \end{pmatrix}.$$

There is nothing special about these two matrices; they happened to be the first pair that I thought up. It is quite rare for two matrices to commute!

Now we could do as we did for polynomials, and examine the proofs of those laws that do hold to see what properties of the matrix entries are being used. We find that these are just the same properties as we are proving for the matrices (commutativity and associativity for addition, associativity for multiplication, distributive laws). In this way, we prove general theorems for matrices with entries from a whole range of number systems.

For completeness, we define addition and multiplication for matrices of arbitrary size. Let A and B be matrices; let A have (i, j) entry a_{ij} (this is shorthand for the entry in the ith row and jth column), and let B have (i, j) entry b_{ij}.

The (i, j) entry of $A + B$ is just the sum $a_{ij} + b_{ij}$ of the corresponding entries of A and B. In order for this to make sense, we need that for each entry of A there is an entry in the same position in B and vice versa; that is, A and B must have the same size. Hence

> $A + B$ is defined if and only if A and B are both $m \times n$ matrices, for some m and n; then $A + B$ is also $m \times n$.

The (i, j) entry of AB is worked out as we described above. We select the ith row of A and the jth column of B; then we multiply each element a_{ik} of the first by the corresponding element b_{kj} of the second, and add all these terms, obtaining the formula

$$\sum_k a_{ik} b_{kj}$$

for the (i, j) entry of AB. For this to work, the number of elements in each row of A (which is the number of columns of A) must be equal to the number of elements in each column of B (which is the number of rows of B). So we have

> AB is defined if and only if A is $m \times n$ and B is $n \times p$ for some m, n, p; then AB is $m \times p$.

(And, in the above sum, k runs from 1 to n.)

In particular, if A and B are both square matrices (this means that they have the same number of rows as columns) of the same size, then both the sum $A + B$ and the product AB are defined. We will see later that, in this case,

$$\det(AB) = \det(A) \cdot \det(B).$$

However, it is not true that $\det(A + B) = \det(A) + \det(B)$.

1.25 Linear equations. The theory of matrices grew from the problem of solving equations, specifically linear equations in several variables. Such a system, involving m equations in n variables, can be written in the form

$$\begin{aligned} a_{11}x_1 + a_{12}x_2 + \cdots + a_{1n}x_n &= b_1, \\ a_{21}x_1 + a_{22}x_2 + \cdots + a_{2n}x_n &= b_2, \\ &\vdots \\ a_{m1}x_1 + a_{m2}x_2 + \cdots + a_{mn}x_n &= b_m. \end{aligned}$$

Such a system gives us a **matrix of coefficients**, the table of numbers a_{ij}:

$$A = \begin{pmatrix} a_{11} & a_{12} & \cdots & a_{1n} \\ a_{21} & a_{22} & \cdots & a_{2n} \\ \vdots & & & \\ a_{m1} & a_{m2} & \cdots & a_{mn} \end{pmatrix}.$$

Such a system of equations may have no solution, or a unique solution, or more than one solution. Matrix theory allows us to determine which possibility occurs and gives us tools to calculate the solutions if they exist. Here, without proof, is part of the answer for the case where the numbers of equations and variables are equal. Suppose that the system of equations is

$$\begin{aligned} a_{11}x_1 + a_{12}x_2 + \cdots + a_{1n}x_n &= b_1, \\ a_{21}x_1 + a_{22}x_2 + \cdots + a_{2n}x_n &= b_2, \\ &\vdots \\ a_{n1}x_1 + a_{n2}x_2 + \cdots + a_{nn}x_n &= b_n. \end{aligned}$$

Let A be the matrix of coefficients, and let b be the $1 \times n$ matrix (the column) with entries $b_1, b_2, \ldots, b_n$. Let B_i be the matrix obtained from A by replacing the ith column by the column b.

Theorem 1.25 *With the above notation,*

(a) If $\det(A) \neq 0$, *then the equations have unique solution, given by*

$$x_i = \frac{\det(B_i)}{\det(A)}, \qquad i = 1, \ldots, n.$$

(b) If $\det(A) = 0$, *then either the equations have no solution, or they have more than one solution.*

This result perhaps explains the name of the mysterious 'determinant' function: it determines whether or not the equations have a unique solution. The formula for the solution in part (a) is **Cramer's Rule**.

We will see the proof of this theorem, and give details of how to calculate which possibility occurs, in Chapter 4.

For example, the equations

$$3x + 2y = 22$$
$$4x + 3y = 31$$

have the unique solution

$$x = \frac{\begin{vmatrix} 22 & 2 \\ 31 & 3 \end{vmatrix}}{\begin{vmatrix} 3 & 2 \\ 4 & 3 \end{vmatrix}} = \frac{66 - 62}{9 - 8} = 4,$$

$$y = \frac{\begin{vmatrix} 3 & 22 \\ 4 & 31 \end{vmatrix}}{\begin{vmatrix} 3 & 2 \\ 4 & 3 \end{vmatrix}} = \frac{93 - 88}{9 - 8} = 5.$$

Exercise 1.47 Prove the distributive law for 2×2 matrices. What properties of the matrix entries are used in your proof?

Exercise 1.48 Look at the example of non-commutative matrices in the text. Observe that the sum of the elements on the 'main diagonal' is the same for AB as for BA (that is, $19 + 50 = 23 + 46$). Is this a coincidence?

Exercise 1.49 An $n \times n$ matrix $A = (a_{ij})$ is **upper triangular** if $a_{ij} = 0$ whenever $i > j$. (This means that all elements below the 'main diagonal' $a_{11}, a_{22}, \ldots, a_{nn}$ are zero.) Prove that the sum and product of upper triangular matrices are upper triangular. Is multiplication of upper triangular matrices commutative?

Exercise 1.50 Solve the following equations:

$$\begin{aligned} x + 2y + 3z &= 10, \\ 2x + 5y + 10z &= 26, \\ 3x + 10y + 26z &= 55. \end{aligned}$$

Exercise 1.51 Solve the following equations, for any real number c:

$$\begin{aligned} x + 2y + 3z &= 10, \\ 2x + 5y + 10z &= 26, \\ 3x + 8y + 17z &= c. \end{aligned}$$

Exercise 1.52 Professor Fibonacci buys a pair of newborn rabbits at the beginning of month 0. Assume that rabbits are infertile until they are two months old, from which time each pair produces one pair of offspring every month for ever. Let x_n be the number of pairs of baby rabbits at the start of month n, and y_n the number of pairs of rabbits which are at least one month old. Show that

$$\begin{pmatrix} x_{n+1} \\ y_{n+1} \end{pmatrix} = \begin{pmatrix} 0 & 1 \\ 1 & 1 \end{pmatrix} \begin{pmatrix} x_n \\ y_n \end{pmatrix},$$

and deduce that

$$\begin{pmatrix} x_n \\ y_n \end{pmatrix} = \begin{pmatrix} 0 & 1 \\ 1 & 1 \end{pmatrix}^n \begin{pmatrix} 1 \\ 0 \end{pmatrix}.$$

How many pairs of rabbits does the Professor have after a year?

Appendix: Logic

Reasoning and logic are to each other as health is to medicine, or—better—as conduct is to morality. Reasoning refers to a gamut of natural thought processes in the everyday world. Logic is how we ought to think if objective truth is our goal—and the everyday world is very little concerned with objective truth. Logic is the science of the justification of conclusions we have reached by natural reasoning.

Julian Jaynes (1976).

1.26   Logic and truth tables.   We are usually able to tell without too much trouble whether or not a simple mathematical argument is valid. For example, the argument

All men are mortal; grass is mortal; therefore all men are grass

is not logically valid, although it may express a poetic truth. In more complicated cases, it is good to know that the rules of logic have been codified and can be applied mechanically.

We build up expressions and arguments from basic **propositions** or statements, each of which may be either true (T) or false (F). For the purpose of the argument, it does not matter what these propositions are; the logical validity of an argument should not depend on the meanings of its propositions. We denote propositions by the letters $p, q, r \ldots$

We are allowed to combine propositions with various **connectives**, as follows. For each connective, we give its meaning in words and then in a **truth table**. All we are interested in is which combinations of truth values of the basic propositions make the compound proposition true.

- **Conjunction, 'and'.** The combination 'p and q' is true if both p and q are true, and is false in all other cases. We write 'p and q' as $p \wedge q$ (or sometimes $p \& q$). The truth table is

p	q	$p \wedge q$
T	T	T
T	F	F
F	T	F
F	F	F

- **Disjunction, 'or'.** The combination 'p or q' is true if either p or q is true (or possibly both). This connective is sometimes called 'inclusive or', as distinct from 'exclusive or', which is true only when exactly one of p and q is true. We write 'p or q' as $p \lor q$. The truth table is

p	q	$p \lor q$
T	T	T
T	F	T
F	T	T
F	F	F

- **Negation, 'not'.** The truth value of 'not p' is opposite to that of p: it is false when p is true, and vice versa. Writing 'not p' as $\neg p$, the truth table is

p	$\neg p$
T	F
F	T

- **Implies, 'if ... then ...'.** The truth table for 'p implies q', written $p \Rightarrow q$, is a bit surprising at first. The implication is true in all cases except when p is true and q is false. We return to this at the end of the section. The truth table is

p	q	$p \Rightarrow q$
T	T	T
T	F	F
F	T	T
F	F	T

- **Equivalent, 'if and only if'.** This connective, written $p \Leftrightarrow q$, is true if and only if p and q have the same truth value (both true or both false). The truth table is

p	q	$p \Leftrightarrow q$
T	T	T
T	F	F
F	T	F
F	F	T

Note that $p \Leftrightarrow q$ means the same as $p \Rightarrow q$ and $q \Rightarrow p$.

Every valid rule of logic can be proved using truth tables. For example, consider the technique of 'proof by contradiction'. In order to prove a proposition p, we assume the negation of p, that is, $\neg p$, and deduce a contradiction x (whose truth value is F). In other words, we prove $(\neg p) \Rightarrow x$. It follows from the truth table for implication that, if x is false and $(\neg p) \Rightarrow x$ is true, then $\neg p$ is false; that is, p is true.

Again, consider the 'proof' that all men are grass. We are given $p \Rightarrow r$ and $q \Rightarrow r$, and are supposed to deduce that $p \Rightarrow q$. But if it happens that p and r are true and q is false, then both $p \Rightarrow r$ and $q \Rightarrow r$ are true but $p \Rightarrow q$ is false. So the deduction is not valid.

The rule that if p is false then $p \Rightarrow q$ is true, whatever proposition q may be, is often found perplexing. But it does agree with everyday usage. Suppose that I say to you, 'If it is fine tomorrow, we will go to the Zoo.' If it rains tomorrow then, whether or not we go to the Zoo, I did not lie to you! The only case in which I have lied would be if it is fine and we do not go to the Zoo.

Bertrand Russell was asked about this by a philosopher, who said, 'Is it true that, if $1 + 1 = 3$, then you are the Pope?' Russell improvised the following argument on the spot.

Suppose that $1 + 1 = 3$. Subtracting 1 from each side we obtain $1 = 2$. Now the Pope and I are two; therefore we are one.

1.27   Sets and logic.   An unexpected application of truth tables is to proving complicated identities about sets. Let $A, B, C, \ldots$ be sets, contained in some large set U (the 'universe'), and let p be the proposition asserting that $x \in A$, q the proposition $x \in B$, r the proposition $x \in C$, and so on. Each of $p, q, r, \ldots$ may be true or false, depending on the point x. But, for example, $p \wedge q$ is true if and only if $x \in A \cap B$, so $p \wedge q$ represents the set $A \cap B$. Similarly, $p \vee q$ represents $A \cup B$; $\neg p$ represents $A' = U \setminus A$ (the complement of A in the 'universe'); and other connectives correspond to other combinations. Now suppose that we have to prove an identity, say

$$A \cap (B \cup C) = (A \cap B) \cup (A \cap C).$$

This will follow if we can show that the two propositions $p \wedge (q \vee r)$ and $(p \wedge q) \vee (p \wedge r)$ are equivalent (that is, always have the same truth value). This can be done quite mechanically as follows:

p	q	r	$q \vee r$	$p \wedge (q \vee r)$	$p \wedge q$	$p \wedge r$	$(p \wedge q) \vee (p \wedge r)$
T	T	T	T	T	T	T	T
T	T	F	T	T	T	F	T
T	F	T	T	T	F	T	T
T	F	F	F	F	F	F	F
F	T	T	T	F	F	F	F
F	T	F	T	F	F	F	F
F	F	T	T	F	F	F	F
F	F	F	F	F	F	F	F

The result follows because the entries in the columns labelled $p \wedge (q \vee r)$ and $(p \wedge q) \vee (p \wedge r)$ are identical.

Truth tables thus formalise the arguments using Venn diagrams that we met earlier.

Exercise 1.53 Prove that $(p \Rightarrow q) \vee (q \Rightarrow p)$ is logically valid (that is, true for all combinations of truth values of p and q).

Remark This shows clearly that logical implication does not involve any material connection between propositions!

Exercise 1.54 Explain why the words 'necessary' and 'sufficient' are used as described on page 6.

Exercise 1.55 If p and q are the propositions $x \in A$ and $x \in B$ respectively, show that $x \in A \triangle B$ is the proposition $\neg(p \Leftrightarrow q)$. What set is represented by $p \Rightarrow q$?

Exercise 1.56 Prove, using truth tables, that $(A \cup B)' = A' \cap B'$ for any two sets A and B.

Miscellaneous exercises

Exercise 1.57 Find the least positive integer N such that every integer $n \geq N$ can be written in the form $4a + 7b$ for some choice of integers $a \geq 0$ and $b \geq 0$. Prove that your N has this property.

Exercise 1.58 True or false? Give reasons.

(a) The square of any integer is congruent to 0 or 1 mod 4.
(b) For any natural number $n \geq 3$, $(n^2 + n + 2)/2$ is a prime number.

Exercise 1.59 Let S be the set of all 2×2 real matrices of the form $\begin{pmatrix} \cos\theta & \sin\theta \\ -\sin\theta & \cos\theta \end{pmatrix}$ for real numbers θ.

(a) Let A and B be two matrices in S. Prove that $AB \in S$.
(b) Is multiplication of matrices in S commutative?

Exercise 1.60 Find all complex numbers z satisfying $z^3 = 1$.

Exercise 1.61 State the Binomial Theorem, and use it to evaluate

$$\sum_{r=0}^{n} \binom{n}{r} (-1)^r 2^{n-r}.$$

Exercise 1.62 (∗) (a) Let n be a positive integer with the property that every positive integer $m \leq n/2$ is a divisor of n. Show that n is 1, 2, 3, 4, or 6.
   (∗∗) (b) Let n be a positive integer with the property that every positive integer $m < \sqrt{n}$ is a divisor of n. Show that n is a divisor of 24.

Exercise 1.63 Here is an extract from some lecture notes on Geometry.

> The scalar product is a way of multiplying two vectors to produce a scalar (a real number).
>
> Let $\mathbf{u}, \mathbf{v}$ be non-zero vectors represented by $\vec{AB}$, $\vec{AC}$. We define the *angle* between $\mathbf{u}$ and $\mathbf{v}$ to be the angle θ (in radians), with $0 \le \theta \le \pi$, between $\vec{AB}$ and $\vec{AC}$.
>
> The *scalar product* (or *dot product*) $\mathbf{u} \cdot \mathbf{v}$ is $|\mathbf{u}|\,|\mathbf{v}|\cos\theta$, if $\mathbf{u}$ and $\mathbf{v}$ are non-zero vectors and θ is the angle between them. If $\mathbf{u} = \mathbf{0}$ or $\mathbf{v} = \mathbf{0}$ then we define $\mathbf{u} \cdot \mathbf{v} = 0$.

(a) What is this? Is it a definition, a theorem, a proof, an example, or just some chit-chat?
(b) Your boss asks you to summarise it in a brief bullet point, without using any symbols. What would you say?
(c) Why are the zeros in the last sentence written in different typefaces?

Exercise 1.64 Prove that, for any two positive integers m and n,

$$\gcd(m,n) \cdot \operatorname{lcm}(m,n) = mn,$$

where $\gcd(m,n)$ and $\operatorname{lcm}(m,n)$ are the greatest common divisor and least common multiple of m and n.

Does any similar result hold for three positive integers?

Exercise 1.65 What is the largest possible order of a permutation of $\{1, 2, \ldots, 8\}$? Give an example of such a permutation. Are there any permutations with this order which have odd parity? Are there any which have even parity?

Exercise 1.66 Give an example of a real polynomial of degree 4 which is reducible even though it has no real roots.

2 Rings

Rings and subrings

This chapter is about rings, which are abstract systems in which addition and multiplication are defined. The prototype is the ring $\mathbb{Z}$ of integers, with the ordinary arithmetic operations. This is the example to which we keep coming back: we are interested in seeing how far the familiar properties of integers (as outlined in Chapter 1) can be extended.

2.1 Introduction. A **ring** is a set with two binary operations called **addition** and **multiplication**. We use the same notation for these operations in a general ring as in the integers: addition is represented by an infix $+$, and multiplication by either juxtaposition or an infix $\cdot$. (That is, we denote the sum of a and b by $a + b$, and their product by either ab or $a \cdot b$.)

A ring is defined by a list of axioms, which follows. These are divided into three groups, involving addition, multiplication, and both operations, respectively. These are meant to be familiar properties of $\mathbb{Z}$. I will assume without proof that they all hold in $\mathbb{Z}$.

Axioms for addition
(A0) (**Closure law**): For all $a, b \in R$, $a + b \in R$.
(A1) (**Associative law**): $a + (b + c) = (a + b) + c$ for all $a, b, c \in R$.
(A2) (**Zero law**): There exists $0 \in R$ such that $a + 0 = 0 + a = a$ for all $a \in R$.
(A3) (**Inverse law**): For all $a \in R$, there exists $b \in R$ with $a + b = b + a = 0$.
(A4) (**Commutative law**): $a + b = b + a$ for all $a, b \in R$.
Axioms for multiplication
(M0) (**Closure law**): For all $a, b \in R$, $ab \in R$.
(M1) (**Associative law**): $a(bc) = (ab)c$ for all $a, b, c \in R$.
Mixed axiom
(D) (**Distributive laws**): $(a + b)c = ac + bc$ and $c(a + b) = ca + cb$ for all $a, b, c \in R$.

The two distributive laws are sometimes called the **left** and **right distributive laws** respectively. Please do not try to remember which is which; I will not use these names.

The closure laws (A0) and (M0) are not strictly necessary: when we say that addition and multiplication are operations on R, it follows that the closure laws must hold! We will see the reason for requiring them when we come to look at subrings in Section 2.4.

Just as, when buying a personal computer, you are offered various extra features (more RAM, larger hard disk, etc.), so it is possible to have extra features in your ring if you are prepared to spend a bit more. Some of these are as follows. A ring R is a **ring with identity** if it satisfies

(M2) (**Identity law**): There exists $1 \in R$ with $1 \neq 0$ such that $a1 = 1a = a$ for all $a \in R$.

R is a **division ring** if it satisfies (M2) and also

(M3) (**Inverse law**): For all $a \in R$ with $a \neq 0$, there exists $b \in R$ with $ab = ba = 1$.

R is a **commutative ring** if it satisfies

(M4) (**Commutative law**): $ab = ba$ for all $a \in R$.

Finally, a commutative division ring is called a **field**.

Note that the extra multiplicative axioms are almost exact parallels of the additive axioms that we require in any ring. The exception is that, in the inverse law, we only require that non-zero elements have multiplicative inverses. We will see the reason for this soon.

Many authors make the convention that a ring must have an identity. In other words, they assume axiom (M2) along with (A0)–(A4), (M0), (M1), and (D).

In fact, some go so far as to use the word **rng** (sic) for a structure which I called a ring (that is, satisfying (A0)–(A4), (M0), (M1), and (D), so that they can use the term r**I**ng for 'rng with **I**dentity'.

Of course it is just convention. When different groups of mathematicians use different conventions, you are free to choose the one you like best, but you must accept that other people will do things differently. When we have learned a bit more about rings, I will explain why I took the decision I did on page 79.

The commutative law for addition follows from the other axioms for a ring with identity. The simple argument for this is outlined on page 69. A different proof is outlined in the solution to Exercise 2.7.

Remark   Remember that the qualifying expressions in the terms 'commutative ring' and 'ring with identity' refer to the *multiplication*. The addition in a ring is always commutative, and there is always an identity (or zero) element for addition.

2.2   Examples of rings.

Example 1   Our prototype of a ring is the ring $\mathbb{Z}$ of integers. It is indeed a ring; in fact, it is a commutative ring with identity (but not a field, since, for example, there is no integer x such that $2x = 1$). I assume that all of these properties of integers are familiar to you. To give formal proofs, it is necessary to have a careful definition of the integers. This is done in courses on the Foundations of Mathematics: we will look at the arguments in Chapter 6.

Example 2 Other familiar number systems, such as $\mathbb{Q}$ (the rational numbers), $\mathbb{R}$ (the real numbers), and $\mathbb{C}$ (the complex numbers), are fields.

Example 3: Matrix rings Let R be any ring. Let $M_n(R)$ denote the set of all $n \times n$ matrices with elements in R. We can define addition and multiplication on $M_n(R)$ by rules which look exactly the same as those for matrices of real numbers, which we saw in the last chapter. That is, if $A = (a_{ij})$ and $B = (b_{ij})$ (this means that the element in row i and column j of A is a_{ij}, etc.), then

$$A + B = C = (c_{ij}), \text{ where } c_{ij} = a_{ij} + b_{ij},$$

$$A \cdot B = D = (d_{ij}), \text{ where } d_{ij} = \sum_{k=1}^{n} a_{ik}b_{kj}.$$

(Note that the rule for matrix addition depends on addition of ring elements, while the rule for matrix multiplication involves calculating n products and adding them up. There is a potential problem here, since we can only add two elements at a time: we will see in the next section that it does not matter how we perform the additions.)

It can be shown that $M_n(R)$ is a ring. (See Exercise 2.2 for the case $n = 2$.)

If R has an identity, then so does $M_n(R)$ (the usual identity matrix with 1 on the diagonal and 0 everywhere else). But $M_n(R)$ is not commutative except in trivial cases, and is never a division ring for $n > 1$.

Example 4: Polynomial rings For any ring R, the set $R[x]$ of all polynomials with coefficients in R is a ring. This is a generalisation of the familiar case of real polynomials. We will discuss exactly what a polynomial is, and how addition and multiplication should be defined in general, in Section 2.9.

Example 5: Finite rings A finite ring can be specified by giving operation tables for its addition and multiplication. For obvious reasons, these tables are usually called **addition tables** and **multiplication tables**.

For example, it can be shown that the structure given by

+	0	1		·	0	1
0	0	1		0	0	0
1	1	0		1	0	1

is a field. This can be proved directly, but it takes some work. For example, to verify the associative law (A1) from the tables, we have to substitute all possible values of $a, b,$ and c. There are two possibilities for each of these, so $2^3 = 8$ instances of the law to be checked. Also, of course, eight instances of the associative law for multiplication, four of the commutative law, 16 of the distributive law .... For larger finite rings the situation is even worse. The moral is that, if all else fails, this method can be used; but usually it is better to have a more theoretical proof!

You probably recognised that the ring with the above tables is $\mathbb{Z}_2$, the integers mod 2. In fact, $\mathbb{Z}_m$ is a ring, for any positive integer m. But not all finite rings are of this kind. Here is one which is not.

+	0	1	a	b		$\cdot$	0	1	a	b
0	0	1	a	b		0	0	0	0	0
1	1	0	b	a		1	0	1	a	b
a	a	b	0	1		a	0	a	a	0
b	b	a	1	0		b	0	b	0	b

Example 6: Zero rings   Let R be a set with one binary operation $+$ satisfying axioms (A0)–(A4). (Later, we will see that such a thing is called an **abelian group**.) Is it possible to define a multiplication on R so that it becomes a ring? The answer is yes: it is always possible, by the trivial rule

$$ab = 0 \text{ for all } a, b \in R,$$

where 0 is the zero element given by (A2).

To prove this, we check the remaining axioms:

(M0): For all $a, b \in R$, $ab = 0 \in R$.

(M1): For all $a, b, c \in R$, $(ab)c = 0 = a(bc)$.

(D): For all $a, b, c \in R$, we have $(a + b)c = 0$, while $ac + bc = 0 + 0 = 0$, using property (A2). Similarly the other way round.

A ring constructed in this manner (one in which all products are zero) is called a **zero ring**. Such rings always exist, but they are not very exciting.

Example 7   The set of all even integers is a ring. It is commutative, but does not have an identity. (There is no even integer e such that $ex = x$ for all even integers x.)

Example 8: Boolean rings   Just as Descartes aimed to turn geometry into algebra by setting up coordinates in the Euclidean plane, so Boole attempted to turn set theory (and logic) into algebra, as we see below. The main legacy of his attempt is that his name is familiar to every computer scientist.

Let X be a set, and let R denote the **power set** of X, the set of all subsets of X. (This is sometimes denoted by $\mathcal{P}(X)$.) We define operations on R as follows. For $A, B \subseteq X$, we let $A + B$ be the **symmetric difference** of A and B, the set of all elements lying in either A or B but not both. (This is sometimes written $A \triangle B$.) Also, we let $A \cdot B$ be the intersection $A \cap B$.

Now R is a ring. Let us check the axioms.

(A0): Clear.

(A1): Use a Venn diagram or truth table to show that $(A + B) + C$ and $A + (B + C)$ are both equal to the set of elements which are either in all three of the sets A, B, C or in exactly one of them.

(A2): $A + \emptyset = A$, since nothing is in the empty set; so $\emptyset$ is the zero element.

(A3): $A + A = \emptyset$, since there is no element which lies in A but not in both A and A.(!) So the inverse of A is A.

(A4): Clear.
(M0): Clear.
(M1): $(AB)C$ and $A(BC)$ both consist of the elements lying in all three sets.
(D): Prove this by means of a Venn diagram or truth table.

Now R is a commutative ring. It has an identity, namely the whole set X (since $X \cap A = A$ for any $A \subseteq X$). But it is not a division ring if X has more than one element: the equation $A \cap B = X$ can never hold if A is a proper subset of X.

A ring of this form is called a **Boolean ring.** For example, when $X = \{0, 1\}$, the addition and multiplication tables are as follows:

+	$\emptyset$	$\{0\}$	$\{1\}$	$\{0,1\}$
$\emptyset$	$\emptyset$	$\{0\}$	$\{1\}$	$\{0,1\}$
$\{0\}$	$\{0\}$	$\emptyset$	$\{0,1\}$	$\{1\}$
$\{1\}$	$\{1\}$	$\{0,1\}$	$\emptyset$	$\{0\}$
$\{0,1\}$	$\{0,1\}$	$\{1\}$	$\{0\}$	$\emptyset$

$\cdot$	$\emptyset$	$\{0\}$	$\{1\}$	$\{0,1\}$
$\emptyset$	$\emptyset$	$\emptyset$	$\emptyset$	$\emptyset$
$\{0\}$	$\emptyset$	$\{0\}$	$\emptyset$	$\{0\}$
$\{1\}$	$\emptyset$	$\emptyset$	$\{1\}$	$\{1\}$
$\{0,1\}$	$\emptyset$	$\{0\}$	$\{1\}$	$\{0,1\}$

2.3 Properties of rings. In this section we prove a few basic properties which follow from the ring axioms.

1. In a ring, we can only add elements two at a time. What if we want to add more than two elements? We have to put in brackets to convert the sum into a succession of pairwise additions. However, because of the associative law, the answer is the same no matter how we put in the brackets. For example, consider $a+b+c+d$. There are five possible ways of evaluating this, corresponding to the five bracketings $((a+b)+c)+d$, $(a+(b+c))+d$, $(a+b)+(c+d)$, $a+((b+c)+d)$, and $a+(b+(c+d))$. Now $(a+b)+c = a+(b+c)$, so the first two are equal. Similarly, $(b+c)+d = b+(c+d)$, so the fourth and fifth are equal. Now consider $(a+b)+(c+d)$. Putting $a+b = x$, this is $x+(c+d) = (x+c)+d = ((a+b)+c)+d$; similarly, putting $c+d = y$, it works out to $a+(b+(c+d))$. So all the expressions are equal.

We usually write $a + b + c + d$, leaving out the brackets.

In fact, the sum of any number of elements does not depend on the bracketing used to work it out. This might seem obvious to you as an extension of the above argument. But we can (and should) give a correct formal proof.

Proposition 2.1 *In a ring, the sum $a_1 + \cdots + a_n$ of any number of elements is independent of the bracketing used to work it out.*

Proof The proof is by induction on n. For $n = 1$ and $n = 2$, there is nothing to prove. For $n = 3$, there are just two possible bracketings, namely $(a_1 + a_2) + a_3$ and $a_1 + (a_2 + a_3)$, and the associative law tells us that they are equal. So let us assume that the result holds for sums of fewer than n terms, and prove it for sums of n terms. The induction hypothesis allows us to write $a_1 + \cdots + a_m$ for the sum of m terms whenever $m < n$.

Now consider two bracketings of the sum of $a_1, \ldots, a_n$. In the evaluation of each bracketing, at the last-but-one stage, we will have two expressions, the sum of $a_1, \ldots, a_k$, and the sum of $a_{k+1}, \ldots, a_n$, which are then added at the last stage. By the induction hypotheses, each of these smaller sums is independent of the bracketing, and so we can write the whole expression as

$$(a_1 + \cdots + a_k) + (a_{k+1} + \cdots + a_n),$$

where $0 < k < n$. Similarly, the other expression reduces to

$$(a_1 + \cdots + a_l) + (a_{l+1} + \cdots + a_n),$$

where $0 < l < n$.

If $k = l$, these expressions are clearly equal. So suppose not. We may assume that $k < l$. Now, again using the inductive hypothesis, we can write the first as

$$(a_1 + \cdots + a_k) + ((a_{k+1} + \cdots + a_l) + (a_{l+1} + \cdots + a_n)),$$

and the second as

$$((a_1 + \cdots + a_k) + (a_{k+1} + \cdots + a_l)) + (a_{l+1} + \cdots + a_n).$$

But these have the form $x + (y + z)$ and $(x + y) + z$, where

$$x = a_1 + \cdots + a_k,$$
$$y = a_{k+1} + \cdots + a_l,$$
$$z = a_{l+1} + \cdots + a_n;$$

by the associative law, they are equal. $\qquad\square$

The argument does not depend on the fact that the operation is called 'addition', but only on the associative law. So the same is true, for example, for the operation of multiplication in a ring.

2. Axiom (A2) guarantees that a zero element exists. Could there be more than one? Suppose that z_1 and z_2 are two zero elements in a ring R; that is, for all $a \in R$,

$$a + z_1 = z_1 + a = a = a + z_2 = z_2 + a.$$

Then we have

$$z_1 = z_1 + z_2 = z_2.$$

So *the zero element is unique*. A very similar argument shows that, in a ring with identity, *the identity element is unique*.

3. It is also true that inverses (as given by (A3)) are unique. For suppose that b and c are both inverses of a. This means that

$$a + b = b + a = 0 = a + c = c + a.$$

Now

$$b + (a + c) = b + 0 = b,$$
$$(b + a) + c = 0 + c = c.$$

By the associative law, the left-hand sides are equal; so $b = c$.

We write the inverse of a as $-a$. Also, we abbreviate $a + (-b)$ to $a - b$. Notice that we are falling into mathematical bad habits here: the symbol $-$ is being used in the first place as a unary operator (taking a to $-a$), and in the second as a binary operator (combining a and b to form $a - b$). In fact, people are used to this double use and have no trouble with it, but many calculators have different buttons for the two different uses of $-$.

Similarly, in a division ring, the multiplicative inverse of a non-zero element a is unique (and is written a^{-1}, so that $aa^{-1} = a^{-1}a = 1$).

4. The **cancellation law** holds for addition:

(C) (**Cancellation Law**): If $a + c = b + c$, then $a = b$.

Proof Suppose that $a + c = b + c$. Add $-c$ to each side:

$$(a + c) - c = (b + c) - c,$$
$$a + (c - c) = b + (c - c),$$
$$a + 0 = b + 0,$$
$$a = b.$$

$\square$

5. Here is the proof that the commutative law for addition follows from the other axioms in a ring with identity.

Expand $(1 + 1)(a + b)$ in two ways. We get

$$(1 + 1)(a + b) = (1 + 1)a + (1 + 1)b$$
$$= a + a + b + b,$$

and

$$(1 + 1)(a + b) = 1(a + b) + 1(a + b)$$
$$= a + b + a + b.$$

So $a + a + b + b = a + b + a + b$. Cancelling a from the front and b from the end gives $a + b = b + a$, as required.

6. For any $a \in R$, $a0 = 0a = 0$.

Proof

$$a(0 + 0) = a0 + a0.$$

Also, $0 + 0 = 0$, so

$$a(0 + 0) = a0 = a0 + 0.$$

Now use (C) to cancel $a0$ from the equation to give $a0 = 0$.

The proof that $0a = 0$ is similar.   □

Remark   This is the reason why axiom (M3) for a division ring only requires *non-zero* elements to have multiplicative inverses; for $0 \neq 1$, and $x0 = 0$ for all $X \in R$, so there cannot exist b with $b0 = 1$.

7. For any $a \in R$, $-(-a) = a$.

Proof $-(-a)$ is the inverse of $-a$; that is, it is an element which, when added to $-a$, gives zero. But we know that a added to $-a$ gives zero. Since inverses are unique, these two inverses of $-a$ must be equal.   □

8. For any $a, b \in R$, $-(a + b) = -b - a$.

Proof   We have to show that $-b - a$ is the inverse of $a + b$. So add it to $a + b$:

$$\begin{aligned}
-b - a + (a + b) &= -b + (-a + a) + b \\
&= -b + 0 + b \\
&= -b + b \\
&= 0,
\end{aligned}$$

as required.   □

2.4   Subrings.   Let R be a ring. A **subring** of R is a subset $S \subseteq R$ which itself forms a ring (using the same operations as those in R).

Let us see what checking the axioms involves in this case.

(A0): We require closure, that is, for all $a, b \in S$, $a + b \in S$.

(A1): The associative law automatically holds for all $a, b, c \in S$, since it holds for all a, b, c in the larger set R.

(A2): We require that the zero element of R lies in S.

(A3): For each $a \in S$, we require that $-a \in S$.

(A4): This holds automatically, by the same argument as for (A1).

(M0): We require that S is closed under multiplication.

(M1): This is automatic, as for (A1). The same is true of (D).

We conclude that, of the eight axioms, four are automatically true, just because we are looking at a subset of a ring. (These are the axioms asserting that all elements satisfy some equation.) So we only have to require the two closure axioms, the zero and inverse axioms.

In fact, we can whittle these down to three:

Theorem 2.2 (First Subring Test)   *A non-empty subset S of a ring R is a subring provided that, for all $a, b \in S$, we have $a + b, ab, -a \in S$.*

Proof We are given the closure and inverse axioms; we have to show that $0 \in S$. But S is non-empty, so take any element $a \in S$. Then, by assumption, $-a \in S$, and so $0 = a + (-a) \in S$, as required. □

We can do even better, reducing the number of tests to two: closure under subtraction and multiplication.

Theorem 2.3 (Second Subring Test) *A non-empty subset S of a ring R is a subring provided that, for all $a, b \in S$, we have $a - b, ab \in S$.*

Proof Suppose that S is closed under subtraction and multiplication. To show it is a subring, we verify the conditions of the First Subring Test. Take an element $a \in S$. Then $a - a = 0 \in S$; so $0 - a = -a \in S$, and the inverse law holds. Now take $a, b \in S$. Then $-b \in S$, and so $a - (-b) = a + b \in S$, and we have closure under addition. Closure under multiplication is given; so S is a subring. □

Example We find all the subrings of the ring $\mathbb{Z}$ of integers.

First, we show that, for any integer m, the set $m\mathbb{Z} = \{mx : x \in \mathbb{Z}\}$ of all multiples of m is a subring. Take $a, b \in m\mathbb{Z}$; let $a = mx$, $b = my$, for some integers x, y. Then

$$a - b = m(x - y) \in m\mathbb{Z},$$
$$ab = m(mxy) \in m\mathbb{Z},$$

so $m\mathbb{Z}$ passes the Second Subring Test.

Now we show that every subring of $\mathbb{Z}$ is of this form. So let S be a subring. Certainly $0 \in S$. If $S = \{0\}$, then $S = 0\mathbb{Z}$ is of the required form. So suppose not. If $n \in S$, then also $-n \in S$; so S must contain some positive integer. Let m be the smallest positive integer in S. We will prove that $S = m\mathbb{Z}$. Proving this equality involves showing that each element of one set is in the other and vice versa.

First, take any element of $m\mathbb{Z}$, say mx. If $x = 0$, then $mx = 0 \in S$. If $x > 0$, then $mx = m + m + \cdots + m$ (x terms), and $m \in S$; so $mx \in S$ by closure. If $x < 0$, let $x = -y$. Then $my \in S$ as above, and then $mx = -my \in S$.

Conversely, take any element of S, say n. By the Division Algorithm for integers, we can divide n by m, obtaining a quotient q and remainder r; thus, $n = mq + r$, and $0 \le r < m$. Now $n \in S$ and $mq \in S$, so $r = n - mq \in S$. If $r > 0$, we have a contradiction to the fact that m is the smallest positive integer in S. So, necessarily, $r = 0$ and $n = mq \in m\mathbb{Z}$.

Remark If you are asked to prove that something is a ring, it is usually much easier to recognise that it is a subset of a structure known to be a ring, and then apply one of the subring tests, than it is to check the eight ring axioms directly. Bear this in mind when you tackle Problems 2.1 and 2.3.

Exercise 2.1 Which of the following sets are rings (with the usual addition and multiplication):

(a) the natural numbers;
(b) the real polynomials of degree at most n;
(c) all polynomials with integer coefficients;
(d) all polynomials with integer coefficients and constant term zero;
(e) all polynomials with integer coefficients and degree at most four;
(f) all real polynomials f such that $f(2) = 0$;
(g) all integers divisible by 3;
(h) all non-singular 2×2 real matrices;
(i) all complex numbers of the form $a + bi$ for $a, b \in \mathbb{Z}$;
(j) all real functions of the form $f(x) = ax + b$ for $a, b \in \mathbb{R}$.

Exercise 2.2 Let R be a ring, and let $M_2(R)$ denote the set of all 2×2 matrices with elements from R. Define addition and multiplication of 2×2 matrices by the usual rules:

$$\begin{pmatrix} a & b \\ c & d \end{pmatrix} + \begin{pmatrix} e & f \\ g & h \end{pmatrix} = \begin{pmatrix} a+e & b+f \\ c+g & d+h \end{pmatrix},$$

$$\begin{pmatrix} a & b \\ c & d \end{pmatrix} \cdot \begin{pmatrix} e & f \\ g & h \end{pmatrix} = \begin{pmatrix} ae+bg & af+bh \\ ce+dg & cf+dh \end{pmatrix}.$$

Prove carefully that $M_2(R)$ is a ring.

Exercise 2.3 Which of the following sets of 2×2 matrices over the real numbers are rings (with the addition and multiplication defined in Problem 2.2)? Which are commutative? Which have an identity? Which are division rings?

(a) The set of all symmetric matrices (matrices A satisfying $A^\top = A$).
(b) The set of all skew-symmetric matrices (matrices A satisfying $A^\top = -A$).
(c) The set of all upper-triangular matrices (matrices of the form $\begin{pmatrix} a & b \\ 0 & c \end{pmatrix}$).
(d) The set of all strictly upper-triangular matrices (matrices of the form $\begin{pmatrix} 0 & a \\ 0 & 0 \end{pmatrix}$).
(e) The set of matrices of the form $\begin{pmatrix} a & b \\ -b & a \end{pmatrix}$.

[The **transpose** $A^\top$ of a matrix $A = \begin{pmatrix} a & b \\ c & d \end{pmatrix}$ is given by $A^\top = \begin{pmatrix} a & c \\ b & d \end{pmatrix}$.]

Exercise 2.4 Prove that, in any ring R,

$$(a_1 + a_2 + \cdots + a_m)(b_1 + b_2 + \cdots + b_n) = a_1 b_1 + a_1 b_2 + \cdots + a_1 b_n$$
$$+ a_2 b_1 + \cdots$$
$$+ a_m b_1 + \cdots + a_m b_n.$$

Exercise 2.5 Let R be a ring. For a positive integer n, let $n \cdot x$ denote $x + \cdots + x$ (n terms). Prove that

(a) $(m + n) \cdot x = m \cdot x + n \cdot x$ and $(mn) \cdot x = m \cdot (n \cdot x)$.
(b) if 1 is an identity and $n \cdot 1 = 0$, then $n \cdot x = 0$ for all x.

Exercise 2.6 Let x and y be elements of a ring R, and suppose that $xy = yx$. Prove the **Binomial Theorem** for $n > 0$:

$$(x + y)^n = \sum_{i=0}^{n} \binom{n}{i} \cdot x^{n-i} y^i.$$

[As in the preceding question, $m \cdot x = x + \cdots + x$ (m terms).]

Exercise 2.7 Let R be a ring with identity element 1.

(a) Prove that $(-1) \cdot x = -x$, where -1 and $-x$ are the additive inverses of 1 and x.
(b) Show that $-(x + y) = -y - x$.
(c) Hence show that the commutativity of addition can be deduced from the other axioms for a ring with identity.

Show that this is false if no identity element exists. [*Hint:* Let all products be zero.]

Exercise 2.8 Let R be a ring in which every element x satisfies $x^2 = x$ (where x^2 means xx).

(a) By evaluating $(x + x)^2$, show that $x + x = 0$ for all $x \in R$.
(b) By evaluating $(x + y)^2$, show that R is commutative.

Remark Any Boolean ring satisfies the condition $x^2 = x$ for all $x \in R$. It can be shown that any finite ring satisfying this condition is a Boolean ring.

In the spirit of abstract algebra, we will re-define the term **Boolean ring** to mean a ring R with identity satisfying $x^2 = x$ for all $x \in R$.

Exercise 2.9 Let R and S be rings. Define operations on $R \times S$ (the set of ordered pairs) by the rules

$$(r_1, s_1) + (r_2, s_2) = (r_1 + r_2, s_1 + s_2),$$
$$(r_1, s_1)(r_2, s_2) = (r_1 r_2, s_1 s_2).$$

Prove that $R \times S$ is a ring. Show further that $R \times S$ is commutative if and only if R and S are commutative, and that $R \times S$ has an identity if and only if R and S do. Can $R \times S$ ever be a field?

Remark $R \times S$ is known as the **direct product** or **direct sum** of the two rings R and S.

Exercise 2.10 (a) Let R be a ring in which the elements are the integers, and the addition is the same as in $\mathbb{Z}$. Is it possible that the multiplication is different from that in $\mathbb{Z}$? Can you describe all such rings?

(b) (**) Let R be a ring in which the elements are the integers, and the multiplication is the same as in $\mathbb{Z}$. Is it possible that the addition is different from that in $\mathbb{Z}$?

(Part (b) is much more difficult than part (a); you are not expected to solve it at this stage. We will return to this later.)

Exercise 2.11 (∗) The field $\mathbb{C}$ of **complex numbers** was described in Chapter 1 as the set of all objects $a + bi$, where a and b are real numbers, where addition and multiplication are defined according to 'the usual rules' subject to the extra condition that $i^2 = -1$. Hamilton constructed a larger number system, the **quaternions**, as follows. The elements of $\mathbb{H}$ are all objects of the form $a + bi + cj + dk$. Addition and multiplication are defined according to the usual rules, subject to the condition that multiplication of the elements $1, i, j, k$ works according to the following table:

·	1	i	j	k
1	1	i	j	k
i	i	−1	k	−j
j	j	−k	−1	i
k	k	j	−i	−1

Prove that $\mathbb{H}$ is a non-commutative division ring. [*Hint*: Define the **conjugate** of the element $z = a + bi + cj + dk$ to be $\overline{z} = a - bi - cj - dk$.] Prove that

$$z\overline{z} = (a^2 + b^2 + c^2 + d^2) \cdot 1.$$

Homomorphisms and ideals

Two rings are essentially the same for the purposes of algebra (the technical term is 'isomorphic') if we can match up their elements in such a way that addition and multiplication correspond; that is, if a corresponds to a' and b to b', then $a + b$ corresponds to $a' + b'$ and ab to $a'b'$. In this section, we define a weaker relationship between rings, which merely asserts that they are somewhat alike.

2.5   Cosets. The first topic seems to be a digression, but its relevance will become clear soon. Let S be a subring of the ring R. We will partition R into subsets called **cosets** of S.

Define a relation E on the set R by the rule that $(a, b) \in E$ if $b - a \in S$. I claim that E is an equivalence relation. It is

- *reflexive*, since $a - a = 0 \in S$;
- *symmetric*, since if $b - a \in S$ then $a - b = -(b - a) \in S$;
- *transitive*, since if $b - a \in S$ and $c - b \in S$ then $c - a = (c - b) + (b - a) \in S$.

So it is indeed an equivalence relation.

By the Equivalence Relation Theorem, R is partitioned into equivalence classes $E(a)$, where $E(a) = \{b : (a, b) \in E\}$. These equivalence classes are called the **cosets** of S in R. We examine them a bit more closely, and observe that

$$E(a) = S + a = \{s + a : s \in S\}.$$

To see this, first take $b \in S + a$. Then $b = s + a$ for some $s \in S$, so $b - a = s \in S$, whence $(a, b) \in E$ and $b \in E(a)$. Conversely, if $b \in E(a)$, then $b - a \in S$. Putting $s = b - a$, we have $b = s + a \in S + a$ as required.

Example Let $R = \mathbb{Z}$ and $S = 4\mathbb{Z}$. Then

$$S = S + 0 = \{\ldots, -8, -4, 0, 4, 8, \ldots\},$$
$$S + 1 = \{\ldots, -7, -3, 1, 5, 9, \ldots\},$$
$$S + 2 = \{\ldots, -6, -2, 2, 6, 10, \ldots\},$$
$$S + 3 = \{\ldots, -5, -1, 3, 7, 11, \ldots\},$$

and $S + 4 = S$, at which point the sequence repeats. (Compare the example in Section 1.4.3.)

More generally, if $R = \mathbb{Z}$ and $S = n\mathbb{Z}$, where n is a positive integer, then the coset $n\mathbb{Z} + a$ is the set of all integers congruent to a mod n. Thus, the cosets are the congruence classes mod n, and there are n of them altogether.

The element a is called a **coset representative** for the coset $S + a$. Note that the system is perfectly democratic: any element of a coset can serve as its representative. (Actually, this is very slightly misleading in one case. The subring S is a coset of itself, namely $S + 0$; and while we could use any of its elements as a representative, it is most natural to use the element 0.)

2.6 Homomorphisms and ideals. I introduce these two slightly strange words by means of an example. Suppose that I am short-sighted (actually this is correct), and that when I look at an integer I can only see whether it is even or odd. I will not know much about the integers, but I will know enough to make some consistent statements about addition and multiplication: for example, even plus even equals even. My knowledge can be summarised in tables:

+	even	odd
even	even	odd
odd	odd	even

·	even	odd
even	even	even
odd	even	odd

This looks very much like the two-element ring of Example 5 in Section 2.2. The point is that it is indeed a ring, and captures a little bit of the 'shape' of the ring of integers.

We define a **homomorphism** from a ring R to a ring S to be a function or map $\theta : R \to S$ which satisfies

$$\theta(a + b) = \theta(a) + \theta(b),$$
$$\theta(ab) = \theta(a)\theta(b)$$

for all $a, b \in R$. Note that the addition and multiplication on the left of these equations are the operations in R, while those on the right are the operations in S.

Having said this, I will perversely change notation right away. There are good reasons for writing the result of applying θ to a, not as $\theta(a)$, but as $a\theta$. With this notation, we say that the map θ is written **on the right**. One reason is that, in algebra, we often have to apply a function θ followed by another function ϕ: this is written as $a\theta\phi$, whereas if we wrote functions on the left we would have to say $\phi(\theta(a))$, and always remember to reverse the order. From now on, functions with an algebraic significance, such as homomorphisms, will always be written on the right; while functions with no such significance, such as polynomials, will be written on the left. So $f(x) = x^2 + 1$ is a polynomial, and $f\theta$ is the result of applying to it some homomorphism from the ring of all polynomials to another ring. Confused? Remember that not everybody uses this convention!

Let us rewrite the definition of a homomorphism in the new notation $\theta : R \rightarrow S$ is a homomorphism if

$$(a + b)\theta = a\theta + b\theta,$$
$$(ab)\theta = (a\theta)(b\theta).$$

The word *homomorphism* means 'similar shape'; this is meant to suggest that some of the 'shape' of the ring R is captured by S, as in our example. In this terminology, if S is the ring $\{0, 1\}$ of Example 5 of Section 2.2, then the function $\theta : \mathbb{Z} \rightarrow S$ defined by

$$n\theta = \begin{cases} 0 & \text{if } n \text{ is even} \\ 1 & \text{if } n \text{ is odd,} \end{cases}$$

is a homomorphism.

A homomorphism which is also a bijection (a one-to-one and onto function) is called an **isomorphism**. If there is an isomorphism from R to S, we say that the rings R and S are **isomorphic**. This means that they are 'matched up' by the function θ in such a way that the addition and multiplication are the same. So, from the point of view of abstract algebra, we will regard the two rings as being the same, even if their actual elements are quite different (one ring might consist of matrices and the other of polynomials, say). We denote 'R and S are isomorphic' by $R \cong S$.

Any homomorphism $\theta : R \rightarrow S$ has the additional properties

$$0\theta = 0,$$
$$(a - b)\theta = a\theta - b\theta$$

for all $a, b \in R$. The first equation follows from

$$0\theta + 0\theta = (0 + 0)\theta = 0\theta = 0\theta + 0$$

by using the Cancellation Law in S. The second from the fact that

$$(a - b)\theta + b\theta = (a - b + b)\theta = a\theta.$$

Our main task in this section is to see just how much a homomorphism blurs the structure of a ring, and how much shape is preserved.

Example We will find all homomorphisms from the ring $\mathbb{Z}$ to itself. Let θ be a homomorphism. Suppose that $1\theta = n$. Then $2\theta = (1+1)\theta = n + n = 2n$, and similarly (by induction), $m\theta = mn$ for all positive integers m. Moreover, $0\theta = 0$, and for positive m we have $(-m)\theta = -m\theta = -mn$. So θ multiplies every integer by n.

So far, we have only used the additive property. Now we turn to multiplication, and observe that

$$n = 1\theta = 1\theta \cdot 1\theta = n \cdot n = n^2,$$

so that $n = 0$ or $n = 1$. So there are only two homomorphisms, namely

- θ_0: $x \mapsto 0$ for all x;
- θ_1: $x \mapsto x$ for all x.

In fact, these rules define 'trivial' homomorphisms on any ring R. So our favourite ring $\mathbb{Z}$ is somewhat poor in homomorphisms: the only homomorphisms from $\mathbb{Z}$ to itself are the trivial ones possessed by all rings. (Of course, as we saw, there are homomorphisms from $\mathbb{Z}$ to other rings.)

A homomorphism $\theta : R \to S$ is a function, and so has an image and a kernel in the sense of Section 1.18. As promised there, we simplify the definition of the kernel slightly.

Definition Let $\theta : R \to S$ be a homomorphism of rings. The **image** of θ is

$$\mathrm{Im}(\theta) = \{s \in S : s = r\theta \text{ for some } r \in R\},$$

and the **kernel** of θ is

$$\mathrm{Ker}(\theta) = \{r \in R : r\theta = 0\}.$$

Remark In Section 1.18, the kernel of a function was defined to be the equivalence relation in which two elements are equivalent if they have the same image. So $\mathrm{Ker}(\theta)$ is the equivalence class containing 0 of the relation $\mathrm{KER}(\theta)$.

Proposition 2.4 *Let $\theta : R \to S$ be a ring homomorphism.*

(a) $\mathrm{Im}(\theta)$ is a subring of S.
(b) $\mathrm{Ker}(\theta)$ is a subring of R which has the additional property that, for any $x \in \mathrm{Ker}(\theta)$ and $r \in R$, we have $rx, xr \in \mathrm{Ker}(\theta)$.
(c) Two elements of R are mapped to the same element of S under θ if and only if they lie in the same coset of $\mathrm{Ker}(\theta)$.

Part (c) of this result states that the equivalence classes of the equivalence relation $\mathrm{KER}(\theta)$ are precisely the cosets of the subring $\mathrm{Ker}(\theta)$. This is why we

78 *Rings*

use the simpler definition: we can obtain the entire kernel (in the first sense) from the subring $\mathrm{Ker}(\theta)$. This also shows an important property of cosets.

Proof (a) We apply the subring test. Take $s_1, s_2 \in \mathrm{Im}(\theta)$. Then $s_1 = r_1\theta$ and $s_2 = r_2\theta$, for some $r_1, r_2 \in R$. Then

$$s_1 - s_2 = r_1\theta - r_2\theta = (r_1 - r_2)\theta \in \mathrm{Im}(\theta),$$
$$s_1 s_2 = (r_1\theta)(r_2\theta) = (r_1 r_2)\theta \in \mathrm{Im}(\theta);$$

so $\mathrm{Im}(\theta)$ is a subring of S.

(b) Similarly, take $r_1, r_2 \in \mathrm{Ker}(\theta)$. Then $r_1\theta = r_2\theta = 0$; so

$$(r_1 - r_2)\theta = r_1\theta - r_2\theta = 0 - 0 = 0,$$
$$(r_1 r_2)\theta = (r_1\theta)(r_2\theta) = 0 \cdot 0 = 0;$$

so $r_1 - r_2, r_1 r_2 \in \mathrm{Ker}(\theta)$, and $\mathrm{Ker}(\theta)$ is a subring.

Now we check the extra condition. Suppose that $x \in \mathrm{Ker}(\theta)$ and $r \in R$. Then

$$(rx)\theta = (r\theta)(x\theta) = r\theta \cdot 0 = 0,$$

and so $rx \in \mathrm{Ker}(\theta)$. Similarly, $xr \in \mathrm{Ker}(\theta)$.

(c) Suppose that $r_1\theta = r_2\theta$. Then $(r_2 - r_1)\theta = 0$, so $x = r_2 - r_1 \in \mathrm{Ker}(\theta)$; then $r_2 \in \mathrm{Ker}(\theta) + r_1$, so r_1 and r_2 lie in the same coset of $\mathrm{Ker}(\theta)$. Conversely, if r_1 and r_2 lie in the same coset, say $r_2 = r_1 + x$ with $x \in \mathrm{Ker}(\theta)$; then $r_2\theta = r_1\theta + x\theta = r_1\theta$, since $x \in \mathrm{Ker}(\theta)$. $\qquad\square$

The extra property of the subring $\mathrm{Ker}(\theta)$ is so important that it is given a special name. An **ideal** of a ring R is a subring S of R such that, for any $s \in S$ and $r \in R$, we have $rs, sr \in S$. The term was invented by Kummer, who invented 'ideal numbers' in an attempt to correct a mistake in his attempted proof of Fermat's Last Theorem. Unfortunately, he did not succeed, and we had to wait another hundred years until Fermat's Last Theorem was proved; but the concept of an ideal is crucial for ring theory.

To test for an ideal, we should test for a subring and then check the extra condition. But we can simplify things:

Theorem 2.5 (Ideal Test) *A non-empty subset S of a ring R is an ideal of R if and only if*

(a) for all $s_1, s_2 \in S$, we have $s_1 - s_2 \in S$;
(b) for all $s \in S$ and $r \in R$, we have $rs, sr \in S$.

Proof All that is missing is closure under multiplication; but this is just the special case of (b) corresponding to the case in which $r \in S$. $\qquad\square$

So we can say more briefly:

The kernel of a homomorphism $\theta : R \to S$ is an ideal of R.

Example   We found in Section 2.4 that the subrings of $\mathbb{Z}$ are all the sets of the form $n\mathbb{Z}$ for $n \in \mathbb{Z}$, where $n\mathbb{Z}$ consists of all multiples of n. Now all of these subrings are ideals. For take $r \in \mathbb{Z}$ and $s \in n\mathbb{Z}$, say $s = nx$ for some $x \in \mathbb{Z}$. Then $rs = sr = n(rx) \in n\mathbb{Z}$. So we have found all the ideals of $\mathbb{Z}$.

2.7   Should a ring have an identity?   What are the reasons for not requiring the existence of an identity in a ring?

First, it is convenient that an ideal is a particular kind of subring. But if rings are required to have identities, then all their ideals (with one exception) fail to be subrings! (In the next chapter, we will see a close analogy between groups, subgroups, and normal subgroups on one hand, and rings, subrings, and ideals on the other. This analogy would fail if ideals were not subrings.)

Proposition 2.6   *Let R be a ring with identity element* 1, *and I an ideal containing* 1. *Then $I = R$.*

Proof   For any element $r \in R$, we have $r = 1 \cdot r \in I$.   □

Second, there are several important examples of rings which are used in various branches of mathematics and which do not contain identities. For example,

- $\mathcal{C}_0(\mathbb{R})$, the ring of continuous real-valued functions with bounded support;
- the direct sum of an infinite collection of rings (even if all the factors have identities).

Third, the argument that a more general definition is better. In topology, there was a debate over whether a topological space should be required to satisfy the 'Hausdorff condition' or not; this was resolved in favour of the more general approach.

Finally, the traditional defence: if a ring does not have an identity, we can put one in!

Proposition 2.7   *Let R be a ring. Then there is a ring R^* with identity which contains a subring isomorphic to R.*

Proof   Let

$$R^* = R \times \mathbb{Z} = \{(r,n) : r \in R, n \in \mathbb{Z}\},$$

and define addition and multiplication on R^* by the rules

$$(r_1, n_1) + (r_2, n_2) = (r_1 + r_2, n_1 + n_2),$$
$$(r_1, n_1) \cdot (r_2, n_2) = (r_1 r_2 + n_2 r_1 + n_1 r_2, n_1 n_2),$$

where the product nr of an integer and a ring element is defined as in Exercise 2.5 for positive n (that is, the sum of n copies of r), with $0r = 0$ and $(-m)r = -(mr)$

for $n = -m < 0$. [*Warning*: This is not the direct product of rings defined in Exercise 2.9.]

Now a small amount of checking shows that

- R^* is a ring;
- $(0, 1)$ is the identity of R^*;
- $S = \{(r, 0) : r \in R\}$ is a subring of R^* (in fact, it is an ideal);
- the map $r \mapsto (r, 0)$ is an isomorphism from R to S.

Of course, if R already has an identity element, then this element is no longer the identity of R^*. □

2.8   Factor rings and isomorphism theorems.   For any integer n, the set $n\mathbb{Z}$ of multiples of n is an ideal of $\mathbb{Z}$ (and all ideals are of this form). We saw earlier that the cosets of $n\mathbb{Z}$ are just the congruence classes mod n. But we can do more: we can add and multiply integers mod n. This means that, if we take two congruence classes $n\mathbb{Z} + x$ and $n\mathbb{Z} + y$, then the sum of any integer from the first class and any integer from the second will always lie in the congruence class $n\mathbb{Z} + (x + y)$; and, similarly, the product of integers from these classes will lie in $n\mathbb{Z} + xy$. In this way, we can define operations of addition and multiplication on the **set** of congruence classes mod n (a finite set with n elements). With these operations, the set of congruence classes becomes a ring.

This all works much more generally; for any ideal I of a ring R, it is possible to make the set of cosets of I in R into a ring, called the 'factor ring' of R by I and denoted R/I, as follows.

Definition   Let I be an ideal in the ring R. The **factor ring** or **quotient ring** R/I is the set of cosets of I in R, with operations of addition and multiplication defined by

$$(I + x) + (I + y) = I + (x + y),$$
$$(I + x)(I + y) = I + xy.$$

Theorem 2.8   *The factor ring, as defined above, is indeed a ring.*

Proof   Before we verify the axioms, there is one very important thing to check: that the definition is a good one. On the face of it, the definition depends on the choice of coset representatives. It is not clear that, if x_1 and x_2 are two representatives of a coset of I, and y_1 and y_2 are two representatives of another coset, then $x_1 + y_1$ and $x_2 + y_2$ lie in the same coset (and similarly for multiplication). Suppose that we have such elements. Then $x_2 = x_1 + a$ and $y_2 = y_1 + b$, for some $a, b \in I$. Then

$$(x_2 + y_2) - (x_1 + y_1) = (x_1 + a + y_1 + b) - (x_1 + y_1) = a + b \in I,$$
$$(x_2 y_2) - (x_1 y_1) = (x_1 + a)(y_1 + b) - (x_1 y_1) = x_1 b + a y_1 + ab \in I,$$

where, in the last step, $ab \in I$ since I is a subring, and $x_1b, ay_1 \in I$ since I is an ideal. So the operations on R/I are indeed well defined.

Now the rest of the proof involves verifying the axioms, which is routine. The closure laws need no proof, since we have well-defined operations. For the associative law (A1), we have

$$((I + x) + (I + y)) + (I + z) = (I + (x + y)) + (I + z)$$
$$= I + ((x + y) + z)$$
$$= I + (x + (y + z))$$
$$= (I + x) + (I + (y + z))$$
$$= (I + x) + ((I + y) + (I + z)).$$

The proofs of (A4), (M1), and (D) are very similar. The zero element is $I + 0 = I$, and the inverse of $I + x$ is $I + (-x)$ (which, of course, we write as $I - x$). □

You were warned in the last chapter that a set of sets is difficult to think about, especially if we have to perform operations on its elements. Here is precisely that situation. But it is so important that it is worth taking some trouble to grasp the ideas.

You may recognise that, if R is the ring $\mathbb{Z}$ of integers and I is the set $m\mathbb{Z}$ of all multiples of the positive integer m, then the ring R/I is precisely the structure we called 'the integers mod m' in Chapter 1, and denoted by $\mathbb{Z}_m$. We will use the same notation here. The integers mod m provide a very good example of a factor ring.

The factor ring comes as the image of a natural homomorphism, as follows: Remember that the *elements* of R/I are the *cosets* of I in R. Now define a map $\theta : R \to R/I$ by the rule that $x\theta = I + x$ for all $x \in R$. Checking that θ is a homomorphism and is straightforward (we could say that the definitions of addition and multiplication in R/I were chosen to make this work):

$$(x + y)\theta = I + (x + y) = (I + x) + (I + y) = x\theta + y\theta,$$
$$(xy)\theta = I + xy = (I + x)(I + y) = (x\theta)(y\theta).$$

The image of θ is R/I, since every coset has the form $I + x$ for some $x \in R$. What is the kernel of θ? Since the zero element of R/I is the coset I, we have

$$\mathrm{Ker}(\theta) = \{x \in R : I + x = I\} = \{x \in R : x \in I\} = I.$$

We call the map θ the **canonical homomorphism** from the ring R to its factor ring R/I. Hence we have proved the following:

Theorem 2.9 *The canonical homomorphism $\theta : R \to R/I$ defined by $x\theta = I + x$ for $x \in R$ is indeed a homomorphism; its image is R/I and its kernel is I.*

Armed with the concept of factor rings and the canonical homomorphism, we can return to our analysis of the image and kernel of an arbitrary homomorphism.

Theorem 2.10 (First Isomorphism Theorem)   *Let $\theta : R \to S$ be a ring homomorphism. Then*

(a) $\mathrm{Im}(\theta)$ is a subring of S;
(b) $\mathrm{Ker}(\theta)$ is an ideal of R;
(c) $R/\mathrm{Ker}(\theta) \cong \mathrm{Im}(\theta)$.

Proof   We have already shown (a) and (b). For (c), there is only one reasonable definition of a map ϕ from R/I to S, where $I = \mathrm{Ker}(\theta)$: we must put $(I + x)\phi = x\theta$ for all $x \in R$. As usual, we have to show that this is well defined. So let x_1 and x_2 be representatives of the same coset of I, so that $x_2 = x_1 + a$ for some $a \in I$. Then

$$x_2\theta = (x_1 + a)\theta = x_1\theta + a\theta = x_1\theta,$$

since $a\theta = 0$; so $(I + x_1)\phi = (I + x_2)\phi$, and ϕ is indeed well defined.

To show that ϕ is a homomorphism, we have

$$(I + (x + y))\phi = (x + y)\theta = x\theta + y\theta = (I + x)\phi + (I + y)\phi,$$

$$(I + xy)\phi = (xy)\theta = (x\theta)(y\theta) = ((I + x)\phi)((I + y)\phi).$$

Now ϕ is clearly onto $\mathrm{Im}(\theta)$, since for any $s \in \mathrm{Im}(\theta)$ we have $s = x\theta = (I+x)\phi$. Finally, suppose that $(I + x)\phi = (I + y)\phi$. Then $x\theta = y\theta$, so $(y - x)\theta = 0$; thus $y - x \in \mathrm{Ker}(\theta)$, and x and y represent the same coset of $\mathrm{Ker}(\theta)$.   □

There are two further 'Isomorphism Theorems' relating a ring R to a factor ring R/I.

Theorem 2.11 (Second Isomorphism Theorem)   *Let I be an ideal of R. There is a one-to-one correspondence between the set of subrings of R which contain I and the set of subrings of R/I. Under this correspondence, ideals of R containing I correspond to ideals of R/I.*

Proof   If S is a subring of R containing I, then any coset of I with a representative in S is completely contained in S. (For, if $I \subseteq S$ and $x \in S$, then $I + x \subseteq S$ by closure of S.) Moreover, I is an ideal of S, since it is closed under subtraction and under multiplication by elements of S. So the factor ring S/I is the set of all cosets of I in S, and is a subring of R/I (as it is a ring in its own right). Conversely, let T be a subring of R/I. Then T is a set of cosets of I; the union of all these cosets is a subset $\hat{T}$ of R, which is easily seen to be a subring of R containing I. Hence we have the one-to-one correspondence. The further statement about ideals is an easy exercise.   □

Theorem 2.12 (Third Isomorphism Theorem)   *Let I be an ideal of R and S a subring of R. Then*

(a) $I + S = \{a + s : a \in I, s \in S\}$ is a subring of R containing I;
(b) $I \cap S$ is an ideal of S;
(c) $S/(I \cap S) \cong (I + S)/I$.

Proof All this can be proved directly; but a little trick, based on the natural homomorphism $\theta : R \to R/I$, makes it easier. We let ϕ be the restriction of θ to S. That is, ϕ maps S to R/I, and the value $s\phi$ for $s \in S$ is just $s\theta$. (We simply forget how θ acts on elements outside S.) Clearly, ϕ is a homomorphism: the two conditions in the definition hold for arbitrary elements of R, and so certainly for all elements of S.

(a) What is the image of ϕ? We see that $\mathrm{Im}(\phi)$ consists of all cosets $I + s$ for which the representative is in S. These form a subring of R/I, by Theorem 2.10(a). The union of all these cosets is the set

$$\{a + s : a \in I, s \in S\} = I + S$$

by Theorem 2.11; so $I + S$ is a subring of R which contains I. Incidentally, we see that $\mathrm{Im}(\phi) = (I + S)/I$.

(b) What is the kernel of ϕ? We see that $\mathrm{Ker}(\phi)$ consists of all the elements of S mapped to zero by θ. Since $\mathrm{Ker}(\theta) = I$, we have $\mathrm{Ker}(\phi) = I \cap S$, which is thus an ideal of S, by Theorem 2.10(b).

(c) By Theorem 2.10(c), $S/\mathrm{Ker}(\phi) \cong \mathrm{Im}(\phi)$; that is, $S/(I \cap S) \cong (I + S)/I$, as required. $\qquad \square$

These theorems are quite abstract, and the proofs are very condensed. Here is an example in detail.

Example Let $R = \mathbb{Z}$, and let I be the ideal $4\mathbb{Z}$. The cosets of I are the congruence classes mod 4. For simplicity, we will write the class $4\mathbb{Z}+k$ as k (being careful to distinguish between the integer and the coset). Now the addition and multiplication tables of $\mathbb{Z}_4 = \mathbb{Z}/4\mathbb{Z}$ are as follows (ignore the underlines for the moment):

+	0	1	2	3		·	0	1	2	3
0	0	1	2	3		0	0	0	0	0
1	1	2	3	0		1	0	1	2	3
2	2	3	0	1		2	0	2	0	2
3	3	0	1	2		3	0	3	2	1

Now $2\mathbb{Z}$ is a subring of $\mathbb{Z}$ containing $4\mathbb{Z}$: the corresponding subring of $\mathbb{Z}/4\mathbb{Z}$ is the set of cosets containing even numbers. These are the underlined cosets in the tables above: inspection of the tables shows that we do indeed have closure, so that $\{0, 2\}$ is a subring of $\mathbb{Z}/4\mathbb{Z}$, as it should be by Theorem 2.11. (Indeed, it is an ideal, also in accordance with that Theorem.)

Let $S = 6\mathbb{Z}$, a subring of $R = \mathbb{Z}$, and I the ideal $4\mathbb{Z}$. Then

$$I + S = \{4x + 6y : x, y \in \mathbb{Z}\} = 2\mathbb{Z},$$

the subring of R containing I described above. Also, $4\mathbb{Z} \cap 6\mathbb{Z} = 12\mathbb{Z}$, since an integer is divisible by both 4 and 6 if and only if it is divisible by 12. So Theorem 2.12 asserts that $6\mathbb{Z}/12\mathbb{Z} \cong 2\mathbb{Z}/4\mathbb{Z}$. The second of these factor rings

consists of the underlined elements in the above tables; the first has the following
tables:

+	0	6		·	0	6
0	0	6		0	0	0
6	6	0		6	0	0

(Note that $6 \cdot 6 = 36 \equiv_{12} 0$.) Inspection shows that the two factor rings are
indeed isomorphic, by the correspondence $0 \leftrightarrow 0$, $2 \leftrightarrow 6$.

2.9   Polynomials.   Like sets, polynomials are easy to understand, but difficult to define; we must make the attempt.

Usually, a polynomial is written as a sum of terms, where each term is a
product of a coefficient and a power of an 'indeterminate' x. Traditionally, the
coefficients are real numbers, and a polynomial is regarded as a function from
$\mathbb{R}$ to $\mathbb{R}$. In keeping with the spirit of abstract algebra, we allow the elements of
any ring R as coefficients; and we do not care what a polynomial really is, as
we are only interested in the rules for adding and multiplying polynomials. (In
fact, over some rings, different polynomials define the same function. We saw in
Section 2.2 that $x^2 = x$ for all elements x of a Boolean ring R; so the polynomials
x and x^2 would define the same function.)

Clearly, a polynomial is specified by giving its coefficients. But even these
are not uniquely determined. If a polynomial has degree n, we can add to it an
extra term $0x^{n+1}$ without changing it. Accordingly, we allow a polynomial to
have infinitely many terms, but specify that in all but a finite number of them
the coefficient is zero.

Now we are ready for the formal definition.

A **polynomial** over a ring R is an infinite sequence $(a_0, a_1, a_2, \ldots)$ of elements
of R, indexed by the non-negative integers, with the property that there exists
an integer n such that $a_i = 0$ for all $i > n$. In accordance with the usual notation,
we write the sequence $(a_0, a_1, a_2, \ldots)$ as $a_0 + a_1 x + a_2 x^2 + \cdots$, or (if n is as in
the definition) as $\sum_{i=0}^{n} a_i x^i$.

Addition and multiplication of polynomials are defined by the 'usual' rules
(essentially the ones we saw in Chapter 1):

$$\left(\sum a_i x^i \right) + \left(\sum b_i x^i \right) = \sum c_i x^i, \text{ where } c_i = a_i + b_i,$$

$$\left(\sum a_i x^i \right) \cdot \left(\sum b_i x^i \right) = \sum d_i x^i, \text{ where } d_i = \sum_{j=0}^{i} a_j b_{i-j}.$$

We let $R[x]$ denote the set of polynomials over R, with the above addition and
multiplication.

Theorem 2.13   *For any ring R, $R[x]$ is a ring. It is commutative if and only
if R is commutative; it has an identity if and only if R has an identity; but it is
never a division ring.*

The proof involves checking all the axioms; it will not be given here. The important thing to note is that the closure laws hold: adding and multiplying polynomials produce a sequence which has only finitely many non-zero terms, hence again a polynomial. Indeed, if we define the **degree** $\deg(f)$ of a non-zero polynomial to be the greatest integer n for which the coefficient of x^n is non-zero, then we have

$$\deg(f+g) \leq \max(\deg(f), \deg(g)),$$
$$\deg(fg) \leq \deg(f) + \deg(g).$$

(We do not define the degree of the zero polynomial, since it has no non-zero terms at all. Some people would define its degree to be $-\infty$, while others would say -1; but these are mere conventions.)

A **constant polynomial** is a polynomial $\sum a_i x^i$ with $a_i = 0$ for $i > 0$. In other words, the constant polynomials are the zero polynomial and the polynomials whose degree is zero. They form a subring of $R[x]$ isomorphic to R. Often, we don't distinguish carefully between the ring element r and the constant polynomial $r = \sum a_i x^i$ with $a_0 = r$ and $a_i = 0$ for $i > 0$.

Remarks   1. If we consider all the infinite sequences of elements of R, without imposing the restriction that only finitely many are non-zero, and use the same definitions of addition and multiplication, we obtain another important ring, the **formal power series ring** over R, denoted $R[[x]]$. (The word 'formal' signifies that we do not attempt to sum the power series, and are not concerned with questions of convergence.)

2. Here is another definition of polynomials, which avoids the need to consider infinite sequences at the expense of another complication. Let X denote the set of all *finite* sequences $(a_0, a_1, \ldots, a_n)$ of elements of R. (The number n can take any value. In particular, we include the empty sequence, which has no terms at all.) Now we define a relation E on X by the rule that $(s, t) \in E$ if t can be obtained from s by either adding or deleting any number of zeros from the right-hand end of the sequence. It can be shown that E is an equivalence relation; its equivalence classes are **polynomials**. To add or multiply polynomials f and g, we choose representative sequences $s = (a_0, \ldots, a_n)$ and $t = (b_0, \ldots, b_m)$ from the equivalence classes f and g. We may assume that $m = n$, by adding zeros to the shorter sequence. Now define $f + g$ to be the equivalence class of $s + t$, and fg the equivalence class of st, where addition and multiplication of sequences is defined as before. It can be shown that these operations do not depend on the choice of representatives of the equivalence classes, so that they are well defined; and that, with these operations, the set of equivalence classes is a ring. Furthermore, this ring is isomorphic to the ring of infinite sequences which we defined before.

3. The upshot of this section is that you already understand polynomials, and you should think of them just as you did before; but they can be put on a proper theoretical basis, with some work.

Exercise 2.12 Let I be an ideal of a ring R. Prove that $M_n(I)$ (the set of $n \times n$ matrices with elements in I) is an ideal of $M_n(R)$. (For an easier question, do the case $n = 2$.) Prove also that $M_n(R)/M_n(I) \cong M_n(R/I)$.

Exercise 2.13 Let I be an ideal in a commutative ring R. Prove that $I[x]$ (the ring of polynomials over I) is an ideal in $R[x]$. Prove also that $R[x]/I[x] \cong (R/I)[x]$.

Exercise 2.14 (*) Let R be a ring with identity. Suppose that J is an ideal in $M_n(R)$. Prove that there is an ideal I of R such that $J = M_n(I)$. [*Hint:* let E_{ij} be the **matrix unit** with 1 in row i and column j and 0 everywhere else. Prove that, if $A = (a_{ij})$, then $E_{ki}AE_{jl}$ has entry a_{ij} in the (k,l) position and zeros elsewhere. Now let I be the set of all elements of R which appear as an entry in some matrix of J. Show that, for any $r \in I$, the matrix with r in the top-left corner and 0 elsewhere belongs to J. Hence show that I is an ideal, and that $J = M_n(I)$.]

Exercise 2.15 (a) Show that, in the ring $\mathbb{Z}$, $m\mathbb{Z}$ contains $n\mathbb{Z}$ if and only if m divides n.

(b) How many ideals does the ring $\mathbb{Z}_{60}$ have? How many of these ideals are maximal (in the sense that I is maximal if $I \neq R$ but no ideal J satisfies $I \subset J \subset R$)?

(c) Repeat part (b) for the ring $\mathbb{Z}_n$, where $n = p_1^{a_1} \cdots p_r^{a_r}$ and $p_1, \ldots, p_r$ are distinct primes, and $a_1, \ldots, a_r$ are positive integers.

Exercise 2.16 (a) Prove that the **Gaussian integers**, the complex numbers of the form $a + bi$, where a, b are integers, form a subring of $\mathbb{C}$.

(b) Prove that the **Eisenstein integers**, the complex numbers of the form $a + b\sqrt{-3}$, where either a, b are integers or $a - \frac{1}{2}, b - \frac{1}{2}$ are integers, form a subring of $\mathbb{C}$. (So, for example, $1 + \sqrt{-3}$ and $-\frac{1}{2} + \frac{5}{2}\sqrt{-3}$ are Eisenstein integers but $\frac{1}{2} - \sqrt{-3}$ is not.)

Exercise 2.17 Let R be a commutative ring and $u \in R$. Show that the map $\theta : R[x] \rightarrow R$ defined by 'substituting u for x'; that is, $\sum a_i x^i \mapsto \sum a_i u^i$, is a homomorphism.

Exercise 2.18 Let R be the ring $\mathbb{R}[x]$ of all real polynomials. Define a function $\theta : R \rightarrow \mathbb{C}$ by the rule that $f\theta = f(i)$. Prove that θ is a homomorphism, that its image is $\mathbb{C}$, and that its kernel is the ideal $(x^2 + 1)R$ consisting of all polynomials divisible by $x^2 + 1$. Hence show that
$$\mathbb{R}[x]/(x^2 + 1)\mathbb{R}[x] \cong \mathbb{C}.$$

Exercise 2.19 Construct a homomorphism from $\mathbb{Z}_{mn}$ to $\mathbb{Z}_n$, for any positive integers m, n.

Exercise 2.20 Let Y be a subset of the set X. Let $\mathcal{P}(X)$ and $\mathcal{P}(Y)$ be the Boolean rings of subsets of X and Y, respectively. Show that the map $\theta : \mathcal{P}(X) \rightarrow \mathcal{P}(Y)$ defined by $A\theta = A \cap Y$ is a homomorphism, and find its image and kernel.

Exercise 2.21 Let R be the ring of real upper-triangular 2×2 matrices (those of the form $\begin{pmatrix} a & b \\ 0 & c \end{pmatrix}$ for $a, b, c \in \mathbb{R}$). Let I be the set of strictly upper-triangular matrices (of the form $\begin{pmatrix} 0 & b \\ 0 & 0 \end{pmatrix}$) and S the set of diagonal matrices (of the form $\begin{pmatrix} a & 0 \\ 0 & c \end{pmatrix}$).

(a) Prove that I is an ideal of R.
(b) Prove that S is a subring of R. Is it an ideal?
(c) Prove that R/I is isomorphic to S.

Factorisation

One of the most important properties of the integers is the so-called Fundamental Theorem of Arithmetic, which asserts that any integer can be factorised into primes in an essentially unique way. We want to examine the rings in which such a result could hold.

2.10   Zero-divisors and units. In contrast to the situation in $\mathbb{Z}$, it can happen in an arbitrary ring that the product of two non-zero elements is zero. For example, $2 \cdot 2 = 0$ in $\mathbb{Z}_4$, and $\begin{pmatrix} 1 & 0 \\ 0 & 0 \end{pmatrix}\begin{pmatrix} 0 & 0 \\ 0 & 1 \end{pmatrix} = \begin{pmatrix} 0 & 0 \\ 0 & 0 \end{pmatrix}$ in $M_2(\mathbb{R})$. For much of the rest of this chapter, we are especially interested in rings in which this does not happen. Accordingly, we define it away:

- A **zero-divisor** in a ring R is a non-zero element $a \in R$ such that there exists a non-zero element $b \in R$ with $ab = 0$.
- An **integral domain** is a commutative ring with identity which has no zero-divisors.

Strictly speaking, in the first definition, a is a **left zero-divisor**, and b is a **right zero-divisor**; but, as the second definition suggests, we are mostly interested in commutative rings, and in these, the concepts of left and right zero-divisor coincide.

The condition 'no zero-divisors' can also be stated in the form: *if $ab = 0$, then either $a = 0$ or $b = 0$.* Thus, $\mathbb{Z}$, our prototype of a ring, is also our prototype of an integral domain, as the name would suggest. Integral domains have many nice properties. For example, there is a multiplicative version of the **cancellation law**:

(C′) In an integral domain, if $ab = ac$ and $a \neq 0$, then $b = c$.

For, if $ab = ac$, then $a(b - c) = 0$; in an integral domain, if $a \neq 0$, this implies that $b - c = 0$; that is, $b = c$.

Let R be a ring with identity. The element $a \in R$ is a **unit** if there exists $b \in R$ with $ab = ba = 1$. The element b is called the **inverse** of a, and is written a^{-1}. It is unique; for if b and c are both inverses of a, then

$$c = 1c = (ba)c = b(ac) = b1 = b.$$

You should compare this with the proof of uniqueness of additive inverses in Section 2.3.

Proposition 2.14 *Let R be a ring with identity.*

(a) *The identity is a unit; it is equal to its inverse.*
(b) *If a is a unit, then so is a^{-1}; its inverse is a.*
(c) *If a and b are units, then so is ab; its inverse is $b^{-1}a^{-1}$.*

Proof   (a) $1 \cdot 1 = 1$.
    (b) This is shown by the equations $aa^{-1} = a^{-1}a = 1$.
    (c) We have

$$(ab)(b^{-1}a^{-1}) = a(bb^{-1})a^{-1} = a1a^{-1} = aa^{-1} = 1,$$

and, similarly, $(b^{-1}a^{-1})(ab) = 1$.   □

Here is how Hermann Weyl explains part (c) of this proposition in his book *Symmetry* (1952).

> With this rule, although perhaps not with its mathematical expression, you are all familiar. When you dress, it is not immaterial in which order you perform the operations; and when in dressing you start with the shirt and end up with the coat, then in undressing you observe the opposite order; first take off the coat and the shirt comes last.

Examples   1. A ring with identity is a division ring if and only if every non-zero element is a unit.

2. The units in $\mathbb{Z}$ are 1 and -1.

3. In the next proposition, we find the zero-divisors and units in the ring $\mathbb{Z}_n$ of integers mod n, where $n > 1$.

Proposition 2.15   *An element $x \neq 0$ of $\mathbb{Z}_n$ is a zero-divisor if and only if x and n have greatest common divisor (g.c.d.) greater than 1; it is a unit if and only if x and n have greatest common divisor 1.*

In other words, in $\mathbb{Z}_n$, every non-zero element is either a zero-divisor or a unit (but not both, see Exercise 2.22).

Proof   Suppose that $d = \gcd(x, n)$ is the greatest common divisor of x and n.

(a) If $d > 1$, then (n/d) is a non-zero element of $\mathbb{Z}_n$; and $x(n/d) = (x/d)n \equiv_n 0$.

(b) Suppose that $d = 1$. By the Euclidean algorithm, there are integers p and q such that $xp + nq = 1$. But this means that $xp = px \equiv_n 1$, so that x is a unit (and p is its inverse).

Conversely, if x is a zero-divisor, then x is not a unit, so d is not 1 — that is, $d > 1$ — and similarly for (b).   □

Two elements a, b of the integral domain R are said to be **associates** if there is a unit $u \in R$ such that $b = au$. Note that, by the above Proposition, it follows that being associates is an equivalence relation: it is

- *reflexive*, since $a = a1$;
- *symmetric*, since $b = au$ implies $a = bu^{-1}$;
- *transitive*, since $b = au$ and $c = bv$ imply $c = a(uv)$.

(Here u and v are units; and the proposition shows that 1, u^{-1} and uv are units.)

By the Equivalence Relation Theorem, R is partitioned into equivalence classes, called **associate classes**.

For example, in $\mathbb{Z}$, the associate classes are the sets $\{n, -n\}$ for all non-negative integers n.

2.11   Irreducibles and factorisation.   In this section, we examine the possibility of factorising elements of a ring into 'irreducible' elements (which cannot themselves be further factorised), and look at a special class of rings in which the analogue of the Fundamental Theorem of Arithmetic holds.

First, we will make some simplifying assumptions about the ring R. We always assume that R is commutative, so that we can regard ab and ba as essentially the same factorisation of a ring element. (So, in a factorisation, we do not care about the order of the factors.) Also, we exclude divisors of zero. For, if $ab = 0$, then $ac = a(b+c)$ for any element c, and there is little chance of unique factorisations.

Accordingly, we assume, in this section and the next two, that

$$R \text{ is an integral domain.}$$

Also, units provide another problem. In $\mathbb{Z}$, we regard $2 \cdot 3$ and $(-2) \cdot (-3)$ as 'essentially the same' factorisation of 6. More generally, if $x = ab$, and u is a unit with inverse v, then $x = (au)(vb)$, and we want to think of this as the same factorisation. We note that a and au are associates, as are b and vb. So we think of two factorisations as the same if the factors in one are associates of factors in the other. For the same reason, we do not regard units as counting towards a factorisation, or we could multiply them *ad infinitum*.

This leads us to the appropriate definitions.

Definition   Let R be an integral domain.

- An element $p \in R$ is **irreducible** if p is not zero or a unit, and if, whenever $p = ab$, either a or b is a unit (and the other is an associate of p).
- R is a **unique factorisation domain** or **UFD** if it holds that
  (a) every element other than zero and units can be factorised into irreducibles;
  (b) if $p_1 \cdots p_m = q_1 \cdots q_n$, where the p_i and q_j are irreducibles, then $m = n$, and (possibly after re-ordering) p_i and q_i are associates for $i = 1, \ldots, n$.

Briefly, condition (a) says that factorisations into irreducibles exist, while (b) says that they are 'unique up to order and associates'. (Note that any associate of an irreducible is irreducible.)

So the 'Fundamental Theorem of Arithmetic' says that $\mathbb{Z}$ is a UFD. (We will prove this later, in Section 2.13.) However, things here are a little different from our first view of the FTA. Instead of factorising positive integers into positive primes, we factorise arbitrary integers into arbitrary (positive or negative) primes—remember that p and $-p$ are associates.

In a trivial way, a field is a UFD: it does not have any elements which are not zero or units!

One of the most substantial results about UFDs is the following:

Theorem 2.16 (Gauss' Lemma)   *If R is a UFD, then $R[x]$ is a UFD.*

The proof of this result will be given in Chapter 7.

In particular, if F is a field, then $F[x]$ is a UFD. This will be proved in Section 2.13, since it uses techniques similar to those for $\mathbb{Z}$.

An important property of UFDs is that 'greatest common divisors exist'. In the case of the integers, we interpret the word 'greatest' in its usual sense for numbers. In general, this is not possible; the greatest common divisor is a common divisor which is divisible by every common divisor, in a sense which the next definition makes precise.

Definition   Let R be a commutative ring.

- For $a, b \in R$, we say that a **divides** b (in symbols, $a \mid b$) if $b = ac$ for some $c \in R$.
- The element d is a **greatest common divisor** or **g.c.d.** of a and b if
    (a) d divides a and d divides b;
    (b) for any $e \in R$, if e divides a and e divides b then e divides d.

Thus, the greatest common divisor is not necessarily greatest in any absolute sense. In an arbitrary ring, two elements may have no greatest common divisor at all.

Theorem 2.17   *(a) In an integral domain, if a divides b and b divides a, then a and b are associates.*

(b) In an integral domain, if a and b have a greatest common divisor, then any two g.c.ds are associates.

(c) In a unique factorisation domain, every two elements have a greatest common divisor.

Proof   (a) If $a = 0$ then $b = 0$ and there is nothing to prove. So suppose not. Let $b = ac$ and $a = bd$, then $a = acd$, so $a(1 - cd) = 0$. Since $a \neq 0$ and R is an integral domain, $cd = 1$; so c and d are units, and a and b are associates.

(b) If d_1 and d_2 are both g.c.ds of a and b, then (by part (b) of the definition) each divides the other; so they are associates.

(c) Assume that a and b are non-zero and not units. (Can you deal with the remaining cases?) Factorise a into irreducibles. Then, up to associates, every divisor of a is a product of some of the irreducibles in the factorisation of a. (For suppose that $a = xy$. Factorise x and y into irreducibles. Combining these gives a factorisation of a, which must be equal to the given one, up to order and associates.) So we find the g.c.d. of a and b by factorising both elements into irreducibles, and taking all the irreducibles which (up to associates) occur in both factorisations. □

If this sounds somewhat complicated, it is just a generalisation of the argument which says, for example, that the g.c.d. of $2^4.3^2.5^2.7$ and $2^3.3^3.5.11$ is $2^3.3^2.5$.

This method finds the greatest common divisor, in principle. But it is not really an algorithm, since it depends on finding the factorisations of a and b, and we do not know how to do this in an arbitrary UFD (only that it can be done).

We conclude this section with an example of failure of the unique factorisation property.

Example Let $R = \{a + b\sqrt{-5} : a, b \in \mathbb{Z}\}$. Then R is a ring, with the usual definition of addition and multiplication of complex numbers. Moreover, R is an integral domain.

We first find the units of R. Let $a + b\sqrt{-5}$ be a unit; suppose that

$$(a + b\sqrt{-5})(x + y\sqrt{-5}) = 1.$$

Taking the square of the modulus of this equation (and using the fact that $|a + b\sqrt{-5}|^2 = a^2 + 5b^2$), we obtain

$$(a^2 + 5b^2)(x^2 + 5y^2) = 1.$$

Since a, b, x, y are integers, the only possibility is $b = y = 0$, $a^2 = x^2 = 1$, so that $a = \pm 1$. So the units are 1 and -1, and the associates of an element r are r and $-r$.

Consider the equation

$$6 = 2 \cdot 3 = (1 + \sqrt{-5})(1 - \sqrt{-5}).$$

We claim that all of the factors $2, 3, 1 + \sqrt{-5}, 1 - \sqrt{-5}$ are irreducible. Then certainly the factorisations are not the same up to order and associates!

To show that 2 is irreducible, suppose that

$$2 = (a + b\sqrt{-5})(x + y\sqrt{-5}).$$

Taking the norm squared as before, we obtain

$$4 = (a^2 + 5b^2)(x^2 + 5y^2).$$

As before, this implies that $b = y = 0$, so that $a = \pm 1$ or ± 2, and $x = \pm 2$ or ± 1. So one factor is a unit, and the other is an associate of 2. So 2 is irreducible. By a very similar argument, all the other factors are irreducible too.

So R is not a unique factorisation domain.

2.12   Principal ideal domains or PIDs. In the ring $\mathbb{Z}$, every ideal consists of all multiples of a fixed integer. This is a very important property, which we now study in general.

Definition Let $a_1, \ldots, a_n$ be elements of a ring R. The ideal **generated** by $a_1, \ldots, a_n$, denoted by $\langle a_1, \ldots, a_n \rangle$, is the smallest ideal containing these elements. ('Smallest' has the sense of inclusion; it is a subset of every ideal that contains $a_1, \ldots, a_n$.) Be aware that this is often written as $(a_1, \ldots, a_n)$. However, this risks confusion with the n-tuple $(a_1, \ldots, a_n)$. Angle brackets as used here are very common in mathematics to convey the idea of generation.

From the definition, it is not obvious that such an ideal exists. It can be shown that it does exist in any ring. But in a special case, it is easy to describe:

Proposition 2.18 *Let R be a commutative ring with identity, and let $a_1, \ldots, a_n$ be elements of R. Then*

$$\langle a_1, \ldots, a_n \rangle = \{x_1 a_1 + x_2 a_2 + \cdots + x_n a_n : x_1, \ldots, x_n \in R\}.$$

Proof Let $I = \{x_1 a_1 + \cdots + x_n a_n : x_1, \ldots, x_n \in R\}$, the set of all linear combinations of $a_1, \ldots, a_n$. We have to show that I is an ideal, that I contains $a_1, \ldots, a_n$, and that any ideal containing $a_1, \ldots, a_n$ necessarily contains all of I.

I is an ideal: (a) if $a, b \in I$, say $a = x_1 a_1 + \cdots + x_n a_n$ and $b = y_1 a_1 + \cdots + y_n a_n$, then

$$a - b = (x_1 - y_1)a_1 + \cdots + (x_n - y_n)a_n \in I.$$

(b) If $a = x_1 a_1 + \cdots + x_n a_n \in I$ and $r \in R$, then

$$ar = ra = (rx_1)a_1 + \cdots + (rx_n)a_n \in I.$$

So I passes the Ideal Test.

I contains $a_1, \ldots, a_n$: for $1 \leq i \leq n$, we have

$$a_i = 0a_1 + \cdots + 0a_{i-1} + 1a_i + 0a_{i+1} + \cdots + 0a_n \in I.$$

Any ideal containing $a_1, \ldots, a_n$ contains I: Let J be an ideal of R containing $a_1, \ldots, a_n$. For any $x_1, \ldots, x_n \in R$, we have $x_i a_i \in J$, and hence $x_1 a_1 + \cdots + x_n a_n \in J$ (using the fact that J is an ideal, and so is closed under addition and under multiplication by elements of R). So every element of I is in J, which means that $I \subseteq J$. $\square$

An ideal of R is **principal** if it is generated by a single element. By the proposition, if R is a commutative ring with identity, then a principal ideal is of the form $\langle a \rangle = aR = \{ax : x \in R\}$. In other words, $\langle a \rangle$ consists of all elements divisible by a. In an integral domain, principal ideals have some further nice properties:

Proposition 2.19 *Let R be an integral domain.*

(a) For $a, b \in R$, if $\langle a \rangle = \langle b \rangle$, then a and b are associates.

(b) For $a, b \in R$, if $\langle a, b \rangle = \langle d \rangle$, then d is a greatest common divisor of a and b.

Proof (a) Suppose that $\langle a \rangle = \langle b \rangle$. Then $a \in \langle b \rangle = bR$, so b divides a, and similarly a divides b. By Theorem 2.17, a and b are associates.

(b) Suppose that $\langle a, b \rangle = \langle d \rangle$. Then $a, b \in \langle d \rangle$, so (as above) d divides both a and b. On the other hand, $d \in \langle a, b \rangle$, so $d = ax + by$ for some $x, y \in R$. Let e be any common divisor of a and b. Then $a = eu$ and $b = ev$ for some u, v. Then we have $d = ax + by = eux + evy = e(ux + vy)$; that is, e divides d. So d is a greatest common divisor. $\square$

Definition A principal ideal domain or **PID** is an integral domain with the property that every ideal is principal.

Proposition 2.20 (a) Let R be a PID. Then any two elements $a, b \in R$ have a greatest common divisor d, which can be written in the form $d = ax + by$ for some $x, y \in R$.

(b) $\mathbb{Z}$ is a PID.

Proof (a) The ideal $\langle a, b \rangle$ is principal, and hence has the form $\langle d \rangle$ for some d. Now apply part (b) of the previous proposition.

(b) We found that every ideal of $\mathbb{Z}$ has the form $n\mathbb{Z}$, that is, $\langle n \rangle$ in the present notation. $\square$

What has all this to do with factorisation? The following important result holds:

Theorem 2.21 *Every principal ideal domain is a unique factorisation domain.*

Proof Let R be a PID. We have to show two things: that elements of R can be factorised into irreducibles; and that the factorisation of an element is unique (up to order and associates). The first part is quite substantial; the proof is deferred until Chapter 7. I will show here that factorisations are unique. This depends on the following fact.

Proposition 2.22 *Let p be an irreducible element in a PID R. If p divides ab, then p divides a or p divides b.*

Proof Suppose that p divides ab but p does not divide a. Then the greatest common divisor of p and a is 1. (Remember that g.c.ds exist in a PID.) So there exist $x, y \in R$ such that $px + ay = 1$. Multiplying this equation by b, we obtain $pxb + aby = b$. Now p clearly divides pxb; and p divides aby (since p divides ab by assumption); so p divides b. The result is proved. $\square$

It follows that if p is irreducible and p divides $a_1 \cdots a_n$, then p divides a_i for some i.

This property fails in the ring $R = \{a + b\sqrt{-5} : a, b \in \mathbb{Z}\}$ discussed in the preceding section. For 2 divides $6 = (1 + \sqrt{-5})(1 - \sqrt{-5})$, but 2 does not divide either factor.

Now we return to the proof that a PID is a UFD. Suppose that we have two factorisations of an element of the PID R, say

$$a = p_1 p_2 \cdots p_m = q_1 q_2 \cdots q_n,$$

where the p_i and q_j are irreducible. Now p_1 divides $q_1 \cdots q_n$, so by the remark following the proposition, p_1 divides q_j for some j. By re-ordering the product, we can assume that p_1 divides q_1. Since p_1 and q_1 are irreducible, they must be associates, say $p_1 = q_1 u$ for some unit u. Then we have

$$p_1 p_2 \cdots p_m = (q_1 u)(u^{-1} q_2) \cdots q_n.$$

By the Cancellation Law,

$$p_2 \cdots p_m = q_2' \cdots q_n,$$

where $q_2' = u^{-1} q_2$, an associate of q_2. Continuing in this manner we find that $m = n$ and that p_i and q_i are associates for all i (after suitable re-ordering), and we are done.   □

We close this section with an example of a UFD which is not a PID. This is the ring $\mathbb{Z}[x]$ of polynomials over the integers. It is a UFD by Gauss' Lemma. The g.c.d. of the elements 2 and x is obviously 1; but there do not exist polynomials f and g such that $2f(x) + xg(x) = 1$, since the constant term of the left-hand side is even. Said otherwise, the ideal $\langle 2, x \rangle$ generated by 2 and x (which is the set of all polynomials whose constant term is even) cannot be generated by a single element.

2.13   Euclidean domains or EDs.   We now look at an even more specialised class of rings (which, however, includes our prototype $\mathbb{Z}$ as well as polynomial rings over fields).

Definition   Let R be a commutative ring with identity.

- A **Euclidean function** on R is a function d from the set of non-zero elements of R to the non-negative integers which satisfies
  (a) $d(ab) \geq d(a)$ for $a, b \neq 0$;
  (b) if $a, b \in R$ with $b \neq 0$, then there exist $q, r \in R$ with $a = bq + r$ and either $r = 0$ or $d(r) < d(b)$.
- R is a **Euclidean domain**, if there exists a Euclidean function on R.

Examples   1. $\mathbb{Z}$ is a ED. Take $d(a) = |a|$ for non-zero $a \in \mathbb{Z}$. If also $b \neq 0$, then clearly

$$d(ab) = |ab| = |a| \cdot |b| = d(a)d(b) \geq d(a),$$

since $d(b) \geq 1$. For (b), suppose that $b \neq 0$. If $b > 0$, divide a by b to obtain a quotient q and remainder r; then $a = bq + r$ with $0 \leq r < b$; that is, $r = 0$ or $d(r) < d(b) = b$. If $b < 0$, divide a by $-b$ instead.

2. For any field F, the polynomial ring $F[x]$ is a ED. In this case, we take the Euclidean function to be $d(f) = \deg(f)$, the degree of the polynomial f. (Recall that we did not define $\deg(f)$ if $f = 0$, but we do not need a value for $d(0)$ either.)

(a) If f, g are non-zero, then $d(fg) = d(f) + d(g) \geq d(f)$.

(b) Suppose that $g \neq 0$; we wish to find q, r with $f = gq + r$ and $r = 0$ or $\deg(r) < \deg(g)$. The proof is by induction on the degree of f. Let $m = \deg(f)$, $n = \deg(g)$. If $m < n$, we can take $q = 0$ and $r = f$. So suppose that $m \geq n$. Let

$$f(x) = a_m x^m + \text{lower terms},$$

$$g(x) = b_n x^n + \text{lower terms},$$

where a_m and b_n are non-zero. Put

$$f_1(x) = f(x) - (a_m b_n^{-1}) x^{m-n} g(x).$$

(This is defined since the coefficients form a field and $b_n \neq 0$.) The coefficient of x^m in f_1 is $a_m - (a_m b_n^{-1}) b_n = 0$, and clearly there are no terms of higher degree. So $\deg(f_1) < m = \deg(f)$. By the induction hypothesis, we have $f_1 = gq_1 + r_1$, where $r_1 = 0$ or $\deg(r_1) < \deg(g)$. Then

$$f = g(a_m b_n^{-1} x^{m-n} + q_1) + r_1;$$

so we can take $q = a_m b_n^{-1} x^{m-n} + q_1$, $r = r_1$.

The reason for the term 'Euclidean domain' is that this is the class of rings in which the Euclidean Algorithm for finding the greatest common divisor of two elements can be made to work. We met the Euclidean Algorithm for integers in Chapter 1. The general case is exactly the same. I will present it here in a way influenced by computer programming, as a recursive algorithm. But you do not need to know anything about computers in order to follow this. Just remember that an algorithm takes some data as input and produces some other data as output; we must specify what the algorithm is expected to do, and then we must prove that the algorithm really does what is claimed.

Euclidean Algorithm Let R be a Euclidean domain.

   Input: Two elements $a, b \in R$.

   Output: An element $c \in R$ which is a greatest common divisor of a and b. We write $c = \gcd(a, b)$ for this output.

   Operation: If $b = 0$, then set $\gcd(a, b) = a$.

   Otherwise, choose $q, r \in R$ such that $a = bq + r$ with either $r = 0$ or $d(r) < d(b)$; set $\gcd(a, b) = \gcd(b, r)$.

It is not clear that we have defined anything: why should $\gcd(b, r)$ be easier to calculate than $\gcd(a, b)$? Imagine that we are given two elements a and b, and are trying to find $\gcd(a, b)$. If $b = 0$, we obtain immediately the result a. Suppose not. Now observe that either $r = 0$, in which case we finish at the second step, or we have to calculate $\gcd(b, r)$ with $d(r) < d(b)$. During the calculation, the second alternative can only occur finitely often, since the value of d on the second argument of the function is strictly smaller at each instance, and a strictly decreasing sequence of non-negative integers cannot continue for ever. So, after a finite number of steps, the algorithm does terminate and produce a result.

Now, we have to show that it gives the correct result. The proof is by induction on $d(b)$ (taking the base case of the induction to be $b = 0$). In order to do this, we need to show two things:

(a) the greatest common divisor of a and 0 is a;
(b) if $a = bq + r$, then the greatest common divisor of a and b is equal to the greatest common divisor of b and r (up to associates).

The first fact should be clear if $a \neq 0$; you should think about it and convince yourself that it also holds if $a = 0$. (Use the definition of greatest common divisor, rather than any prejudices about greatest integers, etc.)

For the second point, observe that any divisor of a and b also divides $r = a - bq$, while any divisor of b and r also divides $a = bq + r$. So the set of all common divisors of a and b is the same as the set of common divisors of b and r, and the greatest common divisors must be associates, as required.

As we saw in case of the integers, the Euclidean Algorithm has another feature; it can be used to express the g.c.d. of a and b as a linear combination of these two elements. This is also true in general:

Enriched Euclidean Algorithm   Let R be a Euclidean domain.
   Input: Two elements $a, b \in R$.
   Output: An element $c \in R$ which is a greatest common divisor of a and b, together with two elements x and y such that $c = ax + by$.
   Operation: If $b = 0$, then we set $c = a$, $x = 1$, $y = 0$.
   Otherwise, we write $a = bq + r$ with $r = 0$ or $d(r) < d(b)$ as usual, and apply the algorithm to b and r. Suppose that the output is c', x', and y'. Then we put $c = c'$, $x = y'$, and $y = x' - y'q$.

The proof that this algorithm terminates, and finds the g.c.d. correctly, is exactly as for the original version. We have to show that $c = xa + yb$. This is obvious in the first case of the algorithm. In the second case (arguing, as before, by induction), we may assume that $c' = bx' + ry'$. Then

$$c = c' = bx' + (a - bq)y' = ay' + b(x' - qy'),$$

as required.

Now we turn to some theoretical properties of Euclidean domains.

Proposition 2.23   *(a) A Euclidean domain is an integral domain.*

(b) If R is a Euclidean domain and a, b are non-zero elements with $a \mid b$ and $d(b) = d(a)$, then a and b are associates.

Proof (a) If a and b are non-zero, then $d(ab) \geq d(a)$, so $ab \neq 0$.

(b) By condition (b), $a = bq + r$, where $r = 0$ or $d(r) < d(a)$. Suppose that $r \neq 0$. Then a divides b, so a divides $r = a - bq$; by condition (a), $d(r) \geq d(a)$, a contradiction. So we must have $r = 0$, whence $a = bq$. So each of a and b divides the other, and these elements are associates. □

Now we come to the main result:

Theorem 2.24 *(a) A Euclidean domain is a principal ideal domain.*
(b) A Euclidean domain is a unique factorisation domain.

Proof (a) Let R be a Euclidean domain. Take any ideal $I \in R$: we have to show that I is a principal ideal. The argument is similar to our determination of the ideals in $\mathbb{Z}$. First, $0 \in I$; and, if I consists only of 0, then $I = \langle 0 \rangle$, and so I is principal. So we may suppose that I contains some non-zero elements. Choose an element $a \in I$ such that $d(a)$ is as small as possible. (This depends on the fact that the values of d are non-negative integers, so there is necessarily a smallest one.) We claim that $I = \langle a \rangle$. As usual, we have to show that any element of either set is contained in the other. First, take $x \in \langle a \rangle$; then x is of the form $x = ar$ for some $r \in R$, and so $x \in I$ (since $a \in I$ and I is an ideal). Conversely, take $x \in I$. By part (b) of the definition of a Euclidean function, we write $x = aq + r$, where $r = 0$ or $d(r) < d(a)$. Now $x \in I$ and $aq \in I$, so $r = x - aq \in I$; and, since a was chosen as an element of I with $d(a)$ as small as possible, it cannot happen that $d(r) < d(a)$, so we must have $r = 0$, and $x = aq \in \langle a \rangle$. Thus indeed $I = \langle a \rangle$.

(b) If we had proved Theorem 2.21, this would immediately follow from (a). Since we didn't do that, we have some work to do. We showed that factorisation is unique (up to order and associates) in any PID; so we only have to do the other part, to prove that any element (other than zero and units) has a factorisation in R. So take $a \in R$ with $a \neq 0$ and a not a unit. We show by induction on $d(a)$ that a has a factorisation. In other words, we assume that any element b with $d(b) < d(a)$ has a factorisation.

If a is irreducible, then we have a factorisation (with only one factor!), so suppose that $a = bc$, where neither b nor c is a unit. Now by condition (a), $d(b) \leq d(a)$ and $d(c) \leq d(a)$. If $d(b) < d(a)$ and $d(c) < d(a)$, then by the inductive hypothesis, both b and c have factorisations; combining these gives a factorisation of a. So we can suppose that $d(b) = d(a)$. But then a and b are associates, by part (b) of the Proposition: a contradiction. The proof is finished. □

Let us summarise our findings. We have three classes of integral domains:

- unique factorisation domains;
- principal ideal domains;
- Euclidean domains.

Theorems 2.21 and 2.24 show that

$$ED \Rightarrow PID \Rightarrow UFD.$$

We see the increasing strength of these conditions by looking at the facts about greatest common divisors:

- In a UFD, any two elements have a g.c.d.;
- In a PID, any two elements a, b have a g.c.d. d, and $d = xa + yb$ for some x, y;
- In a ED, any two elements a, b have a g.c.d. d, and $d = xa + yb$ for some x, y; moreover, d, x, y can be found by using the Euclidean Algorithm.

You might expect here an example of a PID which is not a ED. Such rings do exist; T. S. Motzkin showed that the ring

$$R = \{a + b\sqrt{-19} : \text{ either } a, b \in \mathbb{Z} \text{ or } a - \tfrac{1}{2}, b - \tfrac{1}{2} \in \mathbb{Z}\}$$

is an example. But the proof is more difficult. (You can read it in volume 55 of the *Bulletin of the American Mathematical Society*, starting on page 1142.)

Exercise 2.22 Show that, in a commutative ring with identity, no element can be both a zero-divisor and a unit.

Exercise 2.23 Write down the associate classes in the ring $\mathbb{Z}_{12}$.

Exercise 2.24 (∗) Find all positive integers m with the property that every unit a in $\mathbb{Z}_m$ satisfies $a^2 = 1$.

Exercise 2.25 Let R be an integral domain. Show that the units of $R[x]$ are precisely the constant polynomials which are units of R.

Exercise 2.26 (a) Let F be a field. Show that a matrix $A \in M_n(F)$ is a unit if and only if $\det(A) \neq 0$; and that a non-zero matrix A is a zero-divisor if and only if $\det(A) = 0$.

(∗) (b) More generally, let R be a commutative ring with identity. Prove that $A \in M_n(R)$ is a unit if and only if $\det(A)$ is a unit in R. [*Hint:* If $\det(A)$ is a unit, use the 'cofactor formula' to find an inverse of A. Conversely, if $AB = BA = I$, take determinants to show that $\det(A)$ is a unit.]

Exercise 2.27 Let x be an element in a ring with identity, and suppose that $x^n = 0$ for some positive integer n. Prove that $1 + x$ is a unit. [*Hint:* $(1+x)(1-x+x^2-x^3+\cdots) = 1$.]

Exercise 2.28 (a) Find the greatest common divisor of the real polynomials $f(x) = x^2 + 3x + 2$ and $g(x) = x^5 + 2x^4 + 5x^3 + 6x + 2$.

(b) Give a simple description of the ideal $\langle f(x), g(x) \rangle$ of $\mathbb{R}[x]$.

Exercise 2.29 (a) Show that the ring $R = \{x + yi : a, b \in \mathbb{Z}\}$ of Gaussian integers is a Euclidean domain, with Euclidean function $d(x + yi) = x^2 + y^2$. [*Hint:* For (b), take

$a, b \in R$ with $b \neq 0$, and write $a/b = u + v\mathrm{i}$ in $\mathbb{C}$, where u and v are rational numbers. Then choose $m, n \in \mathbb{Z}$ such that

$$|(u + v\mathrm{i}) - (m + n\mathrm{i})| \leq 1/\sqrt{2},$$

by considering lattice points with integer coordinates in the complex plane.]

(∗) (b) Show that the ring of Eisenstein integers (Exercise 2.16(b)) is a Euclidean domain, by using the triangular lattice similarly.

Exercise 2.30 (a) Describe the units in the ring of Gaussian integers.

(∗) (b) Show that the irreducibles in the ring of Gaussian integers are of two types: primes $p \in \mathbb{Z}$ which cannot be written as the sum of two integer squares; and elements $x + y\mathrm{i}$, where $x^2 + y^2$ is a prime in $\mathbb{Z}$. (For example, 3 is a 'Gaussian prime'; 5 is not, since $5 = (2 + \mathrm{i})(2 - \mathrm{i})$, but the two factors are both Gaussian primes.)

Remark A theorem of Number Theory asserts that a prime p can be expressed as the sum of two squares if and only if $p = 2$ or $p \equiv_4 1$.

Fields

Recall that a field is a commutative ring with identity in which division is possible by non-zero elements. Rings are easy to build: we have seen polynomial rings, matrix rings, Boolean rings, cartesian products. Almost always, these rings turn out not to be fields. In fact, there are only two standard methods of constructing fields: applied to the integers, they produce the rationals, and the integers mod p.

2.14 Field of fractions. The first method involves going from a ring to its 'field of fractions', which generalises the construction of the rational numbers from the integers.

Let R be an integral domain. A field F is a **field of fractions** of R if

(a) R is a subring of F;
(b) Any element of F can be written in the form ab^{-1} for some $a, b \in R$ (where b^{-1} is calculated in F).

For example, the rational numbers $\mathbb{Q}$ form the field of fractions of the integers $\mathbb{Z}$. Any field is its own field of fractions.

Theorem 2.25 *Any integral domain has a field of fractions.*

Proof Let R be an integral domain. We let S be the set of all ordered pairs (a, b) for $a, b \in R$, $b \neq 0$. We intend that the ordered pair (a, b) will represent the element ab^{-1}. But, of course, $ab^{-1} = cd^{-1}$ if (and only if) $ad = bc$; so we want the ordered pairs (a, b) and (c, d) to represent the same element of F if this condition holds. Accordingly, we define an equivalence relation $\sim$ on S by the rule

$$(a, b) \sim (c, d) \text{ if and only if } ad = bc.$$

We prove that this really is an equivalence relation. First, $(a, b) \sim (a, b)$, since $ab = ba$ (R is commutative). Then, if $(a, b) \sim (c, d)$, then $ad = bc$, and so $cb = da$; this means $(c, d) \sim (a, b)$. Finally, suppose that $(a, b) \sim (c, d)$ and $(c, d) \sim (e, f)$. Then $ad = bc$ and $cf = de$. So

$$adf = bcf = bde,$$

and by the cancellation law (since $d \neq 0$) we deduce that $af = be$; so $(a, b) \sim (e, f)$. So $\sim$ really is an equivalence relation.

Now we let $[a, b]$ denote the equivalence class of the ordered pair (a, b): $[a, b] = \{(c, d) : ad = bc\}$. Let F be the set of equivalence classes. We define addition and multiplication on F by the following rules:

$$[a, b] + [c, d] = [ad + bc, bd],$$
$$[a, b] \cdot [c, d] = [ac, bd].$$

(To see where these definitions come from, work out what you would expect $(ab^{-1}) + (cd^{-1})$ and $(ab^{-1})(cd^{-1})$ to be, by the usual rules for fractions.)

We have to check that these operations are well defined, and that they do indeed make F a field. All of this is straightforward checking. Finally, the map that takes a to the equivalence class $[a, 1]$ is a one-to-one homomorphism from R to F, so we can regard R as a subring of F. Moreover, the inverse of the element $[b, 1]$ is $[1, b]$ if $b \neq 0$; and $[a, b] = [a, 1][b, 1]^{-1}$. So, if we identify R with its image in F under this embedding, we see that F is indeed a field of fractions of R. $\square$

In fact, the field of fractions is unique (up to isomorphism). This is another way of saying that the only possible way to construct a field of fractions is the way we actually did it.

2.15   Maximal ideals and fields.   The second method of constructing fields generalises the passage from the integers to the integers modulo a prime: the field is constructed as a factor ring. To study this, first we need a different test for when a ring is a field.

Proposition 2.26   *Let R be a commutative ring with identity. Then R is a field if and only if the only ideals in R are $\{0\}$ and R itself.*

Proof   For the forward implication, suppose that R is a field. Take any ideal I of R. Suppose that $I \neq \{0\}$; we have to show that $I = R$, that is, that every element of R is in I. Certainly, some non-zero element is in I, say $a \in I$. Now, for any $x \in R$, we have $x = (xa^{-1})a \in I$. So $I = R$.

For the converse, let R be a commutative ring with identity whose only ideals are $\{0\}$ and R. We have to show that all its non-zero elements have inverses. So take $a \in R$ with $a \neq 0$. Let I be the ideal $(a) = aR$. Then $I \neq \{0\}$, since $a \in I$; so $I = R$. Thus, $1 \in I = aR$, so there exists $b \in R$ with $ab = ba = 1$, as required. $\square$

We say that the ideal I of the ring R is a **maximal ideal** of R if $I \neq R$ but there is no ideal J properly between I and R; that is, if J is an ideal with $I \subseteq J \subseteq R$, then $J = I$ or $J = R$.

Theorem 2.27 *Let R be a commutative ring with identity, and I an ideal of R. Then R/I is a field if and only if I is a maximal ideal of R.*

Proof This follows immediately from the proposition and the correspondence between ideals of R/I and ideals of R containing I given by the Second Isomorphism Theorem (Theorem 2.11). □

How do we recognise maximal ideals?

Proposition 2.28 *Let R be a principal ideal domain, and take $a \in R$ with $a \neq 0$. Then $\langle a \rangle$ is a maximal ideal of R if and only if a is irreducible.*

Proof In an integral domain, we have $\langle a \rangle \subseteq \langle b \rangle$ if and only if b divides a, and $\langle a \rangle = \langle b \rangle$ if and only if a and b are associates. In particular, $\langle a \rangle = R$ if and only if a is a unit (an associate of 1: note that $\langle 1 \rangle = R$). Hence $\langle a \rangle$ is maximal if and only if every element b which divides a is either an associate of a or a unit; but this is exactly the condition that a is irreducible. Finally, if R is a principal ideal domain, then there are no other ideals to spoil the maximality of (a). □

Example $R = \mathbb{Z}$. We see that $\mathbb{Z}_n$ is a field if and only if n is prime. (In fact we knew this already. For we showed that m is a unit in $\mathbb{Z}_n$ if and only if m and n are coprime; and this holds for all non-zero residues mod n if and only if n is prime.)

We will apply this result to polynomial rings in the next section.

2.16 Field extensions, finite fields. The standard procedure for constructing the complex numbers from the real numbers is to 'adjoin' a square root of -1; that is, an element i satisfying $\mathrm{i}^2 + 1 = 0$. We will now describe this procedure, 'adjoining the root of a polynomial', in more detail.

Theorem 2.29 *Let F be a field, and f a polynomial which is irreducible in $F[x]$. Then there is a field K containing F and an element α satisfying $f(\alpha) = 0$.*

Proof The construction is simple. We set $K = F[x]/\langle f \rangle$. This is a field by the results of the last section: $F[x]$ is a principal ideal domain and we are given that f is irreducible, so $\langle f \rangle$ is a maximal ideal in $F[x]$, and $F[x]/\langle f \rangle$ is a field.

We have to show that

(a) K contains (a field isomorphic to) F;
(b) K contains a root α of f.

(a) Set $I = \langle f \rangle$. For $a \in F$, let $\overline{a}$ denote the coset $I + a$, and let $\overline{F}$ be the set of all such cosets. We show that the map $a \mapsto \overline{a}$ is an isomorphism from F to $\overline{F}$. It is one-to-one, since if $\overline{a} = \overline{b}$ then $b - a \in \langle f \rangle$, so $b - a = 0$ (as any non-zero element of $\langle f \rangle$ has degree at least as great as that of f). The homomorphism property is clear.

(b) We take α to be the coset $I + x$. Let

$$f(x) = a_n x^n + \cdots + a_1 x + a_0.$$

Then

$$
\begin{aligned}
f(\alpha) &= \overline{a_n}(I + x)^n + \cdots + \overline{a_1}(I + x) + \overline{a_0}(I + 1) \\
&= (I + a_n x^n) + \cdots + (I + a_1 x) + (I + a_0) \\
&= I + (a_n x^n + \cdots + a_1 x + a_0) \\
&= I + f(x) \\
&= I
\end{aligned}
$$

and we are done, since I is the zero element of $F[x]/I$.   □

Our proof shows that a field with the required properties exists. However, it is defined as a factor ring, which is not the most convenient form for calculation. As always, calculation in a factor ring is very much easier if we can make a good choice of coset representatives!

Let f be an irreducible polynomial of degree $n > 0$ over the field F. We claim that every coset of the ideal $\langle f \rangle$ in $F[x]$ has a unique representative r satisfying $r = 0$ or $\deg(r) < n$. Such an r exists because of the Euclidean property of $F[x]$. (If g is any polynomial, and $g = fq + r$, then g and r differ by a multiple of f, and so $\langle f \rangle + g = \langle f \rangle + r$.) If r_1 and r_2 are two representatives of the same coset with $r_i = 0$ or $\deg(r_i) < n$ for $i = 1, 2$, then $\langle f \rangle + r_1 = \langle f \rangle + r_2$, so $r_1 - r_2 \in \langle f \rangle$; this means that f divides $r_1 - r_2$. But since $\deg(f) = n$, this implies that $r_1 - r_2 = 0$, so $r_1 = r_2$.

Moreover, a simple argument (similar to the one in the above proof) shows that the coset $\langle f \rangle + r$ is equal to $r(\alpha)$, where α is the coset $\langle f \rangle + x$.

This means that

> Every element of $K = F[x]/\langle f \rangle$ can be uniquely expressed in the form
> $$c_0 + c_1 \alpha + c_2 \alpha^2 + \cdots + c_{n-1} \alpha^{n-1},$$
> where $c_0, c_1, \ldots, c_{n-1} \in F$.
> The addition and multiplication in K are given by the usual arithmetic rules, with the added condition that $f(\alpha) = 0$.

The construction of $\mathbb{C} = \mathbb{R}[x]/\langle x^2 + 1 \rangle$ is a familiar example: every complex number is uniquely expressible as $c_0 + c_1 i$, where $i^2 = -1$.

For another example, let us construct a finite field of order 4.

We start with the field $F = \mathbb{Z}_2$, with elements 0 and 1. Consider the polynomial $x^2 + x + 1 \in F[x]$. This polynomial is irreducible, since the only possible factorisation would be into two linear factors, which would imply that the polynomial has a root in F; but $0^2 + 0 + 1 \neq 0$ and $1^2 + 1 + 1 \neq 0$. Let

$K = F[x]/\langle x^2+x+1\rangle$, and let α be the coset $\langle x^2+x+1\rangle+x$, so that $\alpha^2+\alpha+1 = 0$. The field K has four elements: $K = \{0, 1, \alpha, \alpha + 1\}$. (This is our canonical representation above.) Letting $\beta = \alpha+1 = \alpha^2$ (noting that $x = -x$ in the field K), we obtain the following tables:

+	0	1	α	β		$\cdot$	0	1	α	β
0	0	1	α	β		0	0	0	0	0
1	1	0	β	α		1	0	1	α	β
α	α	β	0	1		α	0	α	β	1
β	β	α	1	0		β	0	β	1	α

If p is a prime, and n a positive integer, then a finite field of order p^n can be constructed in the same way if an irreducible polynomial of degree n over $\mathbb{Z}_p$ can be found. Galois showed that this is always possible:

Theorem 2.30 *For any prime number p and any positive integer n, there is an irreducible polynomial of degree n over $\mathbb{Z}_p$, and hence a finite field of order p^n.*

Exercise 2.31 Show that the polynomial $x^2 + 1$ is irreducible over $\mathbb{Z}_3$, and hence construct a field of order 9.

Exercise 2.32 Show that the polynomials x^3+x+1 and x^3+x^2+1 are both irreducible over $\mathbb{Z}_2$. Are the corresponding fields of order 8 isomorphic?

Exercise 2.33 Show that, if F is a field with q elements and f an irreducible polynomial of degree n over F, then the field $K = F[x]/\langle f\rangle$ has q^n elements.

Exercise 2.34 Prove that any two fields of fractions F_1 and F_2 of an integral domain R are isomorphic, where the isomorphism $\theta : F_1 \to F_2$ can be chosen so that its restriction to the subring R of F_1 is the identity map.

Exercise 2.35 A subset X of a ring R is called **multiplicatively closed** if $a, b \in X$ implies $ab \in X$.

(a) Prove that R is an integral domain if and only if the set of non-zero elements of R is multiplicatively closed.

*(b) Let R be a commutative ring with identity, and let X be a multiplicatively closed subset of R containing 1 but not 0. Define an equivalence relation $\sim$ on $R \times X$ by the rule that $(a, b) \sim (c, d)$ if and only if $ad = bc$. Define operations of addition and multiplication on the set F of equivalence classes of $\sim$ as in Section 2.14. Prove that F is a ring containing R, in which every element of X has an inverse, and every element of F can be written as ab^{-1}, where $a \in R$ and $b \in X$.

Appendix: Miscellany

We end with some miscellaneous topics.

2.17 Cage on zero. The American composer John Cage wrote the following. What is he talking about? (Think about this before reading the following discussion.)

> Curiously enough, the twelve-tone system has no zero in it. Given
> a series: 3, 5, 2, 7, 10, 8, 11, 9, 1, 6, 4, 12 and the plan of obtaining
> its inversion by numbers which when added to the corresponding
> ones of the original series will give 12, one obtains 9, 7, 10, 5, 2,
> 4, 1, 3, 11, 6, 8 and 12. For in this system 12 plus 12 equals 12.
> There is not enough of zero in it.
>
> John Cage (1968).

I contend that Cage is confusing two different zeros, the zero element of the
real numbers and the zero element of the integers mod 12.

Real numbers   Cage was very much attracted to the Zen concept of emptiness.
One of his most famous compositions, entitled 4′33″, involves a pianist sitting
at the keyboard of a piano for 4 minutes and 33 seconds without striking a note;
the audience notices the background noise (since no emptiness is truly empty).
The real numbers represent sound intensity, so zero is the absence of sound.

Integers mod 12   Musical notation is based on the fact that notes an octave
apart (that is, when the frequency of one is double that of the other) have a very
similar subjective effect in melodic terms. So we regard such notes as 'equivalent'.
More generally, two notes are equivalent if they are a whole number of octaves
apart.

In Western music, only a discrete set of notes is used. The octave is divided
into 12 intervals called **semitones**. Thus, the semitones appear (on a keyboard,
say), stretching to infinity in both directions like the integers. As above, two
semitones are equivalent if they differ by a whole number of octaves; that is,
if (as integers) they are congruent mod 12. So the musical scale, for thematic
purposes, has the structure of the integers mod 12. Various musical operations
fit into this framework. For example, transposition just involves adding a fixed
constant to each note. Inversion involves replacing each equivalence class by its
negative. (This is what Cage describes.)

Two kinds of zeros   The equivalence classes referred to are the congruence
classes mod 12, that is, the cosets of $12\mathbb{Z}$ in $\mathbb{Z}$. We can make any choice of coset
representatives we like. Mathematicians usually use $0, 1, 2, \ldots, 11$. Musicians use
12 instead of 0 as the representative of the class $12\mathbb{Z}$, so that their semitones are
labelled $1, 2, 3, \ldots, 12$.

Now Cage's arithmetic checks, since $-3 = 9$, $-5 = 7$, and so on, in $\mathbb{Z}_{12}$ (the
integers mod 12). The mathematician says $-0 = 0$, the musician $-12 = 12$; it is
exactly the same, just involving a different choice of coset representative.

So, contrary to what Cage says, there is a zero in the twelve-tone scale (but
musicians call it 12); and it has nothing to do with the real number zero, the
zero of intensity or absence of sound when the pianist is not striking the keys.

Footnote The title of Cage's piece mentioned above itself blurs the categories between different kinds of numbers. Four minutes and thirty-three seconds make 273 seconds; and -273 is the temperature of absolute zero in the Celsius scale. If he had used the Fahrenheit scale, Cage would presumably have titled his piece $7'39''$; but this duration may have taxed the patience of the audience too far!

2.18   Solution to Exercise 2.10.   Exercise 2.10 asks whether it is possible to have a ring whose elements are the integers, with (a) the same addition as $\mathbb{Z}$ but different multiplication, or (b) the same multiplication but different addition.

Part (a) is easy if you remember the definition of a zero ring from Section 2.2: if we are given the operation of addition satisfying axioms (A0)–(A4), and we define multiplication by $ab = 0$ for all a, b, we obtain a ring. (For a more challenging question, try to describe all the possible definitions of multiplication which would give a ring.)

Part (b) is much harder. If you tried this question, you probably attempted to write down an explicit rule for addition which would make R into a ring. I do not know how to do that. Instead, I will give here a solution which is non-constructive, and is an illustration of the concept of factorisation, which we discussed in Sections 2.10–2.13.

Let R be a ring in which the set of elements is $\mathbb{Z}$ and the multiplication is the same as that in $\mathbb{Z}$. We start by making a list of properties of R. Any property which is defined purely in terms of multiplication, which holds in $\mathbb{Z}$, will hold in R. Thus, we have the following:

(a) R is commutative.
(b) R has an identity element 1.
(c) R has no divisors of zero. (Thus, R is an integral domain.)
(d) R has just two units, 1 and z, where $z^2 = 1$. (In fact, in $\mathbb{Z}$, we have $z = -1$, but -1 depends on the addition, so we cannot assert that $z = -1$ here. Confusingly, z is the integer whose name is -1, but we do not know that it is the additive inverse of 1.)
(e) R has infinitely many irreducibles (by the theorem of Euclid).
(f) R is a unique factorisation domain.

In fact, these properties determine the multiplication in R completely. For, if they hold, then any non-zero element can be uniquely written as $u p_{k_1}^{a_1} \cdots p_{k_r}^{a_r}$, where $u = 1$ or z, $p_1, p_2, \ldots$ are the irreducibles (one from each associate class), and $a_1, \ldots, a_r$ are positive integers. Now the rule for multiplying these elements is clear.

This means that, if we can find a ring S different from $\mathbb{Z}$ having properties (a)–(f), then S will have the same multiplication as $\mathbb{Z}$, and so $R = S$ is a solution to the problem.

The simplest example of such a ring is the polynomial ring $F[x]$, where $F = \mathbb{Z}_3$ is the field of integers mod 3. This is a UFD (since it is a Euclidean domain); its units are the two non-zero constants; and Euclid's proof holds virtually unchanged to show that there are infinitely many irreducibles. (If there were only finitely many irreducibles, say $f_1, \ldots, f_r$, then the polynomial $f_1 \cdots f_r + 1$ would not be irreducible, but would not be divisible by any irreducible, a contradiction.)

For example, we might let the irreducible polynomials x, $x + 1$, $x - 1$, $x^2 + 1$, $\ldots$ correspond to the prime numbers 2, 3, 5, 7, $\ldots$. Then, using $\oplus$ for the new addition, we have $1 \oplus 1 = -1$, $3 \oplus 5 = -2$, $4 \oplus 1 = 7$, $7 \oplus 1 = 15$, and so on.

2.19   Ideals in matrix rings.   A commutative ring R with identity, whose only ideals are the trivial ones (namely, $\{0\}$ and R), is necessarily a field (see Proposition 2.24). This is false if we do not assume commutativity. The ring $M_n(F)$ of all $n \times n$ matrices over the field F has only the trivial ideals, as we shall see; but it is not a division ring for $n > 1$, since there are non-zero singular matrices.

Theorem 2.31   *Let R be a commutative ring with identity, and n a positive integer.*

(a) If S is an ideal of R, then $M_n(S)$ is an ideal of $M_n(R)$.

(b) Every ideal of $M_n(R)$ is of this form.

Proof   (a) If S is an ideal of R, then it is a ring, and so $M_n(S)$ is a ring, so (by definition) a subring of $M_n(R)$. Now take $A = (a_{ij}) \in M_n(S)$, and $X = (x_{ij}) \in M_n(R)$. The (i, j) entry of AX is

$$\sum_{k=1}^{n} a_{ik} x_{kj}.$$

Now $a_{ik} x_{kj} \in S$, since $a_{ik} \in S$ and S is an ideal. Summing over k then gives an element of S. So $AX \in M_n(S)$. Similarly $XA \in M_n(S)$. So $M_n(S)$ is an ideal of $M_n(R)$.

(b) Suppose that T is an ideal of $M_n(R)$. Let S be the set of elements of R which occur as entries in matrices in T. We show that S is an ideal of R and that $T = M_n(S)$.

Let E_{ij} denote the matrix with 1 in row i and column j, and 0 in all other positions. Also, let S' be the set of all elements $x \in R$ such that $xE_{11} \in T$. (Here xE_{11} is the matrix with x in the top left-hand corner and all other entries zero.)

Step 1 $S' = S$.

For clearly $S' \subseteq S$. Let $x \in S$; then there is a matrix $A = (a_{ij}) \in T$ such that $a_{pq} = x$. Now it is easily checked that

$$E_{1p} A E_{q1} = a_{pq} E_{11} = x E_{11}.$$

Since T is an ideal, $xE_{11} \in T$, so $x \in S'$.

Step 2 S is an ideal of R.

Take $x, y \in S$ and $r \in R$. Then $xE_{11}, yE_{11} \in T$. Then, for any $r \in R$, we have

$$(x + y)E_{11} = xE_{11} + yE_{11} \in T,$$
$$(rx)E_{11} = (rE_{11})(xE_{11}) \in T,$$
$$(xr)E_{11} = (xE_{11})(rE_{11}) \in T,$$

since T is an ideal of $M_n(R)$. So $x + y, rx, xr \in S$, and S is an ideal of R.

Step 3 $T = M_n(S)$.

By definition, $T \subseteq M_n(S)$. Suppose that $A = (a_{ij}) \in M_n(S)$. Then

$$A = \sum_{i,j=1}^{n} E_{i1}(a_{ij}E_{11})E_{1j} \in T,$$

since $a_{ij}E_{11} \in T$ by Step 1 and T is an ideal. $\qquad\square$

Thus, for example, the ring of 2×2 matrices over the ring of integers mod 4 has just three ideals:

- the zero ideal;
- the ideal consisting of matrices with every entry 0 or 2;
- the whole ring.

A ring R is defined to be **simple** if the only ideals in R are $\{0\}$ and R. Thus, any field (or, indeed, any division ring) is simple. From Theorem 2.31 we immediately conclude:

Corollary 2.32 *Let F be a field, and n a positive integer. Then $M_n(F)$ is a simple ring.*

Exercise 2.36 Recall the definition of the **direct product** $R \times S$ of two rings R and S: the elements of $R \times S$ are all ordered pairs (r, s), where $r \in R$ and $s \in S$; and the operations are componentwise, that is,

$$(r_1, s_1) + (r_2, s_2) = (r_1 + r_2, s_1 + s_2), \qquad (r_1, s_1) \cdot (r_2, s_2) = (r_1 r_2, s_1 s_2).$$

Let $R' = \{(r, 0) : r \in R\}$ and $S' = \{(0, s) : s \in S\}$. Prove that R' and S' are ideals of $R \times S$ isomorphic to R and S respectively.

Now suppose that T is a ring which contains ideals R and S having the property that every element of T can be written uniquely in the form $r + s$, where $r \in R$ and $s \in S$. Prove the following assertions:

(a) $R + S = T$ and $R \cap S = \{0\}$.
(b) If $r \in R$ and $s \in S$, then r and s commute (that is, $rs = sr$).
(c) The map θ from $R \times S$ to T given by $(r, s)\theta = r + s$ is an isomorphism.
(d) T is isomorphic to $R \times S$.

Exercise 2.37 Let R be a ring, and a an element of R. Remember that $\langle a \rangle$ means the **ideal generated by** a, which by definition is the smallest ideal of R containing a.

(a) Prove that $\langle a \rangle$ is the set of all elements of R of the form

$$na + sa + at + \sum_{i=1}^{m} s_i a t_i,$$

where m and n are integers, $m \geq 0$, and s, t, s_i, t_i are elements of R (and na has its usual meaning).

(b) Suppose that R has an identity. Show that the terms na, sa, and at can be dropped from the expression above.

(c) The element a is said to be **central** if it commutes with every element of R. Show that, if R has an identity and a is central, then

$$\langle a \rangle = aR = \{ar : r \in R\}.$$

(d) Give a description of $\langle a \rangle$ in the case where a is central but R does not necessarily have an identity.

Exercise 2.38 Let F be a field and n a positive integer. Let R be the ring $M_n(F)$ of $n \times n$ matrices over F. Let $a = E_{11}$ be the matrix with entry 1 in the first row and column, and all other entries zero. By Theorem 2.31, we know that $\langle a \rangle = R$. So, by part (b) of the preceding exercise, every element of R can be written in the form $\sum_{i=1}^{m} s_i a t_i$, for some elements $s_i, t_i \in R$. Show that there are elements of R which cannot be expressed as the sum of fewer than n terms of the form $s_i a t_i$, for $s_i, t_i \in R$.

Exercise 2.39 An element e of a ring is said to be an **idempotent** if $e^2 = e$.

(a) Let e be an idempotent in a ring with identity. Show that $1 - e$ is also an idempotent.

(b) Let R and S be rings with identity. Show that the elements $(1, 0)$ and $(0, 1)$ are central idempotents of the direct product $R \times S$, whose sum is the identity of $R \times S$.

(c) Conversely, suppose that T is a ring with identity and e is a central idempotent of T with $e \neq 0, 1$. Prove that $T \cong R \times S$, where $R = eT$ and $S = (1 - e)T$.

Exercise 2.40 An element r of a ring R is said to be **nilpotent** if $r^n = 0$ for some positive integer n.

(a) Prove that a non-zero nilpotent element is a zero divisor. Is the converse true?

(b) Prove that, in a commutative ring, the set of nilpotent elements is an ideal.

(c) Let n be a positive integer. Find all nilpotent elements of the ring $\mathbb{Z}_n$ of integers mod n. (Since, by the previous part they form an ideal, they must consist of all multiples of n^* for some integer n^* dividing n. Your job is to calculate n^* in terms of n.)

Exercise 2.41 Let R be a ring with identity. Prove that, if the element $r \in R$ is nilpotent, then $1 + r$ is a unit.

Exercise 2.42 Prove that, in a matrix ring $M_n(R)$, any strictly upper triangular matrix is nilpotent.

3 Groups

We now turn to the study of groups. A group is a set with a single binary operation. So groups are much less structured than rings. We will see that this gives a different flavour to the subject.

Groups and subgroups

3.1 Introduction. Groups resemble rings in many ways. The main difference is that the conditions defining a group are less stringent (only one operation and four axioms, half as many as for rings), so that examples of groups are more numerous and varied. But also, there is no 'canonical example' of a group, corresponding to the ring $\mathbb{Z}$, on which to base our definitions and test our intuition.

First, there is a problem of notation. Many examples of groups are based on number systems, as we will see. Sometimes, the operation is addition, and the identity element is 0; at other times, the operation is multiplication, and the identity is 1. So we define a group using terminology different from both of these, not carrying the freight of associations of plus or times.

A **group** is a set G with a binary operation $\circ$ satisfying the following laws:

(G0) (**Closure law**): For all $g, h \in G$, $g \circ h \in G$.
(G1) (**Associative law**): $g \circ (h \circ k) = (g \circ h) \circ k$ for all $g, h, k \in G$.
(G2) (**Identity law**): There exists $e \in G$ such that $g \circ e = e \circ g = g$ for all $g \in G$.
(G3) (**Inverse law**): For all $g \in G$, there exists $h \in G$ with $g \circ h = h \circ g = e$.

After we defined rings, we gave some extra conditions that select special classes of rings. We do the same here, but there is only one such class to be defined. We say that a group is **abelian**, or **commutative**, if it satisfies:

(G4) (**Commutative law**): $g \circ h = h \circ g$ for all $g, h \in G$.

(The term 'abelian' is much more common than 'commutative' in this context; it commemorates the mathematician N. Abel. In fact, many things are named after mathematicians who had some involvement in their discovery; but the ultimate accolade is that the word has passed so much into common usage that we use a lower-case letter for it.)

At this point, you should stop and compare the axioms (G0)–(G4) with the first five ring axioms (A0)–(A4). You will see that they are exact translations: we have put 'group G' in place of 'ring R', $\circ$ for $+$, e for the zero element 0, and used letters g, h, k instead of a, b, c. This brings us naturally to our first examples of abelian groups....

3.2 Examples of groups.

Example 1 It follows from our observation at the end of the last section that, if R is any ring, then $(R, +)$ (meaning the *set* R with the *operation* $+$, where we forget entirely about the multiplication), is an abelian group. So every ring gives us an abelian group, called the **additive group** of the ring.

Conversely, we saw in Section 2.2 that, given any abelian group R, where the group operation is written as $+$ and the identity as 0, we obtain a ring (a **zero ring**) by defining the ring multiplication by the rule

$$ab = 0 \text{ for all } a, b \in R.$$

So abelian groups are exactly the same as additive groups of rings. For this reason, in the study of abelian groups, it is customary to write the group operation as $+$ rather than $\circ$, and the identity as 0.

Example 2: Groups of units Here is another construction of a group from a ring, this time using the multiplication. It is not true that the ring with the operation of multiplication forms a group. However, we do find a group as follows:

Let R be a ring with identity. Let $U(R)$ be the set of units of R. (Recall that u is a unit if there exists $v \in R$ such that $uv = vu = 1$.) Now $(U(R), \cdot)$ is a group. To show this, we have to check the axioms. But most of the work is done for us.

Just after the definition of units in Section 2.10, we proved Proposition 2.14, with three parts:

- *The product of two units is a unit.* Thus, the units satisfy (G0).
- *The identity is a unit.* Thus, the units satisfy (G2).
- *The inverse of a unit is a unit.* Thus, the units satisfy (G3).

This leaves us with just the associative law to check; but the associative law holds for units because it holds for all elements of the ring, by (M1). So the claim is proved.

The group $U(R)$ is called the *group of units* of R.

Here are a few examples of this construction.

1. $U(\mathbb{Z}) = \{+1, -1\}$. So this set is a group, having just two elements, under the operation of multiplication.
2. If F is a field, then every non-zero element is a unit. Thus, $F \setminus \{0\}$ is a group under multiplication. This group is called the **multiplicative group** of the field, often written as F^*.
3. Again let F be a field, and let n be a positive integer. Then the set $M_n(F)$ of $n \times n$ matrices over F is a ring. A matrix A is a unit if and only if it is **invertible** (or **non-singular**); that is, $\det(A) \neq 0$. Thus, the group

$U(M_n(F))$ of units of F consists of all the invertible $n \times n$ matrices. This group is referred to as $\mathrm{GL}(n, F)$, and called the **general linear group**. (It is *linear* because matrices are the topic of Linear Algebra, and *general* because it is the largest possible group we could make out of $n \times n$ matrices with the operation of matrix multiplication.)

Example 3: Permutations   Let Ω be a set. A **permutation** of Ω is a one-to-one and onto function (a bijection) $\pi : \Omega \to \Omega$.

We write $x\pi$ for the image of the element x under the permutation π, rather than $\pi(x)$ as you might expect. The reasons are somewhat similar to those that we discussed in the last chapter for writing homomorphisms 'on the right'.

We define the operation of **composition** of permutations as follows: if π_1 and π_2 are permutations, their composition $\pi_1 \circ \pi_2$ is given by $x(\pi_1 \circ \pi_2) = (x\pi_1)\pi_2$. In other words, apply π_1, then π_2.

This shows the reason for writing permutations on the right of their arguments. If we wrote them on the left, then 'first π_1, then π_2' would be $\pi_2(\pi_1(x))$, and we would either have to set $(\pi_1 \circ \pi_2)(x) = \pi_2(\pi_1(x))$, or else redefine composition to mean 'first π_2, then π_1'. Life is complicated enough without that! But you are hereby warned: some people do exactly that.

How do we describe a permutation? In the case where Ω is finite, there are two commonly used notations, which we met in Chapter 1. Take Ω to be the set $\{1, 2, \ldots, n\}$.

Two-line notation for the permutation π: We write the elements $1, 2, \ldots, n$ in the top row of an array. Below the element x we write its image $x\pi$ under the permutation. So, for example,

$$\pi = \begin{pmatrix} 1 & 2 & 3 & 4 & 5 & 6 \\ 2 & 5 & 3 & 6 & 1 & 4 \end{pmatrix}$$

is the permutation which maps 1 to 2, 2 to 5, 3 to 3, and so on.

This notation enables us to count the number of permutations. The image of 1 (the number written under 1) can be any of the n elements of Ω. When it is chosen, the image of 2 can be any of the remaining $n - 1$ elements, the image of 3, any of the remaining $n - 2$, and so on. So the number of permutations is

$$n \cdot (n - 1) \cdot (n - 2) \cdots 2 \cdot 1 = n!,$$

the product of the numbers from 1 to n (or 'n factorial').

Cycle notation is more compact. Given the permutation π, choose a point 1; open a bracket and write 1, followed by its image under π, followed by its image, and so on, until the next step would return us to 1; then close the bracket. Then pick the smallest number not used so far, and repeat the procedure with it; continue until all numbers have been used. The sequences in brackets are called the **cycles** of π.

For example, the permutation π described above in two-line notation would be written as $(1,2,5)(3)(4,6)$. The calculation goes, '1 maps to 2, to 5, back to 1; then 3 maps to itself; then 4 maps to 6, back to 4'. So π has three cycles.

Note that the point 3, which is fixed by the permutation, lies in a cycle with just one element (a cycle of length 1). By convention, we simply omit cycles of length 1. So π would be written as $(1,2,5)(4,6)$. There is one exception to this convention. If we applied it to the identity, then everything would be omitted, and we would write nothing at all! Usually we put in a token cycle of length 1 and write the identity as (1).

Now let $\mathrm{Sym}(\Omega)$, the **symmetric group** on Ω, be the set of all permutations of Ω, with the operation of composition. We claim that it is a group.

(G0) If π_1 and π_2 are permutations, then so is $\pi_1 \circ \pi_2$.
- *Proof that it is one-to-one*: suppose that $x(\pi_1 \circ \pi_2) = y(\pi_1 \circ \pi_2)$. Then $(x\pi_1)\pi_2 = (y\pi_1)\pi_2$. Since π_2 is one-to-one, we see that $x\pi_1 = y\pi_1$; then since π_1 is one-to-one, $x = y$.
- *Proof that it is onto*: Given $x \in \Omega$, there exists $y \in \Omega$ such that $y\pi_2 = x$, since π_2 is onto; then there exists $z \in \Omega$ such that $z\pi_1 = y$, since π_1 is onto. Then $z(\pi_1 \circ \pi_2) = x$.

(G1) $x((\pi_1 \circ \pi_2) \circ \pi_3) = (x(\pi_1 \circ \pi_2))\pi_3 = ((x\pi_1)\pi_2)\pi_3$,
$x(\pi_1 \circ (\pi_2 \circ \pi_3)) = (x\pi_1)(\pi_2 \circ \pi_3) = ((x\pi_1)\pi_2)\pi_3$.
In other words, both $(\pi_1 \circ \pi_2) \circ \pi_3$ and $\pi_1 \circ (\pi_2 \circ \pi_3)$ say 'apply π_1, then π_2, then π_3'.

(G2) The identity permutation ϵ defined by $x\epsilon = x$ for all x (leaving everything where it is) satisfies $\epsilon \circ \pi = \pi = \pi \circ \epsilon$ for all permutations π.

(G3) If π is a permutation, it is a one-to-one and onto function, and hence has an inverse function σ, where $x\sigma = y$ if and only if $y\pi = x$. Then $\pi \circ \sigma = \sigma \circ \pi = \epsilon$.

If $\Omega = \{1, 2, \ldots, n\}$, we write the symmetric group $\mathrm{Sym}(\Omega)$ as S_n for brevity. Thus, S_n is a group with $n!$ elements. For example, S_3 consists of the six elements (in cycle notation) (1) (the identity), $(1,2,3)$, $(1,3,2)$, $(1,2)$, $(2,3)$, and $(1,3)$. It turns out that S_3 is the smallest non-abelian group.

Example 4: Automorphism groups Let R be a ring. An **automorphism** of R is an isomorphism $\theta : R \to R$; in other words, it is a permutation of R which happens also to be a homomorphism satisfying

$$(x + y)\theta = x\theta + y\theta, \quad (xy)\theta = (x\theta)(y\theta),$$

using the notation of Example 3.

Let $\mathrm{Aut}(R)$ be the set of all automorphisms of R. Then $\mathrm{Aut}(R)$ is a group, the **automorphism group** of R.

(G0) If θ_1 and θ_2 are automorphisms, so is $\theta_1 \circ \theta_2$: for

$$
\begin{aligned}
(x + y)(\theta_1 \circ \theta_2) &= ((x + y)\theta_1)\theta_2 \\
&= (x\theta_1 + y\theta_1)\theta_2 \\
&= (x\theta_1)\theta_2 + (y\theta_1)\theta_2 \\
&= x(\theta_1 \circ \theta_2) + y(\theta_1 \circ \theta_2),
\end{aligned}
$$

and similarly for multiplication.

(G1) As in Example 3.

(G2) The identity permutation is an automorphism of R.

(G3) The inverse of an automorphism is an automorphism.

Remark There is absolutely nothing special about rings here, unlike the situation in Examples 1 and 2. If $\mathcal{X}$ is any class of mathematical objects for which we can formulate the notion of homomorphism (or isomorphism), then $\mathrm{Aut}(X)$ is a group for any $X \in \mathcal{X}$.

For example, we will be able to talk about the automorphism group of a group, once we have defined group homomorphisms.

Example 5 Since a group is a set with a binary operation, a finite group can be specified by its operation table, or **Cayley table**, as it is usually called in this case, after Arthur Cayley, who pioneered the use of such tables.

For example, here is the Cayley table of a group with four elements:

$\circ$	e	a	b	c
e	e	a	b	c
a	a	e	c	b
b	b	c	e	a
c	c	b	a	e

You may recognise this as the additive group of the field with four elements which we constructed in Section 2.16, or of the Boolean ring of subsets of $\{1, 2\}$ in Section 2.2. It is an important enough group to have a name: it is the **Klein group**, or the **four-group**. The German translation of the latter name, *Vieregruppe*, gives rise to the notation V_4 for the group. (The first name commemorates the mathematician Felix Klein, not the fact that it is quite a small group.)

We will see further examples later on.

3.3 Properties of groups. Before proceeding further, we will change our notation for groups. As we saw, abelian groups are the same as additive groups of rings, and are usually written with the symbol $+$ for the group operation and 0 for the identity. However, most other groups have more in common with multiplicative systems. Accordingly, we will use juxtaposition instead of $\circ$ for the group operation. Often, we refer to gh as the *product* of g and h. We also use 1 instead of e for the identity element of a group, and g^{-1} for the inverse of the element g. (We will see the uniqueness of identity and inverses shortly.)

Many of these properties will remind you of similar properties for rings. Usually the proofs are almost identical; but they will be repeated here.

1. Products   The product $g_1 \cdots g_n$, strictly speaking, requires the insertion of brackets so that it can be evaluated, and potentially has many values according to how the brackets are inserted. However, in Proposition 2.1, we showed that the value of the sum of n elements of a ring is independent of the bracketing used to work it out. We remarked that the proof uses only the associative law. So the same is true for the product of n elements of a group.

2. Uniqueness of identity   The identity element of a group is unique. For suppose that e and f are two identities in a group. Then

$$e = ef = f.$$

We will denote the unique identity element by 1.

3. Uniqueness of inverses   The inverse of any group element is unique. For, if h and k are both inverses of g, then

$$h = h1 = h(gk) = (hg)k = 1k = k.$$

We will denote the inverse of g by g^{-1}.

Now our notation for a group is consistent with the notation for the group of units of a ring introduced in Section 2.10.

4. Properties of inverses   (a) $(gh)^{-1} = h^{-1}g^{-1}$. [For

$$(gh)(h^{-1}g^{-1}) = g(hh^{-1})g^{-1} = g1g^{-1} = gg^{-1} = 1,$$

and similarly the other way around.]
(b) $1^{-1} = 1$, clearly.
(c) $(g^{-1})^{-1} = g$. [For the equations $gg^{-1} = g^{-1}g = 1$ show that g has the properties of the inverse of g^{-1}; and inverses are unique.]

5. Cancellation Laws

(C1) (**Left Cancellation Law**): If $gx = gy$, then $x = y$.
(C2) (**Right Cancellation Law**): If $xg = yg$, then $x = y$.

For suppose that $gx = gy$. Then

$$x = 1x = (g^{-1}g)x = g^{-1}(gx) = g^{-1}(gy) = (g^{-1}g)y = 1y = y.$$

The proof of the right cancellation law is similar.

6. Exponents   Now we define g^n for any element $g \in G$, and any integer n. If $n > 0$, we define g^n to be the product of n factors equal to g. (As we noted in Point 1 above, this is well defined.) For $n = 0$, we set $g^0 = 1$, the identity. Finally, for $n < 0$, say $n = -m$, we define $g^{-m} = (g^m)^{-1}$. Now we have **laws of exponents**, as in elementary algebra:

(a) $g^m \cdot g^n = g^{m+n}$.
(b) $(g^m)^n = g^{mn}$.
(c) If $gh = hg$, then $(gh)^n = g^n h^n$.

Note that the third of these laws does not hold for arbitrary elements g, h. (We say that g and h **commute** if $gh = hg$. Like regular commuters, if g and h commute, then they can move back and forth in any product of gs and hs. Every time we find a factor $\cdots hg \cdots$ in the product, we can replace it by $\cdots gh \cdots$. In this way, we can bring the gs in front of the hs. In particular, the product $(gh)^n = ghgh \cdots gh$ is equal to $gg \cdots ghh \cdots h = g^n h^n$.) This proves the assertion for positive n. The remaining cases are left as an exercise.

3.4   Subgroups.   Let G be a group. A **subgroup** of G is a subset of G which, using the same operation as in G, is itself a group. We write $H \leq G$ to indicate that H is a subgroup of G (as opposed to $H \subseteq G$, which just means that H is a subset of G). If H is a subgroup of G, which is not the whole of G, we write $H < G$.

Let us consider the group axioms for H, a subset of G.

(G0) Closure requires that the product of any two elements of H is in H.
(G1) Since the associative law holds for any elements of G, it certainly holds for any elements of the subset H.
(G2) We require that the identity 1 of G should belong to H.
(G3) We require that the inverse h^{-1} of any element $h \in H$ should also belong to H.

So the associative law comes free, and we only have to check the other three axioms. If H is non-empty, we can dispense with the identity. For assume the closure and inverse laws, and take any $h \in H$. Then $h^{-1} \in H$ by the inverse law, and so $hh^{-1} = 1 \in H$ by the closure law. Our conclusion is as follows:

Theorem 3.1 (First Subgroup Test)   *Let H be a non-empty subset of the group G. Then H is a subgroup of G if and only if*

(a) for all $h_1, h_2 \in H$, we have $h_1 h_2 \in H$;
(b) for all $h \in H$ we have $h^{-1} \in H$.

Just as for rings, we can replace these two tests by a single one:

Theorem 3.2 (Second Subgroup Test)   *Let H be a non-empty subset of a group G. Then H is a subgroup if and only if, for all $h_1, h_2 \in H$, we have $h_1 h_2^{-1} \in H$.*

Proof Clearly the condition holds if H is a subgroup. So assume that it does hold. Take any $h \in H$. Then $hh^{-1} = 1 \in H$; and so $1h^{-1} = h^{-1} \in H$. So we have condition (b) of the First Subgroup Test. Now take $h_1, h_2 \in H$. Then $h_2^{-1} \in H$, and so $h_1(h_2^{-1})^{-1} = h_1 h_2 \in H$. So condition (a) holds too. $\square$

Remark If the group operation is written as $+$, then $h_1 h_2^{-1}$ would be written as $h_1 - h_2$. So this condition is the exact counterpart of the similar one in the Second Subring Test. We see this in the following example:

Example Let G be the additive group of the ring $\mathbb{Z}$. Find all subgroups of G.

We already found that any subring, and any ideal, is of the form $n\mathbb{Z}$ (all multiples of n) for some integer n. It turns out that the subgroups are exactly the same.

First, we use the Second Subgroup Test to show that $n\mathbb{Z}$ is a subgroup. Clearly it is non-empty. Take two elements of $n\mathbb{Z}$, say nx and ny. Their difference is $nx - ny = n(x - y) \in n\mathbb{Z}$. So $n\mathbb{Z}$ passes the test.

Now let H be an arbitrary subgroup. We follow the strategy we used for subrings. If H consists just of 0, then $H = 0\mathbb{Z}$; so suppose not. The inverse of a negative number is positive, so H must contain positive numbers; let n be the smallest positive number in H. Then any positive multiple of n can be obtained by adding n the appropriate number of times, and so is in H. Then $n(-x) = -nx$ for positive x, so negative multiples are in H; and clearly $0 \in H$. So H contains $n\mathbb{Z}$.

Conversely, take any number $m \in H$; we wish to show that n divides m. Divide m by n; that is, write $m = nq + r$, where $0 \le r < n$. By subtracting n q times from m (or adding it $-q$ times, if q is negative), we see that $r \in H$. Now n is the smallest positive number in H, and $r < n$; so necessarily $r = 0$, and n divides m, as required. So $H = n\mathbb{Z}$.

Exercise 3.1 Which of the following structures $(G, \circ)$ are groups?

(a) $G = \mathcal{P}(X)$, $A \circ B = A \triangle B$ (symmetric difference);
(b) $G = \mathcal{P}(X)$, $A \circ B = A \cup B$;
(c) $G = \mathcal{P}(X)$, $A \circ B = A \setminus B$ (difference);
(d) $G = \mathbb{R}$, $x \circ y = xy$;
(e) G is the set of positive real numbers, $x \circ y = xy$;
(f) $G = \{z \in \mathbb{C} : |z| = 1\}$, $x \circ y = xy$;
(g) G is the interval $(-c, c)$,

$$x \circ y = \frac{x + y}{1 + xy/c^2}$$

[this example describes the addition of velocities in Special Relativity];
(h) $G = \{a, b\}$, $a \circ a = a \circ b = a$, $b \circ a = b \circ b = b$;
(i) $G = \{a, b\}$, $a \circ b = b \circ a = a$, $a \circ a = b \circ b = b$.
   [In (a)–(c), $\mathcal{P}(X)$ is the set of all subsets of X, where X is a set with at least two elements.]

Exercise 3.2 (a) Show that the symmetric group S_3 is non-abelian, by finding two of its elements which do not commute.
(b) Show that S_n is non-abelian for any $n \geq 3$.

Exercise 3.3 Show that the following set of six matrices is a group:

$$\left\{ \begin{pmatrix} 1 & 0 \\ 0 & 1 \end{pmatrix}, \begin{pmatrix} 0 & 1 \\ -1 & -1 \end{pmatrix}, \begin{pmatrix} -1 & -1 \\ 1 & 0 \end{pmatrix}, \begin{pmatrix} 1 & 0 \\ -1 & -1 \end{pmatrix}, \begin{pmatrix} -1 & -1 \\ 0 & 1 \end{pmatrix}, \begin{pmatrix} 0 & 1 \\ 1 & 0 \end{pmatrix} \right\}.$$

Is it an abelian group?

Exercise 3.4 If A is a subset of a group G, we let $A^{-1} = \{a^{-1} : a \in A\}$. Also, for $A, B \subseteq G$, we let $AB = \{ab : a \in A, b \in B\}$. Prove that A is a subgroup of the group G if and only if $AA^{-1} \subseteq A$.

Exercise 3.5 Let $U(R)$ be the group of units of the ring $R = \{a + b\sqrt{2} : a, b \in \mathbb{Z}\}$. Is $U(R)$ finite or infinite?

Exercise 3.6 Let R be a commutative ring with identity element 1. Let S be the set of all solutions of $x^2 = 1$ in R. Show that S, with the operation of multiplication, is an abelian group.

Exercise 3.7 (a) Show that $(gh)^2 = g^2h^2$ if and only if $gh = hg$.
(b) Show that $(gh)^{-1} = g^{-1}h^{-1}$ if and only if $gh = hg$.
*(c) Show that, if there exists a number m such that the equation $(gh)^n = g^n h^n$ holds for $n = m$, $n = m+1$ and $n = m+2$, then $gh = hg$.

[Since $(gh)^0 = 1 = g^0 h^0$ and $(gh)^1 = gh = g^1 h^1$, we see that part (c) is 'best possible'—the equation holding for two consecutive values does not suffice to make g and h commute—and also that (a) and (b) are special cases of (c), taking $m = 0$, $m = -1$ respectively.]

Exercise 3.8 Let R be a ring. Show that $\mathrm{Aut}(R)$ is a subgroup of $\mathrm{Sym}(R)$.

Exercise 3.9 ($*$) Let G be a set with a binary operation $\circ$. (As usual, this presupposes that the closure law (G0) holds.) Suppose that in addition the following three axioms hold:

(a) the associative law (G1), that is, $(g \circ h) \circ k = g \circ (h \circ k)$ for all $g, h, k \in G$;
(b) there exists $e \in G$ such that $e \circ g = g$ for all $g \in G$;
(c) for any $g \in G$, there exists $h \in G$ such that $h \circ g = e$ (where e is as in (b)).

Prove that G is a group.

Exercise 3.10 ($*$) Let G be a set with a binary operation $\circ$. Suppose that g satisfies conditions (a) and (b) of Question 2, and also the following:

(c') for any $g \in G$, there exists $h \in G$ such that $g \circ h = e$ (where e is as in (b)).

Show that G need not be a group. [*Hint*: Take the operation defined by $g \circ h = h$ for all $g, h \in G$.]

Exercise 3.11 Prove the laws of exponents in a group (point 6 of Section 3.3).

Exercise 3.12 (∗∗) A group which contains elements a, b, c, d, e (none of them equal to the identity) such that

$$ab = c, \quad bc = d, \quad cd = e, \quad de = a, \quad ea = b.$$

Find the orders of the elements a, b, c, d, e.

Subgroups and cosets

3.5   Cosets.   Given a subgroup of a group, we can partition the group into cosets, much as we did for subrings of a ring. But there is a complication: because of the non-commutativity, a subgroup has two different kinds of cosets, *left cosets* and *right cosets*.

Let H be a subgroup of a group. We define two relations $\sim_L$ and $\sim_R$ on G, as follows:

- $g_1 \sim_L g_2$ if and only if $g_1^{-1} g_2 \in H$;
- $g_1 \sim_R g_2$ if and only if $g_2 g_1^{-1} \in H$.

Each of these is an equivalence relation. Here is the proof for $\sim_L$; try $\sim_R$ for yourself.

(Eq1) $g_1^{-1} g_1 = 1 \in H$, so $g_1 \sim_L g_1$; $\sim_L$ is reflexive.
(Eq2) If $g_1^{-1} g_2 \in H$, then

$$(g_1^{-1} g_2)^{-1} = g_2^{-1}(g_1^{-1})^{-1} = g_2^{-1} g_1 \in H,$$

so $g_1 \sim_L g_2$ implies $g_2 \sim_L g_1$; $\sim_L$ is symmetric.
(Eq3) If $g_1^{-1} g_2 \in H$ and $g_2^{-1} g_3 \in H$, then

$$(g_1^{-1} g_2)(g_2^{-1} g_3) = g_1^{-1}(g_2 g_2^{-1})g_3 = g_1^{-1} g_3 \in H,$$

so $g_1 \sim_L g_2$ and $g_2 \sim_L g_3$ imply $g_1 \sim_L g_3$; $\sim_L$ is transitive.

The equivalence classes of $\sim_L$ are called **left cosets**, while those of $\sim_R$ are called **right cosets**. We now give a more usable description of these cosets.

Proposition 3.3 *Any left coset of the subgroup H of G has the form $gH = \{gx : x \in H\}$, while any right coset has the form $Hg = \{xg : x \in H\}$.*

Proof We prove this for right cosets; the argument for left cosets is very similar.
Let X be the equivalence class of $\sim_R$ containing the element $g \in G$, so that

$$X = \{y \in G : g \sim_R y\} = \{y \in G : yg^{-1} \in H\}.$$

If $y \in X$, then $yg^{-1} = x \in H$, so $y = xg \in Hg$; and conversely. □

For example, let $G = S_3$, the symmetric group on the set $\{1, 2, 3\}$, and let H be the subgroup $S_2 = \{(1), (1, 2)\}$. The left cosets of H are

$$H = \{(1), (1, 2)\},$$
$$(1, 2, 3)H = \{(1, 2, 3), (2, 3)\},$$
$$(1, 3, 2)H = \{(1, 3, 2), (1, 3)\};$$

while the right cosets are

$$H = \{(1), (1, 2)\},$$
$$H(1, 2, 3) = \{(1, 2, 3), (1, 3)\},$$
$$H(1, 3, 2) = \{(1, 3, 2), (2, 3)\}.$$

[In more detail: The left coset $(1, 2, 3)H$ consists of the two elements $(1, 2, 3)(1)$ and $(1, 2, 3)(1, 2)$. the first is $(1, 2, 3)$, since (1) is the identity. To work out the second, remember that we compose permutations from left to right. So the composite maps 1 to 2, back to 1; 2 to 3 which is then fixed; and 3 to 1 to 2. The result is the permutation $(2, 3)$.]

Note that the left and right cosets are not the same in this case. However, there are equally many cosets of each type. This is not an accident.

Theorem 3.4 *Let H be a subgroup of the group G. Then there is a bijection between the left cosets and the right cosets of H in G; so there are equally many of each.*

Proof For any set X of elements of G, we put $X^{-1} = \{x^{-1} : x \in X\}$. Then $(X^{-1})^{-1} = X$. We show that, if X is a right coset, then X^{-1} is a left coset, and vice versa. So the correspondence $X \mapsto X^{-1}$ is the required bijection.

First note that $H^{-1} = H$. For H contains the inverse of each of its elements, so $H^{-1} \subseteq H$. Now, taking the inverse of both sides, we find that $H = (H^{-1})^{-1} \subseteq H^{-1}$. So equality holds.

Now let $X = Hg = \{hg : h \in H\}$ be a right coset. Then

$$X^{-1} = \{g^{-1}h^{-1} : h \in H\} = g^{-1}H^{-1} = g^{-1}H$$

is a left coset. The reverse implication is similar.   □

In the example, we see that the inverses of the first, second, and third left cosets are the first, third, and second right cosets, respectively.

3.6   Orders; Lagrange's Theorem.   The **order** of a group G, written $|G|$, is the cardinality of the set G, the number of elements in the group. This may be finite or infinite. Thus, the order of the symmetric group S_n is $n!$, while the order of the additive group of $\mathbb{Z}$ is infinite.

If H is a subgroup of G, the **index** of H in G, written $|G : H|$, is the number of right cosets of H in G. (By Theorem 3.4, we could have used left cosets, and the answer would be the same.)

Theorem 3.5 (Lagrange's Theorem) *Let H be a subgroup of the finite group G. Then*

$$|G| = |H| \cdot |G : H|.$$

In particular, the order of H divides that of G.

Proof We know that G can be written as the disjoint union of the right cosets of H. The number of cosets is $|G : H|$. The proof is finished if we can show that the number of elements in each coset is equal to $|H|$.

Define a function $f : H \to Hg$ by $f(h) = hg$. By our characterisation of right cosets, f is onto (this says that every element of Hg is of the form hg for $h \in H$). Now we show that f is one-to-one. Suppose that $f(h_1) = f(h_2)$. Then $h_1 g = h_2 g$. By the right cancellation law, $h_1 = h_2$. So f is indeed one-to-one, and is a bijection. Thus $|Hg| = |H|$, and the proof is complete. □

Remark This gives another proof of Theorem 3.4 in the case of a finite group. For exactly the same argument shows that each left coset has $|H|$ elements, so the numbers of left and right cosets are both equal to $|G|/|H|$.

We now define the order of an element of a group. This is quite a different concept from the order of the group; but we will see that there is a connection.

Let g be an element of a group G. If there exists a positive integer n such that $g^n = 1$, then the least such positive n is called the **order** of g. If no such n exists, then we say that g has **infinite order**.

Theorem 3.6 *(a) Let g be an element of the group G. Then the set*

$$\{g^m : m \in \mathbb{Z}\}$$

is a subgroup of G; its order is equal to the order of g.
(b) The order of any element of a finite group G divides the order of G.
(c) If g has finite order n, then $g^m = 1$ if and only if n divides m.

Proof Let H be the set $\{g^m : m \in \mathbb{Z}\}$. Take two elements of H, say g^p and g^q. Then $g^p(g^q)^{-1} = g^{p-q} \in H$. By the Second Subgroup Test, H is a subgroup.

If g has infinite order, then all of the powers g^m are distinct, since $g^p = g^q$ for $p > q$ implies $g^{p-q} = 1$. So H is infinite.

Suppose, on the other hand, that g has order n. If $m = nk$, then $g^m = (g^n)^k = 1$. Conversely, suppose that $g^m = 1$. Write $m = nq + r$ with $0 \le r < n$. Then $g^r = g^{m-nq} = 1$. We cannot have $r > 0$, since n is the smallest positive integer such that $g^n = 1$. So $r = 0$ and n divides m. This proves (c). Now (a) follows, since the argument shows that any power of g is equal to one of $g^0 = 1, g^1 = g, g^2, \ldots, g^{n-1}$.

Now (b) follows by applying Lagrange's Theorem to the subgroup H. □

Definition We use the notation $\langle g \rangle$ for the subgroup

$$\{g^n : n \in \mathbb{Z}\}$$

used in the preceding proof, and call it the **subgroup generated by** g. [*Note*: This is not the same as the ideal (a) generated by the element a of a ring. If we took all multiples of g by elements of G, we would obtain the whole group!]

3.7 Cyclic groups. A **cyclic group** is a group generated by a single element; that is, a group consisting of all the powers of one of its elements.

Example 1 The additive group of $\mathbb{Z}$ is a cyclic group of infinite order, while the additive group of $\mathbb{Z}_n$ is a cyclic group of finite order n for any positive integer n.

For any positive integer m can be written as $1 + 1 + \cdots + 1$ (m terms); this is the mth power of 1 (but written in additive notation!) The negative integers are the inverses of the positive ones, and so are negative powers of 1; zero is the zero-th power of 1.

Example 2 The set of complex numbers which are nth roots of unity forms a group with the operation of multiplication; this group is cyclic of order n.

For the nth roots of 1 are the complex numbers $e^{2\pi i k/n}$ for $k = 0, 1, \ldots, n-1$; they are all powers of $e^{2\pi i/n}$.

There is only one type of cyclic group of each possible order. (When we have formulated the notion of isomorphism for groups, we will see that any two cyclic groups of the same order are isomorphic.) We will denote the cyclic group of order n by C_n (including the possibility $n = \infty$).

Not every group is cyclic. In the first place, cyclic groups are necessarily abelian: for $g^m g^n = g^{m+n} = g^n g^m$ for all m, n. And not all abelian groups are cyclic. The Klein group V_4 (see Example 5 in Section 3.2) is abelian but not cyclic. Indeed, we can recognise cyclic groups as follows:

Proposition 3.7 *A finite group G of order n is cyclic if and only if it contains an element of order n.*

Proof If g has order n, then the elements $g^0 = 1, g^1 = g, \ldots, g^{n-1}$ of $\langle g \rangle$ are all distinct; since there are n of them, they comprise all of G. The converse is clear. $\qquad\square$

The Klein group has order 4, but all its elements except the identity have order 2. So it is not cyclic.

We can describe completely the subgroups of cyclic groups. For the infinite cyclic group (the additive group of $\mathbb{Z}$), we already did this in Section 3.4. For finite cyclic groups, the following result holds:

Theorem 3.8 *Let $G = \langle g \rangle$ be a cyclic group of order n. Then, for each divisor m of n, there is a unique subgroup of G of order m, which is a cyclic group generated by $g^{n/m}$; and these are all the subgroups of G.*

Proof Let H be a subgroup of G, and let k be the smallest positive integer such that $g^k \in H$. We claim that $g^l \in H$ if and only if k divides l. The proof is along now familiar lines. If $l = kq$, then $g^l = (g^k)^q \in H$. Conversely, suppose $g^l \in H$, and let $l = kq + r$ with $0 \leq r < k$. Then $g^r = g^{l-kq} \in H$, and so $r = 0$.

In particular, $g^n = 1 \in H$, so k divides n. Putting $m = n/k$, we see that H is generated by $g^{n/m}$, and that H has m elements $g^0 = 1, g^k, g^{2k}, \ldots, g^{(m-1)k}$. $\square$

Exercise 3.13 Let H be a subgroup of a group G. Show that any left coset of H is equal to a right coset of some subgroup (not necessarily H).

Exercise 3.14 (a) Let g be an element of a group G. Let $I = \{n \in \mathbb{Z} : g^n = 1\}$. Prove that I is an ideal of $\mathbb{Z}$. Use this to show that (a) and (c) of Theorem 3.6 hold.
(b) Apply a similar idea to the proof of Theorem 3.8.

Exercise 3.15 An **involution** in a group G is an element g having order 2; that is, such that $g^2 = 1$ but $g \neq 1$.

(a) Show that, if G is a group of odd order, then G contains no involutions.
(b) Show that, if G is a group of even order, then G contains at least one involution.
[*Hint*: Pair up the elements of G with their inverses. The only elements which are unpaired (because they are equal to their own inverses) are the identity and the involutions.]

Exercise 3.16 (∗) This exercise generalises part (b) of the preceding one. Prove the following:

Theorem 3.9 (Cauchy's Theorem) *Let G be a finite group, and p a prime number which divides the order of G. Then G contains an element of order p.*

Hint: Let

$$\Omega = \{(g_1, g_2, \ldots, g_p) : g_1 g_2 \cdots g_p = 1\},$$

a subset of the Cartesian power G^p. Let π be the following permutation of Ω:

$$(g_1, g_2, \ldots, g_p)\pi = (g_2, \ldots, g_p, g_1).$$

In other words, π shifts every coordinate back one place and moves the first coordinate to the end. Show that π really is a permutation of Ω, in other words, that if $(g_1, g_2, \ldots, g_p) \in \Omega$, then also $(g_2, \ldots, g_p, g_1) \in \Omega$.

Now decompose Ω into cycles of the permutation π. Show that

(a) if $g^p = 1$, then $(g, g, \ldots, g) \in \Omega$ and this element is fixed by π;
(b) all other elements of Ω belong to cycles of size p.

Show that $|\Omega| = |G|^{p-1}$. It follows that $|\Omega|$ is divisible by p. Since all cycles have size 1 or p, the number of fixed points is also divisible by p. But $(1, 1, \ldots, 1)$ is a fixed point; so there are at least $p - 1$ fixed points of the form $(g, g, \ldots, g)$ where g has order p.

Exercise 3.17 (∗) In this question we use Lagrange's Theorem to prove Fermat's Little Theorem.

Let m be a positive integer. Define $\phi(m)$ to be the number of integers x satisfying $0 \leq x \leq m - 1$ and $\gcd(x, m) = 1$. The function ϕ is called **Euler's totient function**.

(a) Show that the order of the group of units of $\mathbb{Z}_m$ is $\phi(m)$.

(b) Deduce that, if $\gcd(x, m) = 1$, then $x^{\phi(m)} \equiv_m 1$.

(c) If p is a prime number, show that $\phi(p) = p - 1$. Deduce that, if p does not divide x, then $x^{p-1} \equiv_p 1$.

(d) Hence show that, if p is a prime number, then $x^p \equiv_p x$ for all integers x.

Homomorphisms and normal subgroups

3.8 Definitions. The definition of a group homomorphism is almost identical to that of a ring homomorphism; the only difference is the simplification resulting from the fact that there is only one operation. As for rings, we write group homomorphisms 'on the right'.

Let G and H be groups. A **homomorphism** $\theta : G \to H$ is a function θ from G to H that satisfies the condition

$$(g_1 g_2)\theta = (g_1\theta)(g_2\theta)$$

for all $g_1, g_2 \in G$.

It follows from the definition that, if θ is a homomorphism, then

$1_G\theta = 1_H$, where 1_G and 1_H are the identity elements of G and H respectively;
$g^{-1}\theta = (g\theta)^{-1}$ for all $g \in G$.

A homomorphism that is one-to-one and onto is called an **isomorphism**. If there is an isomorphism from G to H, then we say that G and H are **isomorphic**. As is the case for rings, if two groups are isomorphic, then from the point of view of abstract algebra they are the same, even if their elements are completely different.

We have to deal with some unfinished business.

Theorem 3.10 *Two cyclic groups of the same order are isomorphic.*

Proof Let $G = \langle g \rangle$ and $H = \langle h \rangle$ be cyclic groups of the same order; that is, either both are infinite, or both have order n for some positive integer n.

We define a function $\theta : G \to H$ by the rule $g^m\theta = h^m$ for all $m \in \mathbb{Z}$. If G has infinite order, then the powers of g are all distinct, and θ is well defined. This is also true in the finite case. For suppose that $g^k = g^l$. Then $g^{k-l} = 1$, so n divides $k - l$. But then we have $h^{k-l} = 1$, so $h^k = h^l$.

Now θ is trivially a homomorphism, since

$$(g^k g^l)\theta = g^{k+l}\theta = h^{k+l} = h^k h^l = (g^k\theta)(g^l\theta).$$

It is clear that θ is onto. Finally, θ is obviously one-to-one if G is infinite; while, if G has order n, then

$$g^k\theta = g^l\theta \Rightarrow g^{k-l}\theta = 1 \Rightarrow h^{k-l} = 1 \Rightarrow n \mid (k - l) \Rightarrow g^{k-l} = 1 \Rightarrow g^k = g^l.$$

Thus θ is an isomorphism. $\qquad\square$

As mentioned earlier, we denote the cyclic group of order n by C_n.

We saw that a ring homomorphism blurs the structure of a ring; for groups, the situation is very similar.

Example We will find all homomorphisms from the group $\mathbb{Z}$ to itself. Let θ be a homomorphism. Suppose that $1\theta = n$. Then $2\theta = (1+1)\theta = n + n = 2n$, and similarly (by induction), $m\theta = mn$ for all positive integers m. Moreover, $0\theta = 0$, and for positive m we have $(-m)\theta = -m\theta = -mn$. So θ multiplies every integer by n.

So far, this is identical with the situation for rings. But now there is no multiplicative structure to restrict things further: for any m, the function θ_m that multiplies everything by m is a group homomorphism.

A homomorphism $\theta : G \to H$ is a function, and so has an image and a kernel in the sense of Section 1.15. As for rings, we simplify the definition of the kernel.

Definition Let $\theta : G \to H$ be a homomorphism of groups. The **image** of θ is

$$\mathrm{Im}(\theta) = \{h \in H : h = g\theta \text{ for some } g \in G\},$$

and the **kernel** of θ is

$$\mathrm{Ker}(\theta) = \{g \in G : g\theta = 1\}.$$

Proposition 3.11 *Let $\theta : G \to H$ be a group homomorphism. Then:*

(a) $\mathrm{Im}(\theta)$ is a subgroup of H.
(b) $\mathrm{Ker}(\theta)$ is a subgroup of G which has the additional property that, for any $x \in \mathrm{Ker}(\theta)$ and $g \in G$, we have $g^{-1}xg \in \mathrm{Ker}(\theta)$.
(c) Two elements of G are mapped to the same element of H under θ if and only if they lie in the same right coset of $\mathrm{Ker}(\theta)$.

Proof (a) We apply the subgroup test. Take $h_1, h_2 \in \mathrm{Im}(\theta)$. Then $h_1 = g_1\theta$ and $h_2 = g_2\theta$, for some $g_1, g_2 \in G$. Then

$$h_1 h_2^{-1} = (g_1\theta)(g_2\theta)^{-1} = (g_1 g_2^{-1})\theta \in \mathrm{Im}(\theta);$$

so $\mathrm{Im}(\theta)$ is a subgroup of H.

(b) Similarly, take $g_1, g_2 \in \mathrm{Ker}(\theta)$. Then $g_1\theta = g_2\theta = 1$; so

$$(g_1 g_2^{-1})\theta = (g_1\theta)(g_2\theta)^{-1} = 1;$$

so $\mathrm{Ker}(\theta)$ is a subgroup.

Now we check the extra condition. Suppose that $x \in \mathrm{Ker}(\theta)$ and $g \in G$. Then

$$(g^{-1}xg)\theta = (g\theta)^{-1} \cdot 1 \cdot (g\theta) = 1,$$

and so $g^{-1}xg \in \mathrm{Ker}(\theta)$.

(c) Suppose that $g_1\theta = g_2\theta$. Then $(g_2 g_1^{-1})\theta = 1$, so $x = g_2 g_1^{-1} \in \mathrm{Ker}(\theta)$; then $g_2 = xg_1 \in \mathrm{Ker}(\theta)g_1$, so g_1 and g_2 lie in the same right coset of $\mathrm{Ker}(\theta)$.

Conversely, if g_1 and g_2 lie in the same coset, say $g_2 = xg_1$ with $x \in \mathrm{Ker}(\theta)$; then $g_2\theta = (x\theta)(g_1\theta) = g_1\theta$, since $x \in \mathrm{Ker}(\theta)$. □

As in the case of rings, the extra property of the subgroup $\mathrm{Ker}(\theta)$ is so important that it is given a special name. A **normal subgroup** of a group G is a subgroup H of G such that, for any $x \in H$ and $g \in G$, we have $g^{-1}xg \in H$. We write $H \trianglelefteq G$ to indicate that H is a normal subgroup of G. If $H \trianglelefteq G$ and H is not equal to G, we write $H \triangleleft G$.

So we can say more briefly: The kernel of a homomorphism $\theta : G \to H$ is a normal subgroup of G. Normal subgroups play much the same role in group theory that ideals do in ring theory.

There are several equivalent definitions of a normal subgroup.

Theorem 3.12 *Let H be a subgroup of a group G. Then the following are equivalent:*

(a) for all $g \in G$, $x \in H$, we have $g^{-1}xg \in H$;
(b) for all $g \in G$, we have $g^{-1}Hg = H$;
(c) for all $g \in G$, we have $Hg = gH$.

Proof Since, by definition, $g^{-1}Hg$ is the set of all elements $g^{-1}xg$ for $x \in H$, we see that condition (a) can be rewritten as $g^{-1}Hg \subseteq H$. So (a) is implied by (b). Conversely, suppose that (a) holds, so that $g^{-1}Hg \subseteq H$ for all g. Replacing g by g^{-1}, we see that $gHg^{-1} \subseteq H$. Multiply this equation on the left by g^{-1} and on the right by g, to obtain $H \subseteq g^{-1}Hg$. So equality holds, and we have (b).

We get from (b) to (c) by multiplying on the left by g, and back again by multiplying by g^{-1}. □

Part (c) of the theorem says that a subgroup is normal if and only if its left and right cosets are the same. So our earlier example (with $G = S_3$ and $H = S_2$) of a subgroup with different left and right cosets is also an example of a non-normal subgroup.

Here are some simple tests which guarantee that a subgroup is normal.

Proposition 3.13 *Let H be a subgroup of a group G. Each of the following conditions implies that H is a normal subgroup:*

(a) G is abelian;
(b) H is finite and is the only subgroup of G of its order;
(c) H has index 2 in G.

Proof We illustrate the three parts of the preceding theorem.

(a) If G is abelian, then $g^{-1}xg = x$ for all $x, g \in G$. So test (a) applies.

(b) Suppose that H is the only subgroup of G of order m, for some finite m. It is not hard to show that $g^{-1}Hg$ is a subgroup of G, also of order m. So $g^{-1}Hg = H$ for any $g \in G$, and test (b) applies.

(c) The statement that H has index 2 in G implies that it has just two right cosets, one of which is H, so the other coset must be all the rest, namely $G \setminus H$. In the same way, H has just two left cosets, H and $G \setminus H$. Hence the left and right cosets coincide, and test (c) applies. □

3.9 Factor groups and isomorphism theorems. Let N be a normal subgroup of a group G. We are going to define a 'factor group' G/N whose elements are the cosets of N in G (left or right, it's the same, since N is normal). This works in much the same way as for factor rings. We then prove the exact analogues of the isomorphism theorems for rings. Since the proofs are virtually identical to the earlier ones, the discussion will be briefer.

Definition Let N be a normal subgroup in the group G. The **factor group**, or **quotient group**, G/N is the set of (left or right) cosets of N in G, with operation defined by

$$(Ng_1)(Ng_2) = Ng_1g_2.$$

Theorem 3.14 *The factor group, as defined above, is indeed a group.*

Proof First we have to check that the definition is a good one: that is, if g_1' and g_2' represent the same cosets as g_1 and g_2 respectively, then $g_1'g_2'$ represents the same coset as g_1g_2. So suppose that $Ng_1 = Ng_1'$ and $Ng_2 = Ng_2'$. Say $g_1' = xg_1$ and $g_2' = yg_2$, where $x, y \in N$. Then

$$g_1'g_2' = xg_1yg_2 = xzg_1g_2,$$

where the last equality holds because $g_1y \in g_1N = Ng_1$, and so g_1y is equal to zg_1 for some $z \in N$. But then $Ng_1'g_2' = Ng_1g_2$, as required.

Now the rest of the proof involves verifying the axioms, which is routine. The closure law needs no proof, since the operation is well defined. For the associative law (G1), we have

$$((Ng_1)(Ng_2))(Ng_3) = (Ng_1g_2)Ng_3 = N(g_1g_2)g_3,$$
$$(Ng_1)((Ng_2)(Ng_3)) = Ng_1(Ng_2g_3) = Ng_1(g_2g_3),$$

and the right-hand sides are equal, by the associative law for G. The identity of G/N is $N1 = N$, and the inverse of Ng is Ng^{-1}. □

The factor group comes as the image of a natural homomorphism, which (as in the case of rings) is called the **canonical homomorphism**. Remember that the *elements* of G/N are the *cosets* of N in G. Now define a map $\theta : G \to G/N$ by the rule that $g\theta = Ng$ for all $g \in G$. Checking that θ is a homomorphism is straightforward from the definition of the operation in G/N,

$$(g_1\theta)(g_2\theta) = (Ng_1)(Ng_2) = Ng_1g_2 = (g_1g_2)\theta.$$

The image of θ is G/N, since every coset has the form Ng for some $g \in G$. What is the kernel of θ? Since the identity element of G/N is the coset N, we have

$$\text{Ker}(\theta) = \{g \in G : Ng = N\} = \{g \in G : g \in N\} = N.$$

Hence we have proved the following:

Theorem 3.15   *The canonical homomorphism $\theta : G \to G/N$ defined by $g\theta = Ng$ for $g \in G$ is indeed a homomorphism; its image is G/N and its kernel is N.*

Now we return to our analysis of the image and kernel of an arbitrary homomorphism.

Theorem 3.16 (First Isomorphism Theorem)   *Let $\theta : G \to H$ be a group homomorphism. Then:*

(a) $\text{Im}(\theta)$ *is a subgroup of H;*
(b) $\text{Ker}(\theta)$ *is a normal subgroup of G;*
(c) $G/\text{Ker}(\theta) \cong \text{Im}(\theta)$.

Proof   We have already shown (a) and (b). For (c), there is only one reasonable definition of a map ϕ from G/N to H, where $N = \text{Ker}(\theta)$: we must put $(Ng)\phi = g\theta$ for all $g \in N$. As in the ring case, we can show that ϕ is well defined, that it is a homomorphism, that it is onto $\text{Im}(\theta)$, and that it is one-to-one.   □

The second and third 'Isomorphism Theorems' relating a group G to a factor group G/N also work as in rings.

Theorem 3.17 (Second Isomorphism Theorem)   *Let N be a normal subgroup of G. There is a one-to-one correspondence between the set of subgroups of G which contain N and the set of subgroups of G/N. Under this correspondence, normal subgroups of G containing N correspond to normal subgroups of G/N.*

Theorem 3.18 (Third Isomorphism Theorem)   *Let N be a normal subgroup of G and H a subgroup of G. Then:*

(a) $NH = \{nh : n \in N, h \in H\}$ *is a subgroup of G containing N;*
(b) $N \cap H$ *is a normal subgroup of H;*
(c) $H/(N \cap H) \cong (NH)/N$.

3.10   Conjugacy.   There is another equivalence relation defined on a group, which is very important.

Let G be a group. We say that elements x, y of G are **conjugate** (written $x \sim y$) if $y = g^{-1}xg$ for some $g \in G$.

Conjugacy is an equivalence relation, since it is

(Eq1) *reflexive*: $x = 1^{-1}x1$ for any $x \in G$.
(Eq2) *symmetric*: if $y = g^{-1}xg$, then $x = (g^{-1})^{-1}yg^{-1}$.
(Eq3) *transitive*: if $y = g^{-1}xg$ and $z = h^{-1}yh$, then $z = (gh)^{-1}x(gh)$.

Hence, G is the disjoint union of the equivalence classes, which are called **conjugacy classes**.

Conjugacy classes are closely related to normal subgroups:

Proposition 3.19 *The subgroup H of the group G is normal if and only if H is the union of some (possibly all) of the conjugacy classes of G.*

Proof One of the equivalent conditions for normality of H is that $g^{-1}Hg = H$ for all $g \in G$. But this says that $g^{-1}xg \in H$ for all $x \in H$ and all $g \in G$; that is, for every element of H, its entire conjugacy class is contained in H. $\square$

Another important property of conjugacy is the following.

Proposition 3.20 *Conjugate elements of a group have the same order.*

Proof

$$(g^{-1}xg)^n = g^{-1}xg \cdot g^{-1}xg \cdots g^{-1}xg = g^{-1}x^n g,$$

so $(g^{-1}xg)^n = 1$ if and only if $x^n = 1$. $\square$

We can calculate the size of a conjugacy class in terms of another subgroup of G. The **centraliser** of the element $x \in G$, written $C_G(x)$, is the set of all elements of G which commute with x:

$$C_G(x) = \{g \in G : gx = xg\}.$$

Theorem 3.21   *(a) For any element $x \in G$, $C_G(x)$ is a subgroup of G.*

(b) There is a bijection between the conjugacy class of an element x of G and the set of cosets of $C_G(x)$ in G.

*(c) (the **class equation**)*

$$\sum_i \frac{1}{|C_G(x_i)|} = 1,$$

where the elements x_i are representatives of the conjugacy classes.

Proof (a) If g and h commute with x, so do gh and g^{-1}.

(b) The bijection takes the conjugate $g^{-1}xg$ to the coset $C_G(x)g$. To show that it is a bijection, we must show that $g^{-1}xg = h^{-1}xh$ if and only if $C_G(x)g = C_G(x)h$. In fact, both are equivalent to the assertion that $gh^{-1} \in C_G(x)$.

(c) It follows from (b) and Lagrange's Theorem 3.5 that the size of the conjugacy class containing x is equal to $|G|/|C_G(x)|$. Now the sum, over a set of conjugacy class representatives, of the class sizes, is clearly equal to $|G|$. Dividing this equation by $|G|$ gives the result. $\qquad\square$

The **centre** of a group G, written $Z(G)$, is the set of elements of G which commute with every element of G:

$$Z(G) = \{x \in G : xg = gx \text{ for all } g \in G\}.$$

(The letter Z stands for the German *Zentrum*, 'centre'.)

Proposition 3.22 *The centre of a group G is a normal subgroup.*

To show that it is a subgroup, apply the subgroup test: if x and y commute with everything in G, then so does xy^{-1}. Now an element x belongs to $Z(G)$ if and only if x is conjugate only to itself; that is, its conjugacy class is $\{x\}$. Thus, $Z(G)$ is a union of conjugacy classes, and hence a normal subgroup.

We can use these ideas to show that, if the order of G is a prime power, then necessarily G has a non-trivial normal subgroup.

Theorem 3.23 *Let G have order p^n, where p is prime and $n > 0$. Then $Z(G) \neq \{1\}$.*

Proof The sum of the conjugacy class sizes in G is p^n. But each class size is a divisor of p^n, say p^{a_i}, for $i = 1, \ldots, m$. Suppose that k of these class sizes are equal to $p^0 = 1$, so that $|Z(G)| = k$. All of the others are powers of p which are at least p, and hence are divisible by p. So we obtain $k + lp = p^n$. It follows that k is divisible by p. Thus $Z(G)$ is a subgroup of G whose order is divisible by p, hence is not 1. $\qquad\square$

There are other applications of these ideas. For example, we say that two subgroups H and K of the group G are **conjugate** if $K = g^{-1}Hg$ for some $g \in G$. Again, this is an equivalence relation on subgroups. [Check first that, if H is a subgroup, then so is $g^{-1}Hg$.] Now a subgroup is normal if and only if it is conjugate only to itself. Moreover, conjugate subgroups have the same order. So, if H is the only subgroup of G of its order, then it is necessarily normal.

Exercise 3.18 Let F be a field, and let G be the set

$$\left\{ \begin{pmatrix} a & b \\ 0 & 1 \end{pmatrix} : a, b \in F, a \neq 0 \right\}$$

of 2×2 matrices over F. Let N be the set of all matrices in G with $a = 1$, and H the set of all matrices in G with $b = 0$.

(a) Prove that G is a group.
(b) Prove that N is a normal subgroup of G, which is isomorphic to the additive group of F.

(c) Prove that H is a subgroup of G, which is isomorphic to the multiplicative group of F. Is it normal?

(d) Prove that $G/N \cong H$.

Exercise 3.19 (∗) Prove that the group of the preceding exercise, in the case where $F = \mathbb{Z}_3$, is isomorphic to the symmetric group S_3.

Exercise 3.20 Let G be the set of all ordered pairs (a, b), where a and b are real numbers and $a \neq 0$. Define an operation $\circ$ on G by the rule that

$$(a_1, b_1) \circ (a_2, b_2) = (a_1 a_2, b_1 a_2 + b_2).$$

Prove carefully that G with this operation is a group.

Now show that G is isomorphic to the group of all permutations of the real numbers of the form $x \mapsto ax + b$, where $a \neq 0$.

Could you use this information to make the argument for the first part of the question easier?

Exercise 3.21 Let G be a group with the property that $G/Z(G)$ is cyclic. If $G/Z(G)$ is generated by the coset $Z(G)g$, show that every element of G can be written in the form zg^i for some $z \in Z(G)$ and some $i \in \mathbb{Z}$. Deduce that G is abelian (so that in fact, $Z(G) = G$).

Exercise 3.22 (∗) (a) Let G be a group. For any element $g \in G$, let ι_g be the function from G to G defined by $x\iota_g = g^{-1}xg$. Show that ι_g is an automorphism of G. [It is called the **inner automorphism** induced by the element g.]

(b) Show that the set $\{\iota_g : g \in G\}$ is a subgroup of $\mathrm{Aut}(G)$. [This subgroup is called the **inner automorphism group** of G, denoted $\mathrm{Inn}(G)$.]

(c) Show that the map $g \mapsto \iota_g$ is a homomorphism from G to $\mathrm{Aut}(G)$, whose image is the inner automorphism group $\mathrm{Inn}(G)$ and whose kernel is the centre $Z(G)$. Deduce that $\mathrm{Inn}(G) \cong G/Z(G)$.

(d) Show that $\mathrm{Inn}(G)$ is a normal subgroup of $\mathrm{Aut}(G)$. [By definition, the factor group $\mathrm{Aut}(G)/\mathrm{Inn}(G)$ is the **outer automorphism group** of G, denoted $\mathrm{Out}(G)$.]

Exercise 3.23 Find all subgroups of the symmetric group S_3. Which of them are normal subgroups? [S_3 is the group of all permutations of $\{1, 2, 3\}$.]

Exercise 3.24 Show that the group of all real numbers (with the operation of addition) is isomorphic to the group of positive real numbers (with the operation of multiplication).

Exercise 3.25 Let N be a normal subgroup of a group G, and let H be any subgroup. Show that NH is a subgroup of G. [Recall that NH is the set $\{nh : n \in N, h \in H\}$.] Which (if any) of the following statements are true?

(a) If H is a normal subgroup, then NH is a normal subgroup.

(b) If NH is a normal subgroup, then H is a normal subgroup.

Exercise 3.26 Let G_1 be the group of integers (with the operation of addition), and $G_2 = Z_n$, the group of complex nth roots of unity (with the operation of multiplication). Define a function $\theta : G_1 \rightarrow G_2$ by the rule

$$\theta(k) = e^{2\pi i k/n}$$

for $k \in G_1$.

(a) Prove that θ is a homomorphism. What are its image and kernel?

(b) Show that the cosets of the kernel of θ have the form

$$\{x : x \in \mathbb{Z}, x \equiv_n k\}$$

for $k = 0, 1, \ldots, n - 1$. Hence show that the additive group of integers mod n is isomorphic to Z_n.

Exercise 3.27 (a) Prove that C_2 is the only finite group which has just two conjugacy classes. [*Hint*: Use the class equation.]

(b) Find the conjugacy classes in S_3. [There are three of them.]

(c) Prove that no finite group of order greater than 6 can have three conjugacy classes.

**(d) Show that there is a function f such that a finite group with r conjugacy classes has order at most $f(r)$.

Exercise 3.28 Our first example in this chapter was a construction of an abelian group from a ring (the *additive group* of the ring). In this exercise, we reverse the procedure and construct a ring from an abelian group.

Let A be an abelian group. An **endomorphism** of A is a homomorphism from A to A. Let $\text{End}(A)$ be the set of all endomorphisms of A. We define two operations on $\text{End}(A)$, pointwise addition, and composition, as follows:

$$a(\theta + \phi) = a\theta + a\phi,$$
$$a(\theta\phi) = (a\theta)\phi.$$

Prove that $\text{End}(A)$, equipped with these operations, is a ring. (It is called, naturally, the **endomorphism ring** of A.)

Some special groups

3.11   Cayley's Theorem.   Before the rise of the axiomatic method in the late nineteenth century, group theory was already a flourishing subject; but, of course, the meaning of the term 'group' was different. A group always consisted of elements of some special type, with a specified composition law. Most commonly, a group was either a **permutation group** (whose elements are permutations of a set, and whose operation is composition of permutations), or a **matrix group** (whose elements are matrices, and whose operation is matrix multiplication). In modern terminology, we could say that the early group theorists studied subgroups of the symmetric group $\text{Sym}(\Omega)$ or of the general linear group $\text{GL}(n, F)$.

In order that this body of knowledge should not be lost, it is necessary to ensure that the new groups (axiomatically defined) are really the same as the old ones. We already showed in Section 3.2 that the symmetric group and the general linear group are groups in the axiomatic sense, and hence their subgroups are too. The point of Cayley's Theorem is to show the converse of this for permutation groups: that is, every group 'is' a permutation group. Of course this is not literally true, since the elements may not be permutations. In the spirit of abstract algebra, the correct statement goes like this:

Theorem 3.24 (Cayley's Theorem) *Every group is isomorphic to a permutation group (a subgroup of the symmetric group).*

Cayley proved this theorem by means of the Cayley table of the group. Before embarking on the general proof, we will work a particular case: the Klein group V_4. To begin the proof, we rewrite its Cayley table, calling the group elements $\{g_1, g_2, g_3, g_4\}$.

$\circ$	g_1	g_2	g_3	g_4
g_1	g_1	g_2	g_3	g_4
g_2	g_2	g_1	g_4	g_3
g_3	g_3	g_4	g_1	g_2
g_4	g_4	g_3	g_2	g_1

Now we use the columns of this table to define some permutations. Each column is labelled with one of the elements $g_1, \ldots, g_4$. It contains all the elements $g_1, \ldots, g_4$ in some order. We let π_j be the permutation which maps the number i to the index of the element in the ith row and jth column.

For example, in the second column, the elements are (g_2, g_1, g_4, g_3), and so π_2 is the permutation which maps 1 to 2, 2 to 1, 3 to 4, and 4 to 3; that is, in cycle notation, $\pi_2 = (1, 2)(3, 4)$. In the same way, we find the other three permutations: π_1 is the identity; $\pi_3 = (1, 3)(2, 4)$; and $\pi_4 = (1, 4)(2, 3)$.

Now the proof of Cayley's Theorem for this group consists in showing that $\{\pi_1, \pi_2, \pi_3, \pi_4\}$ is a group, and moreover, is isomorphic to the original Klein group (where the isomorphism maps g_i to π_i for $i = 1, 2, 3, 4$).

Proof of Cayley's Theorem Let $G = \{g_1, \ldots, g_n\}$. (This proof presupposes that the group G is finite; but in fact it works in the same way for infinite n, except that the set of indices of the gs will be infinite.) We take $\Omega = \{1, \ldots, n\}$, and find a subgroup of $\mathrm{Sym}(\Omega)$ isomorphic to G.

First, we define permutations $\pi_1, \ldots, \pi_n$ of Ω. For $1 \le i \le n$, let π_i be the function which maps j to k if $g_j g_i = g_k$ holds in G. (This corresponds as above to the ith column of the Cayley table of G.) This function is one-to-one: for, if $j\pi_i = l\pi_i = k$, then $g_j g_i = g_l g_i = g_k$, whence $g_j = g_l$ by the Right Cancellation Law, so $j = l$. Moreover, it is onto, since for any k, if $g_k g_i^{-1}$ is the element g_j, then $g_j g_i = g_k$ and so $j\pi_i = k$. Hence π_i is a bijection, and thus a permutation.

Now we define a map θ from G to $\mathrm{Sym}(\Omega)$ by the rule that $g_i\theta = \pi_i$. We claim that θ is a homomorphism, and that $\mathrm{Ker}(\theta) = \{1\}$.

Take $g_s, g_t \in G$. Suppose that $g_s g_t = g_u$. We have to show that $\pi_s \pi_t = \pi_u$, by applying both sides to an arbitrary element i and checking that the results are equal.

Let $i\pi_s = j$, $j\pi_t = k$, $i\pi_u = l$; we must show that $k = l$. By definition, we have

$$g_i g_s = g_j, \qquad g_j g_t = g_k, \qquad g_i g_u = g_l.$$

Hence

$$g_l = g_i(g_s g_t) = (g_i g_s)g_t = g_j g_t = g_k,$$

and so $l = k$ as required.

Now suppose that $g_i \in \mathrm{Ker}(\theta)$. Then π_i is the identity, so $j\pi_i = j$ for all j. But this means that $g_j g_i = g_j$, whence (by the Left Cancellation Law) $g_i = 1$. So $\mathrm{Ker}(\theta) = 1$.

Thus, by the First Isomorphism Theorem, G is isomorphic to $\mathrm{Im}(\theta)$, a subgroup of $\mathrm{Sym}(\Omega)$. $\qquad\square$

This is not the only way of finding a subgroup of a symmetric group which is isomorphic to a given group. For example, the proof of Cayley's Theorem shows that S_3 (a group of order 6) is isomorphic to a subgroup of S_6; but this group is given to us as a subgroup of S_3. Again, an entirely different way to realise the Klein group inside S_4 is given in Exercise 3.29.

As for matrix groups: Exercise 3.30 shows that every finite permutation group is isomorphic to a matrix group. Hence, by Cayley's Theorem, every finite group is isomorphic to a matrix group. However, not every infinite group is isomorphic to a matrix group.

3.12 Small groups. How many different groups are there? In this section, we will examine groups of small order (up to 8), and verify most of the entries in Figure 3.1, which gives the number of groups of given order (up to isomorphism).

Of course, there is only one group of order 1, since there is no choice about the operation! The next observation settles four of the remaining seven values.

Proposition 3.25 *A group of prime order is cyclic.*

Proof Let G be a group of prime order p. Take any element $g \in G$ which is not the identity. The order of g divides p, and is not 1, hence is p. Now we know that a group G containing an element g whose order is equal to $|G|$ is cyclic. (The p distinct powers of g must comprise the whole of G.) $\qquad\square$

Order	1	2	3	4	5	6	7	8
Number of groups	1	1	1	2	1	2	1	5

Fig. 3.1 Small groups

Of course, we also know that cyclic groups of the same order are isomorphic. This verifies the table entries for orders 2, 3, 5, and 7.

The remaining entries require more work.

Groups of order 4 Let G be a group of order 4. If G contains an element of order 4, then it is cyclic. So suppose not. Then the order of any element of G apart from the identity is equal to 2; in other words, $g^2 = 1$ (or, equivalently, $g^{-1} = g$) for all $g \in G$.

From this, we deduce that G is abelian. For example, take any two elements $g, h \in G$. Then

$$gh = g^{-1}h^{-1} = (hg)^{-1} = hg.$$

Let $G = \{1, a, b, c\}$. We construct the operation in G. We know the product of 1 with anything, and we know that $g^2 = 1$ for any element g. What is ab? It cannot be 1, since this implies $b = a^{-1} = a$; it cannot be a, since this implies $b = 1$ by cancellation; and similarly it cannot be b. So necessarily $ab = c$. In exactly the same way, the product of any two of a, b, c is the third; so the multiplication is determined.

We have shown that there are at most two non-isomorphic groups of order 4, the cyclic group and one other. Since we already know that two different groups exist (namely, the cyclic group C_4 and the Klein group V_4), the verification is complete.

Groups of order 6 We begin with a useful result.

Proposition 3.26 *Let G be a finite group in which every element g satisfies $g^2 = 1$. Then the order of G is a power of 2.*

Proof This follows from Cauchy's Theorem 3.9, proved in Exercise 3.16: for if $|G|$ were not a power of 2, it would be divisible by some odd prime p, and G would contain an element of order p. Here is a direct proof.

We showed above that G must be abelian. Now verify the following:

If H is a subgroup of G and $g \notin H$, then $H \cup Hg$ is a subgroup of G.

We prove this using the First Subgroup Test. Every element of G is equal to its inverse, so $H \cup Hg$ is closed under taking inverses. We have to show that it is closed under multiplication. Take two elements of $H \cup Hg$; each is of the form h or hg, for some $h \in H$. Since G is abelian, we can bring the gs to the end of the product; we find an element of one of the forms $h_1 h_2$, $h_1 h_2 g$, or $h_1 h_2 g^2$. By closure of H and the fact that $g^2 = 1$, this element is in $H \cup Hg$.

Now start with the identity and form an increasing chain of subgroups by applying this result as long as the current subgroup is not the whole of G. Eventually the process must terminate when we reach G. But each subgroup in the chain is twice as large as its predecessor, and so has order a power of 2. $\square$

We return to the matter in hand. Let G be a group of order 6. If G contains an element of order 6, then it is cyclic; so suppose not. Then any element of G has order 1, 2, or 3. Since 6 is not a power of 2, there must be an element a of order 3. Also, since 6 is even, there must be an element b of order 2 (see Exercise 3.15).

We claim that the elements $1, a, a^2, b, ab, a^2b$ of G are all distinct. Certainly the first three are all distinct, since a has order 3, and the last three are all distinct, by the Right Cancellation Law. Moreover, b is different from $1, a, a^2$, since it has order 2. Consider ab. If $ab = 1$, then $b = a^{-1}$; if $ab = a$, then $b = 1$; if $ab = a^2$, then $b = a$. All are impossible. Similarly for a^2b.

So $G = \{1, a, a^2, b, ab, a^2b\}$. It remains to determine the multiplication.

We will know how to multiply any two elements once we know which of the six elements is ba. For the only difficulty will occur when we multiply an element ending with b by an element beginning with a; any other product can be identified by the rules we have already. (For example, $(a^2)(a^2b) = a^4b = ab$.) Now ba is not equal to 1 (or $b = a^{-1}$); it is not a (or $b = 1$); it is not a^2 (or $b = a$); and it is not b (or $a = 1$). So $ba = ab$ or $ba = a^2b$.

If $ba = ab$, then $(ab)^n = a^nb^n$ for all n, by the Exponent Law. Then $(ab)^2 = a^2 \neq 1$, $(ab)^3 = b^3 = b \neq 1$. So the order of ab is not 1, 2, or 3; it must be 6, contrary to our case assumption. So it must be the case that $ba = a^2b$. Then the multiplication is determined, so there is at most one group (up to isomorphism).

Since we already know two groups of order 6 (the cyclic group C_6 and the symmetric group S_3), the entry in the table is verified.

Groups of order 8   This is more difficult (in part because there are five different groups), and we will simply outline the steps. Let G have order 8. If G contains an element of order 8, then G is cyclic; if every element g satisfies $g^2 = 1$, there is just one type of group. So we may assume that every element has order 2 or 4 (except for the identity), and that there is an element a of order 4.

Take any element b which is not a power of a. Then, as above, we find that $G = \{1, a, a^2, a^3, b, ab, a^2b, a^3b\}$. Now we need two pieces of information to determine the multiplication: we have to know which of these eight elements is b^2, and which is ba. We find that $b^2 = 1$ or $b^2 = a^2$, and that $ba = ab$ or $ba = a^3b$. This seems to give us four different groups, which added to the two already found would make six. But two of them are the same. In the group given by $b^2 = 1$ and $ba = ab$, set $b' = ab$; then $(b')^2 = a^2$ and $b'a = ab'$. So the same group arises in two different guises.

It remains to show that all five possibilities really are groups (Exercise 3.34).

3.13   Symmetric and alternating groups.   We have already met the symmetric group S_n, the group of all permutations of $\{1, \dots, n\}$. In this section, we will decide when two elements of S_n are conjugate, and find the normal subgroups of S_n for $n \leq 5$.

The conjugacy test will depend on the cycle notation for permutations, which we met in Section 3.2. To review, let π be a permutation of $\{1, \dots, n\}$. To write cycle notation for π, we do the following: Open a bracket and write any element of $\{1, \ldots, n\}$ (we might as well take 1, but it does not matter.) Then write the image of this point under successive applications of π, until this image returns to its starting point (which it does in a finite number of steps). Then close the bracket. While any elements of $\{1, \ldots, n\}$ remain, open a bracket, and follow the same procedure.

The **cycle structure** of π is the list of the lengths of its cycles. Each number occurs as many times in the list as there are cycles of that length. By convention, we write the cycle lengths in increasing order. Thus, the permutation $(2, 5)(3, 7, 8)(6, 9)$ of $\{1, \ldots, 9\}$ has cycle structure $1, 1, 2, 2, 3$. (Remember that we do not write cycles of length 1).

Here is a first indication of the kind of information that we can read off from the cycle structure.

Proposition 3.27 *The order of a permutation is equal to the least common multiple of the lengths of its cycles.*

Proof Let the cycle lengths of π be $a_1, \ldots, a_r$. If we compose π a_i times, then all points in the cycle of length i return to their starting positions. The same is true if we compose it any number of times which is a multiple of a_i. So, if we evaluate π^m, where m is divisible by all of $a_1, \ldots, a_r$, then every point returns to its starting position; that is, $\pi^m = (1)$. On the other hand, if m is not divisible by some a_i, say $m = a_i q + r$, where $0 < r < a_i$, then points in a cycle of length a_i are shifted r places along by π^m, so $\pi^m \neq (1)$. We conclude that $\pi^m = 1$ if and only if m is a common multiple of $a_1, \ldots, a_r$. So the **order** of π, which is the least positive m such that $\pi^m = 1$, is the least common multiple of $a_1, \ldots, a_r$. $\square$

Recall that two elements x, y of a group G are **conjugate** if $y = g^{-1}xg$ for some $g \in G$.

Theorem 3.28 *Two elements of S_n are conjugate if and only if they have the same cycle structure.*

Proof Suppose first that $y = g^{-1}xg$. Consider any cycle of x, say $(p_1, p_2, \ldots, p_k)$. This means that x maps p_1 to p_2 to $\ldots$ to p_k and back to p_1. Let $q_i = p_i g$ for $i = 1, \ldots, k$. We check that y maps q_1 to q_2 to $\ldots$ to q_k and back to q_1. For $i < k$, the effect of the composition $g^{-1}xg$ on q_i maps it to p_i (for $p_i g = q_i$, so $q_i g^{-1} = p_i$), then to p_{i+1}, then to q_{i+1}. Similarly, q_k goes to q_1. So $(q_1, q_2, \ldots, q_k)$ is a cycle of y. Every cycle arises thus, and the cycle decomposition of y is obtained. So y has the same cycle structure as x.

For the converse, suppose that x and y have the same cycle structure. Write y under x (both in cycle notation) so that cycles of y are under cycles of the same length of x. (Include cycles of length 1 in this step.) Then let g be the permutation that maps each point in the cycle notation for x to the point directly beneath it. The argument of the preceding paragraph shows that $g^{-1}xg = y$, so that x and y are conjugate. $\square$

Example Let $x = (1,4,6)(2,5)(3)$ and $y = (2,5,4)(1,6)(3)$. Then a permutation g such that $g^{-1}xg = y$ is given by $g = \begin{pmatrix} 1 & 2 & 3 & 4 & 5 & 6 \\ 2 & 1 & 3 & 5 & 6 & 4 \end{pmatrix}$ in two-line notation, or $(1,2)(4,5,6)$ in cycle notation. (It is pure coincidence that the cycle structure of g is the same as that of x and y. There is nothing unique about g; if we had written y as $(5,4,2)(6,1)(3)$, we would have obtained $g = \begin{pmatrix} 1 & 2 & 3 & 4 & 5 & 6 \\ 5 & 6 & 3 & 4 & 1 & 2 \end{pmatrix} = (1,5)(2,6)$.)

Let us use this result to compute the conjugacy classes in S_n for $n = 4, 5$, and hence to find the normal subgroups of these groups.

The group S_4 We list the possible cycle structures, and the number of elements of each possible structure, in the following table. (Ignore the column labelled 'Parity' for now.)

Cycle structure	# elements	Parity
$1,1,1,1$	1	E
$1,1,2$	6	O
$2,2$	3	E
$1,3$	8	E
4	6	O
	24	

How do we calculate the numbers? There is a general formula, but in this case it is easier to do it directly. There are $\binom{4}{2} = 6$ permutations of cycle structure $1,1,2$, since we must choose the two points to be transposed from $\{1,2,3,4\}$ and then the other two are fixed. The number for cycle structure $2,2$ is half of this, since each such element is made up of two transpositions, and each transposition occurs once in such a permutation. For type $1,3$, there are four choices of the fixed point, and two ways of permuting the other three in a 3-cycle. Finally, consider 4-cycles. We can take each to start (1, in cycle notation); then there are $3! = 6$ ways to fill in the other three numbers, each giving a different cycle. Of course, there is only one identity. As a check, these numbers add up to the group order.

To find the normal subgroups of S_4, we use the fact that a subgroup is normal if and only if it is a union of some of the conjugacy classes; of course the conjugacy class of the identity must be included. So first we solve the problem: how can we take some of the conjugacy class sizes which add up to a divisor of 24? (This last requirement comes from Lagrange's Theorem.)

If we include 1 but not 8, then (since all other class sizes are multiples of 3) the sum is congruent to 1 mod 3. The only such divisors of 24 are 1 and 4. The first corresponds to the identity only (which is a normal subgroup), and the second to the identity and the three permutations of cycle structure $2,2$ (which form the Klein group, also a normal subgroup).

If we include both 1 and 8, the only possible divisors are $12 = 1 + 8 + 3$, and 24, the sum of all the divisors. The latter corresponds to taking the whole of S_4, which is trivially a normal subgroup. The former case also gives a normal subgroup. This can be checked directly, but it is a special case of something much more general.

We define the **parity** of a permutation π to be the parity of $n - c(\pi)$, where $c(\pi)$ is the number of cycles in the cycle structure of π (including cycles of length 1). It is either even or odd. Check that the parities of elements of S_4 are correctly given in the above table. We also define the **sign** of a permutation π to be $(-1)^{n-c(\pi)}$: so odd and even parity correspond to $-$ and $+$ sign. We saw this definition in Chapter 1; now we will see a remarkable property.

We are going to show the following:

Theorem 3.29 *The map θ that takes a permutation π to its parity is a homomorphism from S_n to the group $\mathbb{Z}_2$ of integers mod 2. Its kernel, the set of all permutations of even parity, is a normal subgroup of S_n having index 2.*

Proof There are several steps to the proof.

Step 1 First we show that, if τ is a transposition (a permutation interchanging two points and fixing the rest), then π and $\pi\tau$ have opposite parity. To see this, we have to count cycles of $\pi\tau$. These are the same as the cycles of π except for the ones containing the two points interchanged by τ, say i and j. Now check that, if i and j are in the same cycle of π, this splits into two cycles of $\pi\tau$, while if they are in different cycles of π, these cycles get 'glued together' in $\pi\tau$.

So $c(\pi\tau) = c(\pi) \pm 1$, and the difference of 1 in either direction changes the parity.

Step 2 Any permutation is a product of transpositions. There are many different expressions, using different numbers of transpositions; but the parity of the number of transpositions needed to express π is equal to the parity of π.

To see that any permutation is a product of transpositions, we only need to check this for one cycle, since we can deal with the cycles separately. Now verify that

$$(1, 2, 3, \ldots, n) = (1, 2)(1, 3) \ldots (1, n).$$

The identity has n cycles of length 1, so its parity is even; and thus, by Step 1, the parity of a product of k transpositions is equal to the parity of k.

Step 3 Parity is a homomorphism. We need to show that the parity of $\pi_1\pi_2$ is the sum (mod 2) of the parities of π_1 and π_2. But this is clear from Step 2, on expressing π_1 and π_2 as products of transpositions. □

The normal subgroup of S_n consisting of permutations with even parity is called the **alternating group**, written A_n. Its order is $n!/2$. (The name comes from a different description of A_n. A function $f(x_1, x_2, \ldots, x_n)$ is called **alternating** if it changes sign when two of its variables are interchanged. The simplest example of such a function is $\prod_{i<j}(x_j - x_i)$, Now the alternating group consists of all permutations which, when applied to the variables $x_1, \ldots, x_n$, leave the value of an alternating function unchanged. The symmetric group is related to 'symmetric functions' in the same way.)

Remark It follows from the theorem that sign is a homomorphism from S_n to the multiplicative group $\{+1, -1\}$.

Now we proceed to find the normal subgroups of S_5. First, we list the conjugacy classes. (For convenience we have given them names.)

Name	Cycle structure	# elements	Parity
C_1	$1,1,1,1,1$	1	E
C_2	$1,1,1,2$	10	O
C_3	$1,2,2$	15	E
C_4	$1,1,3$	20	E
C_5	$2,3$	20	O
C_6	$1,4$	30	O
C_7	5	24	E
		120	

A normal subgroup N of S_5 is a union of conjugacy classes, including the class C_1. If N does not include C_7, then $|N| \equiv_5 1$, and $|N|$ divides 24, whence $|N| = 1$ and $N = \{1\}$. If, however, N does include C_7, then $|N| \geq 25$, whence $|N| = 40, 60,$ or 120, and N also includes C_3. Now

$$(1,2)(3,4) \cdot (1,2)(3,5) = (3,4,5),$$

so if N contains C_3, it also contains at least one element of C_4, whence N contains C_4. This leaves only two possibilities: either $N = C_1 \cup C_3 \cup C_4 \cup C_7 = A_5$, or $N = S_5$. We conclude:

Proposition 3.30 *The only normal subgroups of S_5 are $\{1\}$, A_5, and S_5.*

3.14 Symmetry groups. Some further examples of groups arise geometrically, as groups of symmetries of polygons and polyhedra.

Let $\mathcal{P}$ be a regular polygon with n vertices. Assume that its centre is at the origin. A **symmetry** of the polygon is a transformation which maps vertices of the polygon bijectively to vertices, and edges to edges. Alternatively, we can think of a symmetry as a transformation of the Euclidean space which maps the vertices and edges of the polygon to themselves.

There are two types of symmetries: **rotations** about the origin through multiples of $2\pi/n$ radians; and **reflections** in 'axes of symmetry' of the polygon.

There are clearly n rotations, each of which is obtained by composing the rotation through $2\pi/n$ an appropriate number of times. Hence, the rotations form a cyclic group of order n. Also, there are n reflections. If n is odd, each axis of symmetry joins a vertex to the midpoint of the opposite side. If n is even, however, there are two types of axes of symmetry: one type joins a pair of opposite vertices (there are $n/2$ of these), the other joins the midpoints of a pair of opposite sides (and there are also $n/2$ of these). The full group of symmetries is called the **dihedral group** of order $2n$, written D_{2n} (but note that some people call it D_n).

Figure 3.2 shows the axes of symmetry in the two cases.

Theorem 3.31 *The symmetry group of a regular n-gon is the dihedral group of order $2n$. It has a cyclic normal subgroup of order n consisting of rotations; every element outside this subgroup has order 2.*

How do we represent symmetries? One method is to regard the symmetry group as consisting of permutations of the vertices; we number the vertices from 1 to n and write down the permutations. Another is to think of them (as described above) as Euclidean transformations. Since we chose the polygon to have its centre at the origin, these transformations can be represented by matrices.

For example, consider the square with vertices at $(\pm 1, \pm 1)$, shown in Figure 3.3. Number its vertices anti-clockwise starting from the top left, as in the figure. Then the elements of the group of symmetries are as follows, as either

Fig. 3.2 Axes of symmetry

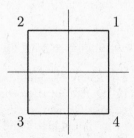

Fig. 3.3 Symmetries of a square

permutations or matrices (rotations first, then reflections):

$$(1), (1,2,3,4), (1,3)(2,4), (1,4,3,2),$$
$$(1,4)(2,3), (1,2)(3,4), (2,4), (1,3)$$

$$\begin{pmatrix} 1 & 0 \\ 0 & 1 \end{pmatrix}, \begin{pmatrix} 0 & 1 \\ -1 & 0 \end{pmatrix}, \begin{pmatrix} -1 & 0 \\ 0 & -1 \end{pmatrix}, \begin{pmatrix} 0 & -1 \\ 1 & 0 \end{pmatrix},$$

$$\begin{pmatrix} 1 & 0 \\ 0 & -1 \end{pmatrix}, \begin{pmatrix} -1 & 0 \\ 0 & 1 \end{pmatrix}, \begin{pmatrix} 0 & 1 \\ 1 & 0 \end{pmatrix}, \begin{pmatrix} 0 & -1 \\ -1 & 0 \end{pmatrix}$$

Note that a rotation has determinant $+1$, while a reflection has determinant -1. The map $A \mapsto \det(A)$ is a homomorphism from the symmetry group of the regular n-gon onto the multiplicative group $\{+1, -1\}$ (cyclic of order 2); the kernel is the rotation group (cyclic of order n).

We have only defined the dihedral group D_{2n} above for $n \geq 3$. However, with the geometrical approach, we can extend the definition to $n = 1$ and $n = 2$ as well, taking rotations through multiples of $2\pi/n$ and corresponding reflections. For $n = 1$, we have the identity rotation and one reflection, giving $D_2 \cong C_2$. For $n = 2$, we have a group of order 4 isomorphic to the Klein group, containing two rotations and two reflections: in matrix terms,

$$D_4 = \left\{ \begin{pmatrix} 1 & 0 \\ 0 & 1 \end{pmatrix}, \begin{pmatrix} -1 & 0 \\ 0 & -1 \end{pmatrix}, \begin{pmatrix} 1 & 0 \\ 0 & -1 \end{pmatrix}, \begin{pmatrix} -1 & 0 \\ 0 & 1 \end{pmatrix} \right\}.$$

In three dimensions, the figures analogous to regular polygons are **regular polyhedra**, which have regular n-gons for faces and regular m-gons for 'vertex figures' (obtained by slicing off a corner). There are only five of these, the so-called **Platonic solids**. To see that there cannot be more than five, recall that the internal angle in a regular n-gon is $\pi(1 - 2/n)$. The angles of the m faces surrounding a vertex must add up to strictly less than 2π. (If the sum was 2π, as for four squares or three hexagons, the figure would lie flat and not fold up; more than 2π would be even further from creating a polyhedron.) So we have

$$m\pi(1 - 2/n) < 2\pi,$$

whence $1/m + 1/n > 1/2$. This inequality has the solutions $(m, n) = (3,3), (3,4),$ $(4,3), (3,5),$ and $(5,3)$.

This shows that not more than five regular polyhedra can exist. But we can construct models of each of the five. In the order described above, they are the tetrahedron, hexahedron (cube), octahedron, dodecahedron, and icosahedron respectively. The Greek prefixes in the names of these figures stand for the total numbers of faces they have, namely 4, 6, 8, 12, and 20 respectively. Figures 3.4 and 3.5 show the regular polyhedra.

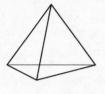
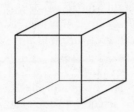
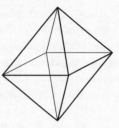

Fig. 3.4 Tetrahedron, cube, and octahedron

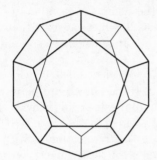

Fig. 3.5 Dodecahedron and Icosahedron

Theorem 3.32 *The properties of the five Platonic polyhedra are given in the following table:*

Name	Faces	Edges	Vertices	Rotation Group	Symmetry Group
Tetrahedron	4	6	4	A_4	S_4
Cube	6	12	8	S_4	$S_4 \times C_2$
Octahedron	8	12	6	S_4	$S_4 \times C_2$
Dodecahedron	12	30	20	A_5	$A_5 \times C_2$
Icosahedron	20	30	12	A_5	$A_5 \times C_2$

Proof You will be greatly helped in following this proof if you have models of the polyhedra available as you read. The models make the first three columns of numbers clear. The other thing to notice is that there is a 'duality' relation between the cube and the octahedron. If we take the six points at the centres of the faces of a cube as vertices, we obtain an octahedron, and vice versa. This explains why the number of faces of the cube is equal to the number of vertices of the octahedron, and vice versa. It also implies that these two figures have the same rotation group and the same symmetry group (thinking of these groups as Euclidean transformations fixing the figures in question). A similar duality relation holds between the dodecahedron and the icosahedron. The tetrahedron is 'self-dual': if we put vertices at its face centres, we obtain another tetrahedron.

Now, in each case, the order of the rotation group is the product of the number of faces and the number of edges of a face. For imagine that we have a frame on the table into which a face of the solid fits. We can specify a rotation by saying which face should go into the frame, and in which orientation. Thus we find that the rotation groups have orders 12, 24, 24, 60, 60 respectively. Also, in all cases the map $A \mapsto \det(A)$ is a homomorphism from the symmetry group onto the group $\{\pm 1\}$ whose kernel is the rotation group; so the symmetry group is twice as large as the rotation group.

It remains to identify the groups.

For the tetrahedron, there are four vertices, which are permuted by any symmetry; since there are $24 = 4!$ symmetries, the symmetry group must be S_4. The rotation group is a normal subgroup of index 2, which must be the alternating group A_4 (either by our determination of all normal subgroups of S_4 in the last section, or by inspection).

For the other figures, the argument is more dependent on the models. A cube has four diagonals (joining opposite vertices) which are permuted by its symmetries. No non-identity rotation can fix all the diagonals [why?], so the rotation group is S_4. However, the reflection represented by the matrix $-I$ (inversion in the centre) fixes all diagonals. In fact, this matrix commutes with all transformations in the group, so $\{\pm I\}$ is the centre of the symmetry group, and is a normal subgroup. From this it can be deduced that the symmetry group is $S_4 \times C_2$ (Exercise 3.40).

In the remaining case, the argument requires a more elaborate model, or very good geometrical intuition! It is possible to 'inscribe' a cube into a dodecahedron, so that the vertices of the cube are eight of the twenty vertices of the dodecahedron, in just five different ways. These five inscribed cubes are permuted among themselves by the rotations of the icosahedron. Thus, the rotation group is a subgroup of order 60 of S_5. Such a subgroup has index 2, and hence is normal, so is necessarily A_5. The proof that the symmetry group is $A_5 \times C_2$ is much as before. (Another approach is given in Exercise 3.41.) $\qquad\square$

Exercise 3.29 Show that the set

$$\{(1), (1,2), (3,4), (1,2)(3,4)\}$$

of permutations also forms a group isomorphic to the Klein group.

Exercise 3.30 Let π be a permutation of $\{1, \ldots, n\}$. Define the **permutation matrix** $P(\pi)$ as follows: $P(\pi)$ is an $n \times n$ matrix whose (i, j) entry is equal to 1 if $i\pi = j$, and is 0 otherwise. So each row or column of $P(\pi)$ contains exactly one entry 1.

Prove that $P(\pi_1 \pi_2) = P(\pi_1) P(\pi_2)$. Hence show that every finite permutation group is isomorphic to a matrix group (over any field).

Exercise 3.31 Let $G = \{g_1, \ldots, g_n\}$ be a group. Show that the Cayley table of G is an $n \times n$ array with entries $g_1, \ldots, g_n$ with the following property: each element g_i occurs exactly once in any row or column of the array. (Arrays with this property are called **Latin squares**. They are used in design of experiments in statistics.)

Show that G is abelian if and only if its Cayley table is symmetric (equal to its transpose).

Exercise 3.32 Let G and H be groups. Let $G \times H$ be the Cartesian product of the sets G and H (the set of all ordered pairs (g, h), with $g \in G$ and $h \in H$). Define an operation on $G \times H$ by the rule

$$(g_1, h_1)(g_2, h_2) = (g_1 g_2, h_1 h_2).$$

Prove that $G \times H$ is a group. (This group is called the **direct product** of the groups g and H.)

Prove that $|G \times H| = |G| \cdot |H|$.

Prove that the direct product of abelian groups is abelian.

Exercise 3.33 (a) Prove that the Klein group V_4 is isomorphic to $C_2 \times C_2$.

(b) Prove that $C_2 \times C_3 \cong C_6$.

(c) Among groups of order 8, we find C_8, $C_4 \times C_2$, and $C_2 \times C_2 \times C_2$. Identify them in the analysis of groups of order 8 given above.

Exercise 3.34 (a) Prove that the eight permutations

$$(1), (1, 2, 3, 4), (1, 3)(2, 4), (1, 4, 3, 2),$$

$$(1, 2)(3, 4), (1, 3), (1, 4)(2, 3), (2, 4)$$

form a non-abelian group. [*Hint*: Construct a Cayley table.]

(b) Prove that the eight matrices

$$\begin{pmatrix} 1 & 0 \\ 0 & 1 \end{pmatrix}, \begin{pmatrix} -1 & 0 \\ 0 & -1 \end{pmatrix}, \begin{pmatrix} i & 0 \\ 0 & -i \end{pmatrix}, \begin{pmatrix} -i & 0 \\ 0 & i \end{pmatrix},$$

$$\begin{pmatrix} 0 & 1 \\ -1 & 0 \end{pmatrix}, \begin{pmatrix} 0 & -1 \\ 1 & 0 \end{pmatrix}, \begin{pmatrix} 0 & i \\ i & 0 \end{pmatrix}, \begin{pmatrix} 0 & -i \\ -i & 0 \end{pmatrix}$$

over the complex numbers form a non-abelian group, not isomorphic to the group in (a). [*Hint*: Count elements of order 2.]

Note These two groups are called the **dihedral group** and the **quaternion group** of order 8, respectively. The quaternion group arises from the **quaternions** discovered by W. R. Hamilton (see Exercise 2.11). If i, j, k are the quaternion 'units', satisfying

$$i^2 = j^2 = k^2 = ijk = -1,$$

then the quaternion group consists of the eight elements

$$\{\pm 1, \pm i, \pm j, \pm k\}.$$

Exercise 3.35 (a) Prove that, if p and q are distinct primes, then the direct product of Z_p and Z_q is isomorphic to Z_{pq}.

(b) Prove that, if p is prime, then the direct product of Z_p with itself is *not* isomorphic to Z_{p^2}.

Exercise 3.36 Show that the alternating group A_4, of order 12, has no subgroup of order 6.

(This shows that the converse of Lagrange's Theorem is false; it is not true in general that, if G is a group of order n, and m divides n, then G has a subgroup of order m.)

Exercise 3.37 Let G be the *dihedral group* of order 8, the group of symmetries of a square. Let z be the symmetry which is a rotation through $180°$. Verify that the centre $Z(G)$ is the subgroup $\{1, z\}$. Now $G/Z(G)$ is a group of order 4: is it the cyclic group or the Klein group?

Exercise 3.38 (*) There are three partitions of the set $\{1, 2, 3, 4\}$ into two sets of size 2, namely,

- $A = \{\{1, 2\}, \{3, 4\}\}$;
- $B = \{\{1, 3\}, \{2, 4\}\}$;
- $C = \{\{1, 4\}, \{2, 3\}\}$.

Any permutation g of $\{1, 2, 3, 4\}$ induces a permutation g^* of the set $\{A, B, C\}$. For example, if g is the cyclic permutation $(1, 2, 3, 4)$, then $g^* = (A, C)(B)$.

Show that the map $\theta : S_4 \to S_3$ defined by $\theta(g) = g^*$ is a homomorphism. Describe the image and kernel of θ, and check that

$$|\operatorname{Im}(\theta)| \cdot |\operatorname{Ker}(\theta)| = |S_4|.$$

Exercise 3.39 (*) Let G be a group having two normal subgroups N and M with the properties that $NM = G$ and $N \cap M = \{1\}$. Show that an element of M and an element of N commute. Show that any element of G can be written uniquely in the form nm, for $n \in N$ and $m \in M$. Hence show that $G \cong N \times M$.

Verify that these conditions are satisfied when G is the symmetry group of a cube, N the rotation group, and $M = \{\pm I\}$.

Exercise 3.40 Calculate the conjugacy classes in the rotation group of the cube (in terms of type of axis of rotation and angle of rotation), and match them up with the conjugacy classes in the symmetric group S_4.

Exercise 3.41 Take a model of either a dodecahedron or an icosahedron; pick it up by one edge, and hold it with this edge horizontal at the 'north pole'. Check that, on the 'equator', there are two horizontal edges at antipodal points (in the direction of the ends of the top edge), and two vertical edges at the intermediate points, while at the 'south pole' there is an edge parallel to the one at the 'north pole'.

This means that any edge belongs to a unique set of six edges in three mutually perpendicular directions. The thirty edges thus define five such sets of six (or 'frames', as we shall call them).

Show that any rotation induces a permutation on the set of five frames, and no rotation except the identity can fix all five frames.

Deduce that the rotation group of the figure is isomorphic to A_5.

Appendix: How many groups?

We have seen that the number of binary operations on a set of n elements is n^{n^2}. These are systems satisfying the axiom (G0).

Even if we count them up to isomorphism, we obtain a very rapidly growing function. For the number of structures on a set of n elements isomorphic to a given one is at most the number $n!$ of bijections of the set; so up to isomorphism the number $F_0(n)$ of binary systems satisfies

$$F_0(n) \geq n^{n^2}/n! \geq n^{n^2}/n^n = n^{n^2-n}.$$

One measure of the strength of the group axioms is to estimate how many structures satisfy various collections of axioms.

It is clear that the identity and inverse laws alone are not very powerful. If we take $A = \{a_1, \ldots, a_n\}$, where a_1 is the identity, then the first row and column of the operation table are determined, but the remaining entries are arbitrary; so there are $n^{(n-1)^2}$ such structures, and at least $n^{(n-1)^2}/(n-1)!$ up to isomorphism. Similarly, we get a lower bound for the number satisfying the inverse law by counting just those for which each element is its own inverse; there are $n^{(n-1)(n-2)}$ of these, so at least $n^{(n-1)(n-2)}/(n-1)!$ up to isomorphism.

A better approach is to consider the cancellation laws. Notice that we used the associative law as well as the identity and inverse laws to prove the cancellation laws in a group. (For example, if $ab = ac$, then $a^{-1}(ab) = a^{-1}ac$; using the associative law, $(a^{-1}a)b = (a^{-1}a)c$, whence $b = c$.) Notice, too, that in a finite structure, the identity and cancellation laws imply the existence of inverses. For the cancellation law implies that the map $x \mapsto ax$ is one-to-one; in a finite structure, this map is also onto, so there exists x such that $ax = 1$. However, we need the associative law to prove that left and right inverses are equal: if $ax = 1 = ya$, then $x = (ya)x = y(ax) = y$.

Accordingly, we define a **quasigroup** to be a set with a binary operation $\circ$ in which the equations $ax = b$ and $ya = b$ have unique solutions x and y for any given a and b. Another way of saying the same thing is that the operation table has the property that each element occurs exactly once in each row or column.

A **Latin square** is an $n \times n$ array containing n different entries, such that each entry occurs exactly once in each row and once in each column. Thus, a set with a binary operation is a quasigroup if and only if its operation table is a Latin square.

In addition, we define a **loop** to be a quasigroup with identity. Taking the identity to be the first element, this requires that the entries in the first row and column of the operation table are equal to the row and column labels.

So we see that the numbers of quasigroups and loops with n elements are each at least $L(n)/n!$, where $L(n)$ is the number of Latin squares of order n. The value of $L(n)$ is not known precisely, but it is known to be at least $(cn)^{n^2}$ for some positive constant c. As before, dividing by $n!$ doesn't have much effect.

By contrast, the number of groups is much smaller:

Theorem 3.33   *The number of non-isomorphic groups of order n does not exceed $n^{n \log_2 n}$.*

Proof   First, we present some terminology. In a group G, the **subgroup generated** by the elements $g_1, \ldots, g_r$ is the smallest subgroup containing these elements; we say that $g_1, \ldots, g_r$ **generate** G if there is no proper subgroup of G which contains them. We write $G = \langle g_1, \ldots, g_r \rangle$ in this case, generalising the notation for cyclic groups. Another characterisation of this is: every element of G can be written as a product of elements chosen from $g_1, \ldots, g_r$ and their inverses.   □

Step 1   A group of order n can be generated by at most $\log_2 n$ elements.

To show this, choose elements $g_1, g_2, \ldots$ so that g_1 is not the identity and g_{i+1} is not in the subgroup H_i generated by $g_1, \ldots, g_i$ for each $i > 1$, as long as possible. Then $H_1 = \langle g_1 \rangle$ is not the identity, so $|H_1| \geq 2$. Also, since H_{i+1} properly contains H_i, its order is a proper multiple of that of H_i by Lagrange's Theorem, and so $|H_{i+1}| \geq 2|H_i|$. By induction, $|H_i| \geq 2^i$ for all i. When the process terminates, we have $H_i = G$, so $n \geq 2^i$, or $i \leq \log_2 n$; and, by construction, G can be generated by i elements.

Step 2   By Cayley's Theorem, there is a subgroup of the symmetric group S_n which is isomorphic to G. Thus, there is an isomorphism θ from G to a subgroup of S_n. If $G = \langle g_1, \ldots, g_r \rangle$, then any element of G is a product of some of $g_1, \ldots, g_r$ and their inverses. So the image of this element under θ is determined by the images of $g_1, \ldots, g_r$, which are r elements of S_n.

Hence the number of groups of order n is not greater than the number of choices of r elements of S_n, where $r = \lfloor \log_2 n \rfloor$ (the integer part of $\log_2 n$). This number is $(n!)^{\log_2 n}$.

Finally, $n!$, which is the product of the n numbers from 1 to n, is not greater than n^n. So the number of groups of order n does not exceed $n^{n \log_2 n}$.   □

In the above proof, the essential ingredients are Lagrange's Theorem and Cayley's Theorem, two of the most basic results about groups. Using much more advanced group theory, this result has been improved. It is known that the number of groups of order n is at most $n^{c(\log n)^2}$ for some positive constant c. In other words, the exponent is reduced from $n \log n$ to the much smaller $c(\log n)^2$.

Even Cayley's Theorem is not essential for this proof. If we are building the Cayley table of G, it is enough to construct the rows corresponding to the generating elements $g_1, \ldots, g_r$, since all other products can then be computed using the associative law. Now the number of $r \times n$ tables does not exceed $n^{nr} \leq n^{n \log_2 n}$.

On the basis of this theorem, we might be tempted to conclude that the associative law is the most powerful of the group axioms. We define a **semigroup** to be a set with a binary operation satisfying the associative law, that is, a structure in which axioms (G0) and (G1) hold.

It is beyond the scope of this book to give an estimate for the number of semigroups with n elements. However, the numbers of small semigroups have been calculated. The table below shows that the number grows much faster than the number of groups. The numbers in the table are up to isomorphism.

So it is more accurate to say that the combination of all the group axioms is more than the sum of its parts.

n	Operations	Quasi groups	Loops	Semi groups	Groups
1	1	1	1	1	1
2	10	1	1	5	1
3	3330	5	1	24	1
4	178981952	35	2	188	2
5	2483527537094825	1411	6	1915	1
6	14325590003318891522275680	1130531	109	28634	2

These numbers are taken from Neil Sloane's *On-Line Encyclopedia of Integer Sequences*, on the web at http://www.research.att.com/~njas/sequences/

Recently, a team of group theorists (Hans Ulrich Besche, Bettina Eick and Eamonn O'Brien) marked the end of the second millennium by counting the groups with order at most 2000 up to isomorphism. There are 49 910 529 484 groups, of which 49 487 365 422 have order $2^{10} = 1024$. By contrast, of course, if p is prime, there is only one group of order p. So the counting function for groups is very erratic.

4   Vector spaces

By the time you reach this point, you will probably have met vector spaces in another course: perhaps matrices, geometry, mechanics, or linear algebra. The treatment here may be somewhat different. A vector space is an algebraic object like a ring or a group, and we will start with a collection of axioms, as we did in the chapters on rings and groups. Then we turn to subspaces and homomorphisms, and find that homomorphisms between vector spaces can be represented by matrices, and indeed by matrices of a particularly simple form. In the last section, we see that for matrices with elements in a Euclidean domain (such as the integers), similar results apply. This result will seem a bit unmotivated; but we will put it to work in the next chapter!

Vector spaces and subspaces

4.1   Introduction.   The notion of a vector space grew from the discovery by Descartes that points in the Euclidean plane can be represented by ordered pairs of real numbers (and points in 3-dimensional space by ordered triples). We are faced with two completely different descriptions: a point in the plane, or a pair of real numbers. Moreover, operations on vectors look quite different according to which description is used. For example, we add vectors by the 'parallelogram law' used in mechanics, and we add pairs of real numbers 'componentwise'; but the result is the same.

Furthermore, we want the possibility to generalise. We want to be able to talk about Euclidean space of n dimensions, for any n. (This is not just an intellectual game, but has important applications in fields as far apart as quantum mechanics, statistics, and signal processing.) Also, we want to be able to use fields other than the real numbers. Computers send information as sequences of binary digits (that is, n-tuples from the field $\mathbb{Z}_2$). So we define the concept of a vector space *over* an arbitrary field F. We call the elements of F **scalars**, to distinguish them from the vectors.

To set up the axioms, we regard two operations on vectors as basic. There is a binary operation of **addition**, written as $+$ as usual. Also, for every field element c, there is a unary operation of **scalar multiplication** by c, written as juxtaposition: the product of the scalar c and the vector v is written as cv (with the scalar on the left).

There is a potential problem here, since we have two different kinds of things, scalars and vectors, both of which can be added; there are two multiplications, one combining two scalars, the other a scalar and a vector; and we will see that there is a vector named 0 as well as a scalar with the same name. Sometimes, this problem is dealt with by using bold type for vectors. This is common in 3-dimensional mechanics, for example, where a vector is a 'geometric' object, unlike a scalar. I want to stress that a vector space, just like a field or a group, is an algebraic object; so I do not adopt this convention. I will usually use letters from near the end of the alphabet (typically u, v, w) for vectors, and letters from the other end (c, d) for scalars. There will be times when no convention can avoid confusion completely!

We now give the formal definition. Let F be a field. A **vector space** over F, or F-**vector space**, is a set V with a binary operation $+$ (addition) and, for each $c \in F$, a unary operation of scalar multiplication by c, satisfying the following axioms:

Addition axioms
(VA0) **(Closure law):** For all $v, w \in V$, we have $v + w \in V$.
(VA1) **(Associative law):** $u + (v + w) = (u + v) + w$ for all $u, v, w \in V$.
(VA2) **(Zero law):** There exists $0 \in V$ such that $v + 0 = 0 + v = v$ for all $v \in V$.
(VA3) **(Inverse law):** For all $v \in V$, there exists $w \in V$ with $v + w = w + v = 0$.
(VA4) **(Commutative law):** $v + w = w + v$ for all $v, w \in V$.
Scalar multiplication
(VM0) **(Closure law):** For all $c \in F$ and $v \in V$, we have $cv \in V$.
(VM1) For all $c \in F$ and $v, w \in V$, we have $c(v + w) = cv + cw$.
(VM2) For all $c, d \in F$ and $v \in V$, we have $(c + d)v = cv + dv$.
(VM3) For all $c, d \in F$ and $v \in V$, we have $(cd)v = c(dv)$.
(VM4) **(Unital law):** For all $v \in V$, we have $1v = v$ (where 1 is the identity of F).

Remark 1. The addition axioms assert that a vector space, with the operation of addition, is an abelian group. This is a convenient way to remember them.

2. It is possible to state the other axioms more briefly, too. If A is an abelian group, then the set $\text{End}(A)$ of all homomorphisms from A to A is a ring, whose addition is defined pointwise, and whose multiplication is composition. (This is the **endomorphism ring** of A: see Exercise 3.28.)

$$a(\phi_1 + \phi_2) = a\phi_1 + a\phi_2,$$
$$a(\phi_1\phi_2) = (a\phi_1)\phi_2.$$

Now let V be a vector space over F, and, for each $c \in F$, let μ_c denote the operation of scalar multiplication by c, the function from V to V given by $v\mu_c = cv$. Then axiom (VM1) says that μ_c is a homomorphism; that is, $\mu_c \in \text{End}(V)$. Next, let θ be the function from F to $\text{End}(V)$ mapping the element c to the homomorphism μ_c. Then axioms (VM2) and (VM3) say that θ is a ring homomorphism, and axiom (VM4) says that θ maps the identity element of F to the identity homomorphism.

So we can reformulate the definition as follows:

> A vector space over F consists of an abelian group V and a ring homomorphism θ from F to $\text{End}(V)$ which maps the identity to the identity.

Perhaps you find this less helpful than the list of ten axioms. But it does point to a close connection between vector spaces, groups and rings, and it is in a form which makes it easier to generalise in a meaningful way.

4.2   Examples.

Example 1 Let $V = F^n$, the set of all n-tuples of elements of F. Define addition and multiplication 'coordinatewise': that is,

$$(a_1, a_2, \ldots, a_n) + (b_1, b_2, \ldots, b_n) = (a_1 + b_1, a_2 + b_2, \ldots, a_n + b_n),$$

$$c(a_1, a_2, \ldots, a_n) = (ca_1, ca_2, \ldots, ca_n).$$

This is a very important example. When we considered rings, we saw that, while there are quite varied examples, the ring of integers was sufficiently typical to act as a prototype. For groups, the examples were so varied that there was no useful prototype. The situation is different here. Not only is F^n the prototype of a vector space; we will see that every 'finite-dimensional' vector space looks exactly like F^n.

For this reason, it is worth stopping to check that the ten vector space axioms do hold for F^n. All the arguments are similar, and straightforward. For example, here is the proof of (VM3):

$$(cd)(a_1, \ldots, a_n) = (cda_1, \ldots, cda_n) = c(da_1, \ldots, da_n) = c(d(a_1, \ldots, a_n)).$$

(We used the associative law for multiplication in F here.)

Example 2 Let Ω be any set, and let V be the set of all functions from Ω to the field F. Define addition and scalar multiplication of functions 'pointwise':

$$(f + g)(x) = f(x) + g(x),$$
$$(cf)(x) = cf(x)$$

for all $x \in \Omega$. Then V is a vector space over F.

In the case where Ω is finite, say $\Omega = \{x_1, \ldots, x_n\}$, we can represent the function f uniquely by giving the list of its values, $(f(x_1), \ldots, f(x_n))$. Any n-tuple of elements of F forms the list of values of a unique function. So we can identify V with the set F^n of all such lists. The addition and multiplication are the same as in Example 1.

Things are more interesting in the case where Ω is infinite. Suppose, for example, that F is the field $\mathbb{R}$ of real numbers, and that Ω is either $\mathbb{R}$ or an interval in $\mathbb{R}$. By restricting to 'interesting' classes of functions, such as continuous functions or differentiable functions, we obtain further vector spaces. This is the subject-matter of Functional Analysis.

Example 3 Let K be a field, containing a subfield F. Now suppose that we are slightly forgetful, and while we can remember how to add elements of K, we can only remember how to multiply them by elements of F. Then with this loss of information, K becomes an F-vector space.

In particular, any field F is a vector space over itself; but this is no surprise, since we can represent F as F^1, the set of all 1-tuples.

A more interesting case is that where $K = F[x]/(f)$, where f is an irreducible polynomial over F of degree n. As we saw in Section 2.16, K can be represented as

$$K = \{c_0 + c_1\alpha + \cdots + c_{n-1}\alpha^{n-1} : c_0, c_1, \ldots, c_{n-1} \in F\},$$

and in this representation we can add elements of K or multiply them by elements of F (coordinatewise) without knowing the precise equation $f(\alpha) = 0$ satisfied by α. So this is just Example 1 again.

Example 4 The first vector space studied by mathematicians was the Euclidean plane. How do we see it as a vector space?

A vector, in elementary geometry or mechanics, has magnitude and direction. If we choose a point in the plane to be the origin, then a vector is thought of as an arrow with its tail at the origin and its head at an arbitrary point of the plane. (The zero vector is a special case. Its head and tail are both at the origin; its length is zero and its direction is not defined.)

Two vectors v, w are added by the **parallelogram law**: construct a parallelogram with one vertex at the origin and two sides corresponding to v and w (so that the heads of these vectors are two more vertices); then the fourth vertex is the head of $v + w$. (This has to be modified if v and w point in the same or opposite directions, since then the parallelogram degenerates into a line.) If c is positive, then to multiply a vector v by c we take a vector with c times the length but in the same direction. If c is negative, we multiply the length by $-c$ and reverse the direction. Finally, we take $0v$ to be the zero vector.

Following Descartes, we represent each point of the plane by a pair (x, y) of real numbers (its **coordinates**). In a similar way, we can represent a vector by coordinates, taking the coordinates of its head. Thus, this labelling identifies the set of vectors with $\mathbb{R}^2$. It can be checked that the rules for addition and multiplication, when expressed in coordinates, are precisely those of Example 1.

Example 5 In communication between computers, data is sent in the form of **binary words**, consisting of n-tuples of zeros and ones (for some fixed word length n). We regard the entries in such a word as being integers mod 2, that is, belonging to the field $\mathbb{F}_2 = \mathbb{Z}/2\mathbb{Z}$. Assume for example that $n = 8$. Now suppose that, during transmission, the signal is distorted by interference, so that, at the receiving end, the second, third, and fifth bits of the transmitted word are received incorrectly. Since changing an element of $\mathbb{F}_2$ can be done simply by adding 1 to it, we see that the effect of the noise is to add to the transmitted word the vector (01101000). In this sense, the received word is 'signal plus noise', the addition being performed in the vector space $\mathbb{F}_2^8$. We will consider error correction further in Chapter 8.

4.3   Properties of vector spaces.   Since a vector space is an abelian group (so far as addition is concerned), we can immediately deduce that all the properties of abelian groups hold. For example,

(a) the sum of n vectors is well defined, independent of bracketing or order;
(b) the zero is unique;
(c) the additive inverse of any element is unique.

Here are a couple more simple properties.

(d) $0v = 0$ for any v;
(e) $(-1)v = -v$ for any v.

Proof   (d) $0v = (0+0)v = 0v + 0v$, by (VM2). Hence $0v = 0$ by the Cancellation Law (which is valid in any abelian group).

(e) $(-1)v + v = (-1)v + 1v = (-1+1)v = 0v = 0$; so $(-1)v$ is the inverse of v.   □

4.4   Subspaces.   What happens next should be fairly familiar now. A subset W of the vector space V (over a field F) is a **subspace** if it forms a vector space in its own right. As we saw for groups and rings, in order that W is a subspace, it is enough to check the various closure properties, since universal laws such as the associative and unital laws will be inherited by W from V. In this case, we have to check closure under addition and scalar multiplication, and that W contains 0 and contains the inverses of its elements. In fact, the first two conditions suffice:

Theorem 4.1 (First subspace test)   *A non-empty subset W of a vector space V is a subspace of V if and only if it is closed under addition and scalar multiplication; that is, $w_1, w_2 \in W$ implies $w_1 + w_2 \in W$, and $c \in F$, $w \in W$ implies $cw \in W$.*

Proof   Closure under scalar multiplcation does the trick for us. For any $w \in W$, we have $0 = 0w \in W$ and $-w = (-1)w \in W$, by the properties proved in the last section.   □

As before, we can combine the two kinds of closure into a single test:

Theorem 4.2 (Second subspace test)   *The non-empty subset W of the vector space V over F is a subspace if and only if, for any $c_1, c_2 \in F$ and $w_1, w_2 \in W$, we have $c_1 w_1 + c_2 w_2 \in W$.*

Proof   If W is a subspace, and c_1, c_2, w_1, w_2 are given, then the closure laws show that $c_1 w_1 + c_2 w_2 \in W$.

Conversely, suppose that this condition holds. Then, choosing $c_2 = 0$, we see that $c_1 w_1 \in W$ for all $c_1 \in F$ and $w_1 \in W$; that is, W is closed under scalar multiplication. Similarly, choosing $c_1 = c_2 = 1$ shows that it is closed under addition. □

4.5 Linear independence and bases. Let V be a vector space over F, and let $v_1, v_2, \ldots, v_n$ be vectors of V. We say that $v_1, v_2, \ldots, v_n$ are **linearly dependent** if there are scalars $c_1, c_2, \ldots, c_n$, *not all zero*, such that

$$c_1 v_1 + c_2 v_2 + \cdots + c_n v_n = 0.$$

Note the importance of the phrase 'not all zero'. If we allowed the coefficients $c_1, \ldots, c_n$ to be all zero, then the equation would be true for any n vectors. Note that if one of the vectors, say v_i, is zero, then they are linearly dependent: take $c_i = 1$ and $c_j = 0$ for $j \neq i$. In a similar manner, if two of the vectors are equal, say $v_i = v_j$, then they are linearly dependent: take $c_i = 1$, $c_j = -1$, and $c_k = 0$ for $k \neq i, j$.

Linear dependence can be formulated in another way. An expression $a_1 w_1 + \cdots + a_m w_m$ is called a **linear combination** of the vectors $w_1, \ldots, w_m$.

Proposition 4.3 *The vectors $v_1, \ldots, v_n$ are linearly dependent if and only if one of them can be expressed as a linear combination of the others.*

Proof Suppose first that the vectors are linearly dependent. That is, we have $c_1 v_1 + \cdots + c_n v_n = 0$, where the coefficients are not all zero. Say that $c_i \neq 0$. Then we have

$$v_i = -(c_1 c_i^{-1}) v_1 - \cdots - (c_{i-1} c_i^{-1}) v_{i-1} - (c_{i+1} c_i^{-1}) v_{i+1} - \cdots - (c_n c_i^{-1}) v_n.$$

In other words, v_i is a linear combination of the others.

Conversely, let v_i is expressed as a linear combination of the other vectors. Then, subtracting v_i from both sides of this equation, we find a linear combination of all the vectors equal to zero, where the coefficient of c_i is $-1 \neq 0$; so the vectors are linearly dependent. □

If the vectors $v_1, v_2, \ldots, v_n$ are not linearly dependent then, naturally enough, they are called **linearly independent**. This is a negative definition, so we reformulate the concept more positively. The vectors $v_1, v_2, \ldots, v_n$ are linearly independent if, whenever an equation

$$c_1 v_1 + c_2 v_2 + \cdots + c_n v_n = 0$$

holds, then we have $c_1 = c_2 = \ldots = c_n = 0$.

Example Show that the vectors $(1, 1, 1)$, $(1, 2, 0)$, and $(0, 1, -3)$ in $\mathbb{R}^3$ are linearly independent.

Solution Suppose that the equation

$$a(1, 1, 1) + b(1, 2, 0) + c(0, 1, -3) = 0$$

holds. In other words,

$$(a + b, a + 2b + c, a - 3c) = (0, 0, 0).$$

This gives us three equations

$$a + b = 0,$$
$$a + 2b + c = 0,$$
$$a - 3c = 0,$$

for the three unknowns a, b, c. Solving these equations, we find that $a = b = c = 0$. So the vectors are linearly independent.

A related concept is the **span** of a list of vectors. Let $v_1, v_2, \ldots, v_n$ be elements of a vector space V over F. The span of these vectors, written $\langle v_1, \ldots, v_n \rangle_F$, is the set of all linear combinations of $v_1, \ldots, v_n$. (If the field F is clear, we omit it from the notation.)

Proposition 4.4 $\langle v_1, \ldots, v_n \rangle_F$ *is a subspace of V.*

Proof We apply the subspace test. Take two vectors in $\langle v_1, \ldots, v_n \rangle_F$, say $w = c_1 v_1 + \cdots + c_n v_n$ and $w' = c'_1 v_1 + \cdots + c'_n v_n$. Then. for any $a, a' \in F$,

$$aw + a'w' = (ac_1 + a'c'_1)v_1 + \cdots + (ac_n + a'c'_n)v_n$$

is a linear combination of $v_1, \ldots, v_n$. $\qquad\qquad\square$

A list $v_1, v_2, \ldots, v_n$ is called a **spanning set** for V if $\langle v_1, \ldots, v_n \rangle = V$. If it is both linearly independent and a spanning set, it is called a **basis**.

Theorem 4.5 *The following conditions for a finite subset X of a vector space are equivalent:*

(a) X is a maximal linearly independent set;
(b) X is a minimal spanning set;
(c) X is a linearly independent spanning set.

Proof We show that each of (a) and (b) is equivalent to (c).

(c) implies (a): If X is a basis for V, then every vector in V is a linear combination of X; so adding any vector to X gives a linearly dependent set, which means that X is maximal.

(a) implies (c): If X is maximal independent, then any vector v added to X gives a linearly dependent set. In a dependence relation, the coefficient of v must be non-zero (else X would be linearly dependent); so we can use the relation to express v as a linear combination of X. So X spans V, and is a basis.

(c) implies (b): If X is a basis, then no element of X is a linear combination of the others, so no proper subset is spanning.

(b) implies (c): If X is a minimal spanning set, then no element of X is a linear combination of the others (or it could be dropped without losing the spanning property); so X is linearly independent. □

Now if a vector space has a basis, we know what it looks like:

Theorem 4.6 *(a) If X is a basis for V, then every element of V has a unique expression as a linear combination of X.*
(b) If V has a basis containing n elements, then V is isomorphic to F^n.

Proof Let $v_1, \ldots, v_n$ be a basis. Any element $v \in V$ has an expression as a linear combination, say $v = c_1 v_1 + \cdots + c_n v_n$. Suppose that there is another such expression, say $v = c'_1 v_1 + \cdots + c'_n v_n$. Then

$$(c_1 - c'_1)v_1 + \cdots + (c_n - c'_n)v_n = 0.$$

Since $v_1, \ldots, v_n$ are linearly independent, the coefficients are all zero; so $c_i = c'_i$ for all i.

Now we map V to F^n by taking $v = c_1 v_1 + \cdots + c_n v_n$ to the n-tuple $(c_1, \ldots, c_n)$. The first part of the theorem shows that it is a bijection; and clearly it preserves addition and scalar multiplication. □

Now we know exactly what a vector space with a basis looks like, except for the possibility that there are bases with different numbers of elements. We now show that this cannot happen.

To do this, we first need a technical result about linear equations.

Let $x_1, \ldots, x_n$ be variables taking values in a field F. A **homogeneous linear equation** in $x_1, \ldots, x_n$ is an equation of the form $c_1 x_1 + \cdots + c_n x_n = 0$, where $c_1, \ldots, c_n$ are given elements of F. A solution is just an assignment of values to the variables so that the equation holds.

Proposition 4.7 *Given m homogeneous linear equations in n variables, with $m < n$, there is a simultaneous solution to the equations with not all the variables equal to zero.*

Proof We prove the result by induction on m. If $m = 0$, there are no equations, and any assignment of values will do. So the induction starts.

Suppose that the result holds for fewer than m equations. Consider one of the given equations, say $c_1 x_1 + \cdots + c_n x_n = 0$. If all the coefficients $c_1, \ldots, c_n$ are equal to zero, then the equation carries no information and can be discarded, reducing the number of equations by one. So suppose that some coefficient is non-zero, without loss the coefficient c_n. Now divide this equation by c_n, and it expresses x_n in terms of the other variables. Substitute this expression in the remaining equations. We end up with $m - 1$ equations in $n - 1$ variables. By induction, these have a non-zero solution. So we obtain a non-zero solution to the original set of equations.

The result is proved. □

Now we return to the properties of linear independence. Let V be a vector space, and let $\mathcal{I}$ denote the set of linearly independent finite subsets of V. By definition, the empty set is linearly independent.

Theorem 4.8 (Properties of linear independence)   *(a) If $X \in \mathcal{I}$ and $Y \subset X$, then $Y \in \mathcal{I}$.*
*(b) **(Steinitz exchange axiom)** Suppose that $X, Y \in \mathcal{I}$ with $|Y| > |X|$. Then there exists $y \in Y \setminus X$ such that $X \cup \{y\} \in \mathcal{I}$.*

Proof   (a) If we have a linear combination of a subset Y of X equal to zero, then we obtain a linear combination of all of X with the same value, by taking the coefficients of the remaining vectors to be zero. Since X is linearly independent, all the coefficients must be zero. So Y is linearly independent.

(b) Suppose that both $\{x_1, \ldots, x_m\}$ and $\{y_1, \ldots, y_n\}$ are linearly independent, with $m < n$. Let us also suppose, arguing for a contradiction, that the set $\{x_1, \ldots, x_m, y_i\}$ is linearly dependent, for any i (with $1 \leq i \leq n$). This means that there is a linear combination of $x_1, \ldots, x_m, y_i$ which is equal to zero, with not all the coefficients zero. The coefficient of y_i must be non-zero, since $x_1, \ldots, x_m$ are linearly independent. Dividing by this coefficient, and taking y_i to the other side of the equation, we find that y_i is a linear combination of $x_1, \ldots, x_m$ for every i. Suppose that

$$y_i = c_{i1}x_1 + \cdots + c_{im}x_m$$

for $i = 1, \ldots, n$. We claim that the vectors $y_1, \ldots, y_n$ are linearly dependent. Could an equation

$$a_1y_1 + \cdots + a_ny_n = 0$$

hold? Substituting the expression for the ys in terms of the xs, we find that the coefficient of x_j is

$$a_1c_{1j} + \cdots + a_nc_{nj} = 0.$$

Regarding these as m linear equations for the n unknowns $a_1, \ldots, a_n$, we see from the proposition that they have a non-zero solution. So $y_1, \ldots, y_n$ are linearly dependent. But this contradicts the fact that they are linearly independent! So the original assumption, that $x_1, \ldots, x_m, y_i$ is linearly dependent for all i must be wrong; so this set is linearly independent for some value of i, as required. This completes the proof. $\square$

From this theorem we can deduce an important property. If there are no bases, then we call the vector space **infinite-dimensional**; it is **finite-dimensional** otherwise. So a space is infinite-dimensional if every linearly independent set can be enlarged to a larger linearly independent set. Infinite-dimensional spaces are important, but here we are only concerned with the finite-dimensional ones.

Theorem 4.9 *If V has a basis, then any two bases have the same number of elements.*

Proof Suppose that X and Y are both maximal. By the Steinitz exchange axiom, if $|X| < |Y|$, then we could obtain a linearly independent set containing X by adding an element of Y, which contradicts the maximality of X. Similarly, the assumption that $|Y| < |X|$ leads to a contradiction. So $|X| = |Y|$. □

If the vector space V has a basis (that is, if it is finite-dimensional), then the number of elements in any basis is called the **dimension** of V.

Theorem 4.10 *Two finite-dimensional vector spaces over the same field F are isomorphic if and only if they have the same dimension.*

Proof Combine the results of Theorems 4.9 and 4.6. □

4.6 Intersection and sum. Let V be a vector space, and let U, W be subspaces of V. The intersection $U \cap W$ is a subspace, and is the largest subspace of V contained in both U and W. See Exercise 4.2 which also asks you to show that the union $U \cup W$ is not usually a subspace. Instead, we define the following:

The **sum** of the subspaces U and W is the set

$$U + W = \{u + w : u \in U, w \in W\}.$$

It is a subspace, and is the smallest subspace of V which contains both U and W.

Proof We apply the Subspace Test. Take two vectors $u + w$ and $u' + w'$ in $U + W$, and two scalars $c, c' \in F$. Then

$$c(u + w) + c'(u' + w') = (cu + c'u') + (cw + c'w') \in U + W,$$

since $cu + c'u' \in U$ and $cw + c'w' \in W$.

Clearly, both U and W are contained in $U + W$. (For example, a vector in U has the form $u + 0$.) Moreover, any subspace which contains both U and W must contain $U + W$, so it is the smallest such subspace. □

In the finite-dimensional case, the following equation connects the dimensions of these spaces:

Theorem 4.11 *If U and W are subspaces of V, then*

$$\dim(U \cap W) + \dim(U + W) = \dim(U) + \dim(W).$$

Proof Let $\dim(U) = m$, $\dim(W) = n$, and $\dim(U \cap W) = k$. Choose a basis $x_1, \ldots, x_k$ for $U \cap W$. This is a linearly independent set in U, and so it can be extended to a basis for U, say $x_1, \ldots, x_k, u_1, \ldots, u_{m-k}$. Similarly, it can be extended to a basis $x_1, \ldots, x_k, w_1, \ldots, w_{n-k}$ for W. We *claim* that all the vectors $x_1, x_k, u_1, \ldots, u_{m-k}, w_1, \ldots, w_{n-k}$ form a basis for $U + W$. Given this, the result follows: for we have

$$\dim(U + W) = k + (m - k) + (n - k) = m + n - k$$
$$= \dim(U) + \dim(W) - \dim(U \cap W).$$

Any vector in $U + W$ can be written as a linear combination of the x_i, u_i, and w_i; so we have a spanning set. Suppose that we have a linear dependence

$$a_1 x_1 + \cdots + a_k x_k + b_1 u_1 + \cdots + b_{m-k} u_{m-k} + c_1 w_1 + \cdots + c_{n-k} w_{n-k} = 0.$$

Transposing, we obtain

$$a_1 x_1 + \cdots + a_k x_k + b_1 u_1 + \cdots + b_{m-k} u_{m-k} = -c_1 w_1 - \cdots - c_{n-k} w_{n-k}.$$

The left-hand expression is in U, and the right-hand expression in W; so both sides lie in $U \cap W$. So they can be expressed in a third form, $d_1 x_1 + \cdots + d_k x_k$. Now this and the left-hand side are two expressions for a vector of U in terms of the basis of U, so they coincide; that is, $a_i = d_i$ for $i = 1, \ldots, k$, and $b_i = 0$ for $i = 1, \ldots, m - k$. Performing the same argument with the right-hand expression gives $0 = d_i$ for $i = 1, \ldots, k$, and $c_i = 0$ for $i = 1, \ldots, n - k$. Combining these, we see that all the coefficients are zero. So the xs, us, and ws are linearly independent, completing the proof. $\qquad \square$

Exercise 4.1 Let V be the vector space of all real-valued functions on the unit interval $[0, 1]$. Show that each of the following is a subspace of V:

(a) the bounded functions;
(b) the continuous functions;
(c) the differentiable functions;
(d) the functions f satisfying $f(0) = f(1)$.

Exercise 4.2 Let W and U be subspaces of the vector space V.

(a) Prove that the intersection $W \cap U$ is a subspace of V.
(b) Prove that the union $W \cup U$ is a subspace if and only if one of W and U contains the other.

Exercise 4.3 Let V be the Euclidean plane, regarded as a real vector space.
(a) Prove that the set $\{0\}$, the whole space V, and any line through the origin, are subspaces of V.
(b) Prove further that every subspace of V is of one of these types.

Exercise 4.4 Show that the set of all n-tuples of elements of F which satisfy a given collection of homogeneous linear equations is a subspace of F^n.

Exercise 4.5 In $V = \mathbb{R}^4$, let $U = \langle(1,2,0,-1),(2,1,1,3),(1,-1,1,2)\rangle$, and let $W = \langle(3,2,0,2),(2,2,0,1)\rangle$. Find a basis for $U \cap W$.

Exercise 4.6 ($**$) The conditions we proved for linearly independent sets in a vector space in Theorem 4.8 have been applied much more widely. In the tradition of abstract mathematics, they are now considered as axioms.

A **matroid** consists of a finite set E and a non-empty family $\mathcal{I}$ of subsets of E, satisfying the two conditions

(Mat1) If $X \in \mathcal{I}$ and $Y \subseteq X$, then $Y \in \mathcal{I}$.
(Mat2) **(Exchange axiom)** If $X, Y \in \mathcal{I}$ and $|X| < |Y|$, then there exists $y \in Y \setminus X$ such that $X \cup \{y\} \in \mathcal{I}$.

Thus, Theorem 4.8 states: A vector space, equipped with the family of its independent subsets, is a matroid. This was the original, motivating example. But there are other important examples in discrete mathematics. Verify the following.

(a) A **graph** consists of a set V of **vertices** and a set E of **edges**, each edge joining a pair of vertices. We allow an edge to join a vertex to itself (such an edge is called a **loop**), and we also allow more than one edge to join the same pair of vertices (such edges are called **parallel**).

A **circuit** in a graph is a sequence $e_1, e_2, \ldots, e_n$ of edges, such that there are distinct vertices $v_1, v_2, \ldots, v_n$ so that e_i joins v_i and v_{i+1} or $i < n$ while e_n joins v_n and v_1. A circuit with $n = 1$ is a loop, and a circuit with $n = 2$ consists of two parallel edges. A set of edges is **acyclic** if it contains no circuit.

Prove that, if $\mathcal{I}$ is the set of acyclic sets of edges, then $(E, \mathcal{I})$ is a matroid.

(b) Let E be a set, and let $\mathcal{X}$ be a family of subsets of E. We say that a subset $\{e_1, e_2, \ldots, e_r\}$ of E is a **partial transversal** for $\mathcal{X}$ if there exist distinct sets $X_1, X_2, \ldots, X_r$ in $\mathcal{X}$ such that $e_i \in X_i$ for $i = 1, 2, \ldots, r$.

Prove that, if $\mathcal{I}$ is the set of all partial transversals for $\mathcal{X}$, then $(E, \mathcal{I})$ is a matroid.

Remark In a matroid $(E, \mathcal{I})$, all maximal members of $\mathcal{I}$ have the same number of elements.

Linear transformations and matrices

4.7 Linear transformations. In this section, we examine homomorphisms of vector spaces. So this corresponds to 'homomorphisms and ideals' of rings, or 'homomorphisms and normal subgroups' of groups. There are two differences. First, the homomorphisms of vector spaces are almost universally called 'linear transformations' or 'linear maps'. Second, in the cases of rings and groups, we saw that kernels of homomorphisms are subrings or subgroups that satisfy some additional property. Here, by contrast, any subspace can be the kernel of a homomorphism.

Let V and W be vector spaces over the same field F. A **linear transformation** from V to W is a function $\theta : V \to W$ which satisfies the following conditions:

(a) For any $v_1, v_2 \in V$, $(v_1 + v_2)\theta = v_1\theta + v_2\theta$.
(b) For any $v \in V$, $c \in F$, $(cv)\theta = c(v\theta)$.

These two conditions are equivalent to the single condition

$$(c_1 v_1 + c_2 v_2)\theta = c_1(v_1 \theta) + c_2(v_2 \theta)$$

for $v_1, v_2 \in V$ and $c_1, c_2 \in F$.

As you would expect, the **image** of θ is

$$\mathrm{Im}(\theta) = \{w \in W : w = v\theta \text{ for some } v \in V\},$$

while its **kernel** is

$$\mathrm{Ker}(\theta) = \{v \in V : v\theta = 0\}.$$

The image is a subspace of W, whose dimension is called the **rank** of θ; the kernel is a subspace of V, whose dimension is the **nullity** of θ.

Let U be a subspace of V. Set $v_1 \sim v_2$ if $v_2 - v_1 \in U$. This is an equivalence relation, whose equivalence classes are the **cosets** of U in V. (This is exactly the same usage as for the cosets of a subgroup of an abelian group, where no distinction between left and right cosets has to be made. In fact, the cosets in the present sense are exactly the cosets of U in the additive group of V.) A coset of U has the form

$$U + v = \{u + v : u \in U\}.$$

Theorem 4.12   *(a)* $\mathrm{Im}(\theta)$ *is a subspace of* W.
(b) $\mathrm{Ker}(\theta)$ *is a subspace of* V.
(c) *Two vectors* $v_1, v_2 \in V$ *satisfy* $v_1 \theta = v_2 \theta$ *if and only if they lie in the same coset of* $\mathrm{Ker}(\theta)$.

Proof   This works in the same way as for groups or rings.

(a) If $w_1, w_2 \in \mathrm{Im}(\theta)$, say $w_1 = v_1 \theta$ and $w_2 = v_2 \theta$, and $c_1, c_2 \in F$, then

$$c_1 w_1 + c_2 w_2 = c_1(v_1 \theta) + c_2(v_2 \theta) = (c_1 v_1 + c_2 v_2)\theta \in \mathrm{Im}(\theta),$$

so $\mathrm{Im}(\theta)$ is a subspace.

(b) If $v_1, v_2 \in \mathrm{Ker}(\theta)$, then $v_1 \theta = v_2 \theta = 0$, so

$$(c_1 v_1 + c_2 v_2)\theta = c_1(v_1 \theta) + c_2(v_2 \theta) = 0,$$

so $c_1 v_1 + c_2 v_2 \in \mathrm{Ker}(\theta)$; so $\mathrm{Ker}(\theta)$ is a subspace.

(c) $v_1 \theta = v_2 \theta$ if and only if $v_2 - v_1 \in \mathrm{Ker}(\theta)$, that is, if and only if $\mathrm{Ker}(\theta) + v_1 = \mathrm{Ker}(\theta) + v_2$. $\qquad\square$

Given a subspace U of a vector space V, we define the **factor space** V/U as follows: its elements are the cosets of U in V, and addition and scalar multiplication are defined in the now-familiar way:

(a) $(U + v_1) + (U + v_2) = U + (v_1 + v_2)$;
(b) $c(U + v) = U + cv$.

Proposition 4.13 *If U is a subspace of V, then the factor space V/U is a vector space. If V is finite-dimensional, then $\dim(V/U) = \dim(V) - \dim(U)$.*

Now the Isomorphism Theorems hold. I state the first one without proof, and leave the others as exercises.

Theorem 4.14 (First Isomorphism Theorem) *Let $\theta : V \to W$ be a linear transformation. Then*

(a) $\operatorname{Im}(\theta)$ is a subspace of W;
(b) $\operatorname{Ker}(\theta)$ is a subspace of V;
(c) $V/\operatorname{Ker}(\theta) \cong \operatorname{Im}(\theta)$.

As a corollary, we have:

Proposition 4.15 (Rank and Nullity Theorem) *If $\theta : V \to W$ is a linear transformation, then $\dim(\operatorname{Im}(\theta)) + \dim(\operatorname{Ker}(\theta)) = \dim(V)$.*

We give another proof of this important result.

Proof Choose a basis $u_1, \ldots, u_r$ for $\operatorname{Ker}(\theta)$, where $r = \dim(\operatorname{Ker}(\theta))$. This is a linearly independent set in V, and so can be extended to a basis for V, say $u_1, \ldots, u_r, v_1, \ldots, v_{n-r}$, where $n = \dim(V)$. We claim that $v_1\theta, \ldots, v_{n-r}\theta$ is a basis for $\operatorname{Im}(\theta)$. From this, it follows that $\dim(\operatorname{Im}(\theta)) = n - r$, and the result is proved.

$v_1\theta, \ldots, v_{n-r}\theta$ *spans* $\operatorname{Im}(\theta)$: choose any vector of $\operatorname{Im}(\theta)$, say $v\theta$. Write v in terms of the basis for V, say

$$v = b_1 u_1 + \cdots + b_r u_r + c_1 v_1 + \cdots + c_{n-r} v_{n-r}.$$

Now apply θ. Since $u_i\theta = 0$ for all i (as these vectors are in the kernel of θ), we have

$$v\theta = c_1(v_1\theta) + \cdots + c_{n-r}(v_{n-r}\theta),$$

as required.

$v_1\theta, \ldots, v_{n-r}\theta$ *are linearly independent*: Suppose that

$$c_1(v_1\theta) + \cdots + c_{n-r}(v_{n-r}\theta) = (c_1 v_1 + \cdots + c_{n-r}v_{n-r})\theta = 0.$$

So $c_1 v_1 + \cdots + c_{n-r}v_{n-r} \in \operatorname{Ker}(\theta)$. Hence this vector can be written in terms of the basis for $\operatorname{Ker}(\theta)$, say

$$c_1 v_1 + \cdots + c_{n-r}v_{n-r} = b_1 u_1 + \cdots + b_r u_r.$$

This implies that

$$-b_1 u_1 - \cdots - b_r u_r + c_1 v_1 + \cdots + c_{n-r}v_{n-r} = 0.$$

But the us and vs are linearly independent, since they form a basis for V. So $b_i = c_i = 0$ for all i, as required. $\qquad\square$

4.8 Matrices. A matrix is simply a rectangular array, whose entries can be anything at all but are most usually numbers (or, more generally, elements of a field). Its purpose may be just to record these numbers (which might be distances between cities, results in a football league, or transition probabilities in a stochastic process); but by far the most important use of matrices is to describe linear transformations of a vector space.

Let U and V be finite-dimensional vector spaces. Choose bases $u_1, \ldots, u_m$ for U, and $v_1, \ldots, v_n$ for V. Then U and V can be identified with F^m and F^n respectively, where, for example, the vector $(x_1, \ldots, x_m) \in F^m$ corresponds to $x_1 u_1 + \cdots + x_m u_m \in U$. Now let $S : U \to V$ be a linear transformation. For $1 \leq i \leq m$, $u_i S$ is a vector in V, so has the form $a_{i1} v_1 + \cdots + a_{in} v_n$, where $a_{ij} \in F$. Then we say that the matrix $A = (a_{ij})$, the $m \times n$ matrix having entry a_{ij} in row i and column j, **represents** S relative to the given bases for U and V.

The choice of bases thus has two effects: the vector spaces are identified with F^m and F^n, and the linear transformation is represented by a matrix. The connection goes deeper:

Proposition 4.16 *Let $S : U \to V$ be a linear transformation. Choose bases in U and V, and let A be the matrix representing S relative to these bases. Then if U and V are identified with F^m and F^n, the action of S is given by $x \mapsto xA$, where $x \in F^m$ is regarded as a $1 \times m$ matrix, and the operation is matrix multiplication.*

Proof We calculate. The m-tuple $(x_1, \ldots, x_m)$ corresponds to $\sum_{i=1}^m x_i u_i$, which is mapped to

$$\sum_{i=1}^m x_i(u_i S) = \sum_{i=1}^m \sum_{j=1}^n x_i a_{ij} v_j.$$

The coefficient of v_j is $\sum_{i=1}^m x_i a_{ij}$, which is the jth coordinate in the matrix product xA. $\qquad\square$

Composition of linear transformations corresponds to matrix multiplication:

Proposition 4.17 *Let $S : U \to V$ and $T : V \to W$ be linear transformations of finite-dimensional vector spaces. Choose bases in the three spaces, and let A and B be the matrices representing S and T respectively. Then the matrix representing ST is AB.*

Proof Let the basis for U be $u_1, \ldots, u_m$, let that for V be $v_1, \ldots, v_n$, and let that for W be $w_1, \ldots, w_p$. Also, let $A = (a_{ij})$ and $B = (b_{jk})$. These matrices are $m \times n$ and $n \times p$ respectively, they can be multiplied.

We have

$$u_i ST = \sum_{j=1}^n a_{ij} v_j T = \sum_{j=1}^n \sum_{k=1}^p a_{ij} b_{jk} w_k.$$

So the coefficient of w_k is $\sum_{j=1}^{n} a_{ij}b_{jk}$, which is precisely the (i,k) entry of AB, by the definition of matrix multiplication. □

These results are not a miracle. Once we agree that the most important purpose of a matrix is to represent a linear transformation, we naturally want to define matrix multiplication so as to reflect the composition of linear transformation. So we should instead regard these results as justifying the rather unintuitive definition of matrix multiplication. Think of the above proposition in these terms: 'I want the matrix product to correspond to composition of transformations. What definition should I use?'

4.9   Change of basis. The matrix representing a linear transformation is defined only relative to bases for the two vector spaces. Change the bases, and the matrix will change. We now have to understand how this change works. Again, matrices are involved.

Let V be a finite-dimensional vector space. Let $v_1, \ldots, v_n$ and $v_1', \ldots, v_n'$ be two bases for V. The **transition matrix** between the two bases is the $n \times n$ matrix $Q = (q_{ij})$, where $v_i' = \sum_{j=1}^{n} q_{ij}v_j$. (The rule is: express the new basis vectors in terms of the old ones, and take the matrix of coefficients.)

Suppose that $v_1'', \ldots, v_n''$ is yet another basis, and let Q' be the transition matrix from the primed to the doubly primed basis. Then the transition matrix from the unprimed to the doubly primed basis is $Q'Q$. (Note the reversal!) For we have

$$v_i'' = \sum_{j=1}^{n} q_{ij}'v_j' = \sum_{j=1}^{n}\sum_{k=1}^{n} q_{ij}'q_{jk}v_k.$$

The coefficient of v_k is thus $\sum_{j=1}^{n} q_{ij}'q_{jk}$, which is the (i,k) entry of $Q'Q$.

The transition matrix from a basis to itself is the identity matrix I. Hence it follows that, if Q is the transition matrix between two bases of V, then the transition matrix between the bases in the reverse direction is Q^{-1}. In particular, we see that a transition matrix is invertible (that is, it has an inverse).

Now we can describe the effect of changes of basis on the matrix of a linear transformation.

Theorem 4.18 *Let $S : U \to V$ be a linear transformation. Choose two bases for U and V; let the transition matrix between the first and second bases in U be P, and the transition matrix between the first and second bases in V be Q. Suppose that the matrix representing S relative to the first bases in U and V is A, and the matrix representing S relative to the second bases is A'. Then*

$$A' = PAQ^{-1}.$$

Proof   Let $u_1, \ldots, u_m$ and $u'_1, \ldots, u'_m$ be the two bases for U, and let $v_1, \ldots, v_n$ and $v'_1, \ldots, v'_n$ be the two bases for V. Then $A' = (a'_{il})$, where

$$u'_i S = \sum_{j=1}^{n} a'_{ij} v'_j.$$

Set $Q^{-1} = R = (r_{ij})$. Then

$$u'_i = \sum_{j=1}^{m} p_{ij} u_j, \qquad v_k = \sum_{k=1}^{n} r_{kl} v'_l.$$

Hence

$$u'_i S = \sum_{j=1}^{m} p_{ij} u_j S = \sum_{j=1}^{m} \sum_{k=1}^{n} p_{ij} a_{jk} v_k = \sum_{j=1}^{m} \sum_{k=1}^{n} \sum_{l=1}^{n} p_{ij} a_{jk} r_{kl} v'_l$$

So $a'_{il} = \sum_j \sum_k p_{ij} a_{jk} r_{kl}$, or $A' = PAR = PAQ^{-1}$. $\qquad \square$

Now, if S is a linear transformation from U to V, there is a good choice of basis which greatly simplifies the form of the matrix. This is the choice that we made in the Rank and Nullity Theorem, with a small twist.

Theorem 4.19   *Let $S : U \to V$ be a linear transformation. Then there is a natural number r and a choice of bases in U and V such that the matrix of S relative to these bases is*

$$\begin{pmatrix} I_r & O \\ O & O \end{pmatrix},$$

where I_r is an $r \times r$ identity matrix and O denotes a zero matrix of the appropriate size.

Proof   Choose a basis for $\mathrm{Ker}(S)$ and extend to a basis for U. Choose the numbering so that, if $u_1, \ldots, u_m$ is the basis for U, then the last $n - r$ basis vectors form a basis for $\mathrm{Ker}(S)$. The Rank and Nullity Theorem tells us that, if $v_i = u_i S$ for $i = 1, \ldots, r$, then $v_1, \ldots, v_r$ are linearly independent, and so can be extended to a basis for V, say $v_1, \ldots, v_n$.

Now we have

$$u_i S = \begin{cases} v_i & \text{if } i \leq r, \\ 0 & \text{otherwise}; \end{cases}$$

so the matrix representing S relative to this basis is as claimed. $\qquad \square$

The matrices P and Q are not unique here; but, whichever ones we choose, the number r (called the **rank** of the matrix A) is the same, since r is the dimension of the image of the linear transformation represented by A.

Sometimes, two matrices A and A' of the same size are called 'equivalent' if they represent the same linear transformation relative to possibly different bases. Another way of saying the same thing is that A and A' are equivalent if there exist invertible matrices P and Q such that $A' = PAQ^{-1}$. Now 'equivalence' is an equivalence relation; and every equivalence class contains a unique matrix of the form $\begin{pmatrix} I & O \\ O & O \end{pmatrix}$. So the number of 'equivalence' classes is equal to the number of possible values of r, which is $1 + \min(m, n)$, since all the values $0, 1, \ldots, \min(m, n)$ can occur. (The matrices $\begin{pmatrix} I & O \\ O & O \end{pmatrix}$ form a set of **canonical forms** for the relation of 'equivalence', since each equivalence class contains just one of them.)

4.10 Elementary operations. Theorem 4.19 describes canonical forms for the relation of 'equivalence', but gives no hint about how to find r, P, Q for a given A. We now consider this problem, and end up giving a different (and algorithmic) proof of the theorem.

We define three types of **elementary row operations** on a matrix A, as follows:

(Er1) Add a scalar multiple of one row to another.
(Er2) Multiply a row by a non-zero scalar.
(Er3) Interchange two rows.

Note Strictly speaking, the operations of type (Er3) are unnecessary, since they can be obtained by a combination of the other types. Given two rows (say the ith and jth rows, R_i and R_j), the sequence of operations:

- add the ith row to the jth;
- subtract the jth row from the ith;
- add the ith row to the jth;
- multiply the ith row by -1;

has the effect

$$(R_i, R_j) \mapsto (R_i, R_j + R_i) \mapsto (-R_j, R_j + R_i) \mapsto (-R_j, R_i) \mapsto (R_j, R_i).$$

However, it is convenient to keep all three types.

For example, consider the matrix $\begin{pmatrix} 2 & 3 \\ 4 & 5 \end{pmatrix}$, and perform elementary operations as follows:

- Subtract twice the first row from the second: we obtain $\begin{pmatrix} 2 & 3 \\ 0 & -1 \end{pmatrix}$.
- Add three times the second row to the first: we obtain $\begin{pmatrix} 2 & 0 \\ 0 & -1 \end{pmatrix}$.

- Multiply the first row by 1/2: we obtain $\begin{pmatrix} 1 & 0 \\ 0 & -1 \end{pmatrix}$.
- Multiply the second row by -1: we obtain $\begin{pmatrix} 1 & 0 \\ 0 & 1 \end{pmatrix}$.

You should compare this procedure with the technique for solving linear equations. Suppose, for example, that we were given the equations

$$2x + 3y = 8,$$
$$4x + 5y = 14.$$

We subtract twice the first equation from the second to eliminate x, giving $-y = -2$. Adding three times this equation to the first gives $2x = 2$. Now these equations imply $x = 1$, $y = 2$.

Not every matrix can be transformed to the identity by performing elementary row operations on it. We now investigate just what can be achieved by these operations.

Definition A matrix A is said to be in **echelon form** if the following two conditions hold:

(Ech1) If a row of A is non-zero, then its first non-zero entry is equal to 1. (This entry is called the **leading** 1 of the row.)

(Ech2) The non-zero rows occur before all the zero rows; and, if the ith and jth rows are both non-zero with $i < j$, then the leading 1 in the jth row occurs to the right of that in the ith row.

It is in **reduced echelon form** if the following condition also holds:

(Ech3) All other entries in the column containing a leading 1 are zero.

The term 'echelon form' is meant to suggest the way geese fly: each goose flies behind and to one side of the one in front (presumably for aerodynamic reasons).

For example, the matrix

$$\begin{pmatrix} 1 & 2 & 0 & 3 & 4 \\ 0 & 0 & 1 & 5 & 6 \\ 0 & 0 & 0 & 0 & 0 \end{pmatrix}$$

is in reduced echelon form. If the entry in the first row and third column had instead been 7, the matrix would be in echelon but not reduced echelon.

Note that it follows from (Ech1) and (Ech2) that, if the matrix A is in echelon form, then all the entries in the column of a leading 1 which lie below that leading 1 must be zero; that is, half of the condition (Ech3) holds automatically.

Theorem 4.20 *Any matrix can be transformed into reduced echelon form by a sequence of elementary row operations.*

Proof The algorithm proceeds in two stages. First, we apply a recursive procedure to bring the matrix into echelon form. Then it is a fairly simple matter to convert echelon to reduced echelon. In fact, the algorithm is just the familiar solution method for systems of linear equations, slightly disguised.

Let A be a matrix. We define the following procedure. If $A = O$, then A is already in echelon form; so we simply report success. Suppose that $A \neq O$. Let m be the number of the leftmost column which contains a non-zero entry, and let k be the number of the topmost row having a non-zero entry in this column. First, if $k > 1$, then interchange the first and kth rows; thus we can assume that $k = 1$. Next, if $a_{1m} \neq 1$, multiply the first row by $(a_{1m})^{-1}$; thus we can assume that $a_{1m} = 1$. Now, for every $i > 1$ for which $a_{im} \neq 0$, subtract a_{im} times the first row from the ith row. This gives us a matrix in which every entry after the first in the mth column is zero.

Let B be the submatrix consisting of all rows except the first. Apply the algorithm recursively to reduce B to echelon form. Since the first row of A is untouched by these operations, and all elements of A in the other rows and the first m columns are zero, the same operations can be applied to A without affecting what we have done already. The resulting matrix is in echelon form, as all the leading 1s in rows after the first occur in columns after the mth. □

The description in terms of linear equations is very natural. If all columns before the mth are zero, then the first $m - 1$ variables do not actually appear in the equations at all (they will end up as free parameters in the solution). We take the first variable x_m which does occur, and divide by its coefficient in some equation. This effectively uses that equation to express x_m in terms of later variables. Subtracting multiples of this equation from the others amounts to using the expression for x_m to eliminate it from the remaining equations. The recursive step says 'solve these equations for $x_{m+1}, \ldots, x_n$'. Then use these solutions to find x_m, and we have finished.

In Stage 2, we have A in echelon form, and wish to convert it to reduced echelon. We work through the non-zero rows from top to bottom. Consider the ith row, and suppose that its leading 1 appears in the m_ith column. As we remarked earlier, all elements of this column below the ith row are already zero. If an earlier row (say, the jth) has a non-zero element in the m_ith column, we remove it by subtracting a_{jm_i} times the ith row from the jth row. Since all entries in the ith row earlier than the m_ith column are zero, earlier entries in the jth column are unaffected, although later ones may change.

This corresponds to massaging the solution we already found for the linear equations. As we saw, columns containing leading 1s correspond to variables which can be expressed in terms of later variables; all other variables appear as independent parameters in the solution. Now converting to reduced echelon ensures that, in the solution, each variable is expressed in terms of the free parameters only; no further substitutions are required to read off the values.

170   *Vector spaces*

All this is easier to understand in terms of a worked example. We take a matrix and apply the algorithm to it, following our progress with the corresponding set of linear equations.

Let

$$A = \begin{pmatrix} 0 & 1 & 2 & 3 & 4 & 5 \\ 0 & 2 & 4 & 7 & 9 & 12 \\ 0 & 0 & 0 & 2 & 5 & 4 \end{pmatrix}.$$

The corresponding system of equations reads

$$x_2 + 2x_3 + 3x_4 + 4x_5 + 5x_6 = 0,$$
$$2x_2 + 4x_3 + 7x_4 + 9x_5 + 12x_6 = 0,$$
$$2x_4 + 5x_5 + 4x_6 = 0.$$

(For simplicity, we have taken the right-hand sides of the equations to be zero.)

The first non-zero elements occur in the second column. The first row already has entry 1 in this column, so we do not need to swap or divide; this entry will be a leading 1. Subtracting twice the first row from the second, we obtain

$$\begin{pmatrix} 0 & 1 & 2 & 3 & 4 & 5 \\ 0 & 0 & 0 & 1 & 1 & 2 \\ 0 & 0 & 0 & 2 & 5 & 4 \end{pmatrix}.$$

In terms of equations, we have expressed $x_2 = -2x_3 - 3x_4 - 4x_5 - 5x_6$, and substituted in the second equation to obtain $x_4 + x_5 + 2x_6 = 0$. The third equation is unaltered.

Now we consider, recursively, the matrix

$$B = \begin{pmatrix} 0 & 0 & 0 & 1 & 1 & 2 \\ 0 & 0 & 0 & 2 & 5 & 4 \end{pmatrix},$$

and the corresponding equations

$$x_4 + x_5 + 2x_6 = 0,$$
$$2x_4 + 5x_5 + 4x_6 = 0.$$

Again, no swap or division is required, and subtracting twice the first row from the second makes the new second row $(0\,0\,0\,0\,3\,0)$. With the equations, we have written $x_4 = -x_5 - 2x_6$ and substituted to obtain $3x_5 = 0$.

At the last step we have $C = (0\,0\,0\,0\,3\,0)$. Dividing by 3 gives $(0\,0\,0\,0\,1\,0)$, which is in echelon form. We have simply deduced that $x_5 = 0$. So we have solved the equations in terms of the free parameters x_1, x_3, x_6; our matrix in echelon form is

$$\begin{pmatrix} 0 & 1 & 2 & 3 & 4 & 5 \\ 0 & 0 & 0 & 1 & 1 & 2 \\ 0 & 0 & 0 & 0 & 1 & 0 \end{pmatrix}.$$

Now, in Stage 2, we remove the 3 above the leading 1 in the second row by subtracting three times the second row from the first, giving a new first row $(0\ 1\ 2\ 0\ 1\ -1)$; and the 1s above the leading 1 in the third row by subtracting the third row from each of the first and second rows. This gives the matrix

$$\begin{pmatrix} 0 & 1 & 2 & 0 & 0 & -1 \\ 0 & 0 & 0 & 1 & 0 & 2 \\ 0 & 0 & 0 & 0 & 1 & 0 \end{pmatrix}$$

in reduced echelon. The corresponding solution of the equations is

$$x_2 = -2x_3 + x_6,$$

$$x_4 = -2x_6,$$

$$x_5 = 0,$$

where x_1, x_3, x_6 are arbitrary.

There is another way we can view this theorem. For each elementary row operation, there is a corresponding **elementary matrix** obtained by performing that operation on the identity matrix. For example, if there are two rows, then the operation 'add twice the first row to the second' corresponds to the elementary matrix $\begin{pmatrix} 1 & 0 \\ 2 & 1 \end{pmatrix}$.

Proposition 4.21 *(a) Let E be the elementary matrix corresponding to a given elementary row operation τ. Then the matrix obtained by performing the operation τ on A is EA.*

(b) The result of applying any sequence $\tau_1, \tau_2, \ldots, \tau_r$ of elementary row operations to a matrix A is $E_r \cdots E_2 E_1 A$.

Proof (a) This is proved by multiplying the matrices to check, for each type of row operation. For example, if τ is the operation of adding λ times the ith row to the jth, then E is the matrix with diagonal entries equal to 1, (j, i) entry λ, and all other entries zero. Then the (k, l) entry of EA is equal to $\lambda a_{il} + a_{jl}$ if $k = j$, and is a_{kl} otherwise; that is, EA is the matrix obtained from A by the operation τ. The argument in the other cases is similar.

(b) This is now obvious. $\qquad\qquad\qquad\qquad\qquad\qquad\qquad\qquad\qquad\square$

Note that any elementary matrix is invertible, since any elementary row operation can be 'undone' by another operation of the same type.

Theorem 4.22 *(a) Any invertible matrix is a product of elementary matrices.*

(b) For any matrix A, there is an invertible matrix P such that PA is in reduced echelon form.

Proof Let A be an invertible $n \times n$ matrix. By the previous result, there exist elementary matrices $E_1, E_2, \ldots, E_r$ so that $B = E_r \cdots E_1 A$ is in reduced echelon form. Now B must be the identity matrix. For B is invertible, so no row of B can be zero. Let the leading 1 in the ith row of B occur in column m_i. Then $1 \leq m_1 < m_2 < \ldots < m_n \leq n$. So we have $m_i = i$ for every i. Now, since every column contains a leading 1, all the elements apart from the leading 1s are zero. So B has entries 1 on the diagonal and 0 elsewhere; that is, $B = I$.

Now it follows that $A = E_1^{-1} E_2^{-1} \cdots E_r^{-1}$ is a product of elementary matrices.

(b) This follows immediately from part (b) of the Proposition together with Theorem 4.20.   □

In order to complete the analysis, we define **elementary column operations** in a similar way to the elementary row operations:

(Ec1) Add a scalar multiple of one column to another.
(Ec2) Multiply a column by a non-zero scalar.
(Ec3) Interchange two columns.

Suppose that the matrix A is in reduced echelon form. Let $A = (a_{ij})$ have its leading 1s in row i and column m_i, for $1 \leq i \leq r$, with $m_1 < m_2 < \cdots < m_r$. Apart from the leading 1s, any non-zero element a_{ij} occurs in a row $i \leq r$ and a column $j > m_i$ which is not of the form m_k for any k. Now column m_i has entry 1 in the ith row and zeros everywhere else; so if we subtract a_{ij} times column m_i from column j, we replace a_{ij} by zero and do not change any other element of the matrix. Thus, by applying a number of operations of type (Ec1), we produce a matrix in which all the elements apart from the leading 1s are zero.

Now use operations of type (Ec3) to swap columns 1 and m_1, 2 and m_2, ..., r and m_r (if necessary). The resulting matrix has its leading 1s in position (i, i) for $i = 1, \ldots, r$. In other words, it has the form $\begin{pmatrix} I & O \\ O & O \end{pmatrix}$, where the identity matrix I has size $r \times r$. This is the canonical form of A under 'equivalence'.

Furthermore, just as elementary row operations can be performed by multiplying on the left by elementary matrices, so elementary column operations can be performed by multiplying on the right (except that it is necessary to transpose the elementary matrices in order to have the correct effect). As before, a product of elementary matrices is invertible.

So we have given an algorithmic proof of Theorem 4.19.

Our algorithm for converting a matrix into the canonical form for equivalence was based on the idea of performing first the row operations required to convert it to reduced echelon, and then finishing with comparatively simple column operations. This is efficient for many purposes. For example, to calculate the rank of a matrix, apply row operations until it is in echelon form (it is not necessary to continue to reduced echelon), and count the number of non-zero rows.

However, if we do not insist on the strict separation between row and column operations, it is possible to reduce a matrix more simply, as we now see. The advantage of this simpler method is that it applies (with suitable modification) in a more general situation, as we see in the next section.

Algorithm for canonical form under equivalence   We are given an $m \times n$ matrix A.

If $A = O$, we are finished.

Otherwise, suppose that a_{ij} is a non-zero element. By swapping the first row with the ith (if $i > 1$) and the first column with the jth (if $j > 1$), we may assume that $i = j = 1$.

Multiplying the first row by a_{ij}^{-1}, we may assume that $a_{11} = 1$.

Now subtracting a_{i1} times the first row from the ith (for $i > 1$) and subtracting a_{1j} times the first column from the jth (for $j > 1$), we may assume that all entries in the first row and column other than a_{11} are zero.

Let B be the matrix obtained by deleting the first row and column of A. Recursively reduce B to canonical form. Then re-insert the first row and column to find the canonical form of A.

4.11   Determinants.   For small matrices, it is very common to define the determinant by writing down the formula and deducing the interesting properties. Thus, for example,

$$\text{if } A = \begin{pmatrix} a & b \\ c & d \end{pmatrix}, \text{then} \det(A) = ad - bc;$$

$$\text{if } A = \begin{pmatrix} a & b & c \\ d & e & f \\ g & h & i \end{pmatrix}, \text{then} \det(A) = aei + bfg + cdh - afh - bdi - ceg.$$

For larger matrices, the formulae become too cumbersome. (We will see that the formula for the determinant of an $n \times n$ matrix has $n!$ terms.) So we give an axiomatic treatment instead.

We define 'a' determinant function to be a function satisfying three axioms. We prove that there is a unique such function. After this is done, we are justified in referring to 'the' determinant.

Definition   A **determinant** function on $n \times n$ matrices over a field F is a function $\det : M_n(F) \to F$ satisfying the following axioms:

(D1) $\det(A)$ is a linear function of the ith row of A (keeping the other rows constant), for each i.

(D2) If two rows of A are equal, then $\det(A) = 0$.

(D3) $\det(I) = 1$, where I is the identity matrix.

Condition (D1) means the following. If A, A', A'' are three matrices which have the same entries in all rows except the ith, and if their ith rows are respectively R_i, R_i', R_i'', where $R_i'' = cR_i + c'R_i'$ for some $c, c' \in F$, then $\det(A'') = c \det(A) + c' \det(A')$. For example,

$$\det \begin{pmatrix} x + 3y & 2x + 4y \\ 5 & 6 \end{pmatrix} = x \det \begin{pmatrix} 1 & 2 \\ 5 & 6 \end{pmatrix} + y \det \begin{pmatrix} 3 & 4 \\ 5 & 6 \end{pmatrix}.$$

Theorem 4.23 *There is a unique determinant on $M_n(F)$, for any positive integer n and field F.*

Proof The key to this theorem is to study the effects of the three types of elementary row operation on the determinant.

Let A be any $n \times n$ matrix over F.

(ED1): *If B is obtained from A by adding c times the ith row to the jth, then* $\det(B) = \det(A)$. For let C be the matrix obtained from A by deleting the jth row and substituting the ith. Then $\det(C) = 0$ by (D2), since C has two equal rows. By (D1),

$$\det(B) = \det(A) + c\det(C) = \det(A).$$

(ED2): *If B is obtained from A by multiplying the ith row by c, then* $\det(B) = c\det(A)$. This is immediate from (D1).

(ED3): *If B is obtained from A by interchanging two rows, then* $\det(B) = -\det(A)$. For we noted, after the three types of elementary operation were defined, that the operation (Er3) (interchanging two rows) can be realised by composing three operations of type (Er1) (which do not change the determinant, by point (ED1)) with one of type (Er2), namely multiplying a row by -1 (which multiplies the determinant by -1, by point (ED2)).

Now there is a sequence of elementary row operations which converts A to reduced echelon form (say B). Thus we have

$$E_r \cdots E_2 E_1 A = B,$$

where $E_1, \ldots, E_r$ are elementary matrices. Since A is square, either B has a zero row, or B is the identity I. In the first case, $\det(B) = 0$ by (D1); in the second, $\det(B) = 1$, by (D3). Moreover, each elementary operation multiplies $\det(A)$ by a factor which depends only on the operation applied. So, as a result, $\det(B) = c_r \cdots c_1 \det(A)$, where $c_1, \ldots, c_r$ are the factors associated with the operations. Thus $\det(A)$ is determined uniquely. $\square$

We have shown that, if there exists a determinant function, then it is unique. It remains to show that there really is such a function. This is somewhat technical, and you may want to skip the next argument at first reading.

Recall that the **sign** of a permutation $g \in S_n$ is equal to $(-1)^p$, where p, the **parity** of g, is equal to the parity of n minus the number of cycles of g. We write it as $\text{sign}(g)$. Now consider the following function $D(A)$ of $n \times n$ matrices $A = (a_{ij})$:

$$D(A) = \sum_{g \in S_n} \text{sign}(g) a_{1\,1g} a_{2\,2g} \ldots a_{n\,ng},$$

a sum of $n!$ terms, each a product of n factors a_{ij} and a sign. We claim that the function D satisfies (D1)–(D3).

(D1): Each term in the sum involves just one entry from the ith row, namely $a_{i\,ig}$, and so is a linear function of the ith row. The same is true of the sum.

(D2): Suppose that the ith and jth rows are equal, that is, $a_{ik} = a_{jk}$ for all k. Let t be the transposition $(i\ j)$. Then $H = \{1, t\}$ is a subgroup of S_n, and so S_n can be partitioned into $n!/2$ right cosets of H. Consider the two terms in $D(A)$ corresponding to the elements g, tg of a coset Hg. The factors taken from rows other than the ith and jth are the same. From the ith and jth rows, we take $a_{i\,ig}a_{j\,jg}$ and $a_{i\,itg}a_{j\,jtg}$. But these are equal, since $it = j$, $jt = i$, and $a_{ik} = a_{jk}$. Moreover, g and tg differ by a transposition, so have opposite signs. Thus, the two terms cancel. Since this holds for each coset, $D(A) = 0$.

(D3): For $A = I$, the only non-zero entries are the diagonal entries a_{ii}, and so the only non-zero term in the expression for $D(A)$ is the one coming from the identity permutation. Since $a_{i\,i1} = 1$ for all i, and the identity has sign 1, the value of $D(I)$ is 1.

In order to prove the important properties of determinants, we look once more at the elementary row operations.

We associated with each such operation τ a factor $c(\tau)$, by which the determinant of A is multiplied when τ is applied to A. We also have an elementary matrix $E(\tau)$, obtained by applying τ to the identity matrix.

Proposition 4.24 $c(\tau) = \det(E(\tau))$.

Proof This is clear from the fact that $\det(I) = 1$. □

Theorem 4.25 *(a) For any $A, B \in M_n(F)$, we have $\det(AB) = \det(A)\det(B)$.*
(b) $\det(A) \neq 0$ if and only if A is invertible.

Proof There is an invertible matrix P (a product of elementary matrices) such that PA is in reduced echelon form; thus, either PA has a zero row (if A is not invertible), or $PA = I$ (if A is invertible). Moreover, if $P = E_r \cdots E_1$, then $c(E_r) \cdots c(E_1)\det(A) = \det(PA)$.

If A is not invertible, then $\det(PA) = 0$, and so $\det(A) = 0$. If A is invertible, then $PA = I$, so $\det(PA) = 1$. Thus $c(E_r) \cdots c(E_1)\det(A) = 1$, and so $\det(A) \neq 0$. This proves (b).

If A is not invertible, then neither is AB, and so $\det(AB) = 0 = 0\det(B) = \det(A)\det(B)$. Suppose that A is invertible. Then, as above,

$$c(E_r) \cdots c(E_1)\det(A) = 1.$$

Now the same sequence of elementary operations, applied to AB, yields the equation $E_r \cdots E_1 AB = PAB = IB = B$, and so

$$c(E_r) \cdots c(E_1)\det(AB) = \det(B).$$

It follows from these two equations that

$$\det(A)^{-1}\det(AB) = c(E_r)\cdots c(E_1)\det(AB) = \det(B),$$

so that $\det(AB) = \det(A)\det(B)$, proving (a). $\qquad\qquad\square$

Theorem 4.26 $\det(A) = \det(A^\top)$, *where $A^\top$ is the transpose of the matrix A.*

Proof A typical term in the formula for $\det(A^\top)$ is

$$\mathrm{sign}(g)a_{1g\,1}a_{2g\,2}\cdots a_{ng\,n}.$$

Now, if h is the inverse of g, then $\mathrm{sign}(h) = \mathrm{sign}(g)$ (since g and h have the same cycle structure); so we can re-order the factors and write this term as

$$\mathrm{sign}(h)a_{1\,1h}a_{2\,2h}\cdots a_{n\,nh}.$$

As σ ranges over the symmetric group, so does its inverse; so we obtain all the terms in $\det(A)$, and conclude that $\det(A^\top) = \det(A)$. $\qquad\qquad\square$

Determinants over rings It is possible to define determinants of matrices whose entries are taken from any commutative ring. There are several ways to do this. One of these is an axiomatic approach similar to the one we used before. It is easier to use what we already know.

If R is an integral domain, then it can be embedded in a field F (its field of fractions, see Section 2.14). Now any matrix over R can be regarded as a matrix over F, and its determinant calculated as before. One advantage of this approach is that our previous results will apply: for example, the axioms (D1)–(D3) all hold.

If R is an arbitrary commutative ring, we can define the determinant of an $n \times n$ matrix over R using the formula we worked out in Theorem 4.23. This time, however, we cannot assume that earlier results apply, but must rework the proofs. As an example, if R is a zero ring, then the determinant of any $n \times n$ matrix over R is zero for $n > 1$.

One final method can be applied when the ring R contains a subring K which is a field. We consider the matrix $X = (x_{ij})$ whose entries are independent indeterminates over K (that is, elements of the polynomial ring $S = K[x_{11}, \ldots, x_{nn}]$). Now S is an integral domain, so we can compute the determinant of X in the field of fractions of S. Now we obtain the determinant of an arbitrary matrix over R by substituting elements of R for the indeterminates. (This substitution is a ring homomorphism from S to R.)

Some special determinants The general formula for a determinant is very cumbersome even for moderate n. It is important that there are various special matrices whose determinants can be evaluated more simply by 'product formulae'. In particular, it is easy to decide whether or not such determinants are zero.

We consider two types here: Vandermonde determinants and circulants.

Definition A **Vandermonde determinant** is one of the form

$$V(a_1, a_2, \ldots, a_n) = \det \begin{pmatrix} 1 & 1 & \cdots & 1 \\ a_1 & a_2 & \cdots & a_n \\ a_1^2 & a_2^2 & \cdots & a_n^2 \\ \cdots & \cdots & \cdots & \cdots \\ a_1^{n-1} & a_2^{n-1} & \cdots & a_n^{n-1} \end{pmatrix}$$

for $a_1, a_2, \ldots, a_n \in F$.

Proposition 4.27 $\displaystyle V(a_1, a_2, \ldots, a_n) = \prod_{1 \le i < j \le n} (a_j - a_i).$

In particular, the determinant is zero if and only if $a_i = a_j$ for some $i \ne j$.

Proof First, we evaluate $V(x_1, x_2, \ldots, x_n)$, where $x_1, x_2, \ldots, x_n$ are indeterminates over F (this is calculated in the field of fractions of $F[x_1, x_2, \ldots, x_n]$). Once the formula of the proposition is established in this situation, it is just a polynomial identity, and remains true if we substitute a_i for x_i, $i = 1, \ldots, n$.

Let us then regard V as a polynomial $f(x_n)$ in x_n with coefficients in the field of fractions of $F[x_1, \ldots, x_{n-1}]$. Under the substitution $x_n = x_i$, for $i < n$, the matrix has two equal columns, and so its determinant is zero. So $(x_n - x_i)$ is a factor of $f(x_n)$. By inspection, f has degree $n - 1$, so $x_1, \ldots, x_{n-1}$ are all its roots. Thus

$$V(x_1, \ldots, x_n) = W(x_1, \ldots, x_{n-1}) \prod_{i<n} (x_n - x_i),$$

where W does not depend on x_n. Repeating this procedure for the other variables, we see that

$$V(x_1, \ldots, x_n) = Z \prod_{i<j} (x_j - x_i),$$

where Z does not depend on any of the variables; that is, $Z \in F$.

To evaluate Z, we consider the term in $x_2 x_3^2 \cdots x_n^{n-1}$ in the expansion of V. In the formula for the determinant as a sum over permutations, this term comes only from the identity permutation, and so its coefficient is $+1$. In the product form, we must take x_n from all $n-1$ factors containing it, then x_{n-1} from all $n-2$ factors which contain it but not x_n, and so on. In other words, from each factor, we take the variable with the greater suffix, which has the positive sign. So the coefficient is $+Z$. We conclude that $Z = 1$, and the proposition is proved. $\square$

Definition A **circulant** $C(a_0, a_1, \ldots, a_{n-1})$ is a determinant of the form

$$C(a_0, a_1, \ldots, a_{n-1}) = \det \begin{pmatrix} a_0 & a_1 & \cdots & a_{n-2} & a_{n-1} \\ a_{n-1} & a_0 & a_1 & \cdots & a_{n-2} \\ \cdots & \cdots & \cdots & \cdots & \cdots \\ a_1 & \cdots & \cdots & a_{n-1} & a_0 \end{pmatrix},$$

where $a_0, \ldots, a_{n-1} \in F$.

Proposition 4.28 *Suppose that F contains a primitive nth root of unity, say ω. Then*

$$C(a_0, a_1, \ldots, a_{n-1}) = \prod_{i=0}^{n-1} (a_0 + a_1\omega^i + \cdots + a_{n-1}\omega^{(n-1)i}).$$

Proof Consider the vector $v_i = (1 \ \omega^{-i} \ \omega^{-2i} \ \ldots \ \omega^{-(n-1)i})$. (We number the coordinates from 0 to $n-1$, and where necessary we take them mod n.) Let A be the $n \times n$ matrix whose determinant is C. Also, let

$$\lambda_i = a_0 + a_1\omega^i + \cdots + a_{n-1}\omega^{i(n-1)}.$$

Then

$$(v_i A)_j = a_j + a_{j-1}\omega^{-i} + \cdots + a_{j-n+1}\omega^{-i(n-1)}.$$

But this is equal to $\lambda_i \omega^{-ij}$, in other words, λ_i times the jth entry of v_i. Thus $v_i A = \lambda_i v_i$.

The vectors $v_0, v_1, \ldots, v_{n-1}$ form a basis for F^n. (There are n of them, and they are linearly independent, since the determinant of the matrix with columns $v_0^\top, v_1^\top, \ldots, v_{n-1}^\top$ is the Vandermonde determinant $V(1, \omega^{-1}, \ldots, \omega^{-(n-1)})$; and the elements $1, \omega^{-1}, \ldots, \omega^{-(n-1)}$ are all distinct, since they are all the powers of a primitive nth root of unity.) So, if P is the matrix whose rows are $v_1, v_1, \ldots, v_{n-1}$, then PAP^{-1} is the matrix representing the same linear transformation as A relative to the new basis. This matrix is diagonal, and has as its diagonal entries $\lambda_0, \lambda_1, \ldots, \lambda_{n-1}$. Hence

$$\det(A) = \det(PAP^{-1}) = \prod_{i=0}^{n-1} \lambda_i,$$

as required. $\qquad\qquad\qquad\qquad\qquad\qquad\qquad\qquad\qquad\qquad\qquad\square$

4.12 Matrices over Euclidean domains. We are going to prove a 'canonical form theorem' for matrices over a Euclidean domain. (In fact, the same result holds more generally, over a principal ideal domain; but the proof needs an extra trick, and is not given here.) This theorem appears to be of no particular use. But we will see that, by applying it to the two most important examples of Euclidean domains (the ring of integers, and the polynomial ring over a field), we obtain two unexpected bonuses: a structure theorem for finitely generated abelian groups, and a canonical form for the matrix of a linear transformation from a vector space to itself.

We start with a quick revision course on Euclidean domains. See Section 2.13 for more details. A **Euclidean domain** is an integral domain R (a commutative ring with identity and no divisors of zero) having a **Euclidean function** d from the set of non-zero elements of R to the set of non-negative integers, satisfying the following conditions:

(a) if a and b are non-zero, then $d(ab) \geq d(a)$;
(b) if $b \neq 0$ and a is arbitrary, then there exist $q, r \in R$ with $a = bq + r$ and either $r = 0$ or $d(r) < d(b)$.

A Euclidean domain is a principal ideal domain (each ideal is generated by a single element), and hence a unique factorisation domain. Moreover, the **Euclidean Algorithm** finds, for any two elements a and b, the greatest common divisor (g.c.d.) d of a and b, and elements x and y such that $d = xa + yb$.

Let R be any commutative ring with identity. We define **elementary row operations** on matrices over R almost exactly as we did over a field. The only difference is that, in order to ensure that any elementary row operation can be undone, we restrict operation (Er2) by allowing only multiplication by units. In detail, the operations are:

(Er1) Add any multiple of one row to another.
(Er2) Multiply a row by a unit of R.
(Er3) Interchange two rows.

We also define the analogous **elementary column operations** (Ec1), (Ec2), and (Ec3).

Theorem 4.29 *Let A be an $m \times n$ matrix over a Euclidean domain R. Then A can be transformed, by means of elementary row and column operations, to a matrix of the form $\begin{pmatrix} D & O \\ O & O \end{pmatrix}$, where D is an $r \times r$ matrix with diagonal entries $d_1, d_2, \ldots, d_r$ and zeros elsewhere, and O is a zero matrix of any appropriate size. Moreover, $d_i \neq 0$ for $1 \leq i \leq r$, and d_i divides d_{i+1} for $1 \leq i \leq r - 1$. The number r is uniquely determined by A, and the elements $d_1, \ldots, d_r$ are unique up to associates.*

This theorem generalises Theorem 4.19, since if R is a field then any non-zero element of R is associate to 1.

The ring elements $d_1, d_2, \ldots, d_r$ are called the **invariant factors** of the matrix A. The matrix $\begin{pmatrix} D & O \\ O & O \end{pmatrix}$ is called the **Smith normal form** of A.

To prove that the required form exists, it is enough to prove the following:

> By performing elementary row and column operations, we can convert A into a matrix B such that the element $d_1 = b_{11}$ divides every element in B and the remaining elements in the first row and column are all zero.

For then B has the form

$$B = \begin{pmatrix} d_1 & O \\ O & C \end{pmatrix},$$

where O is a $1 \times (n-1)$ row or a $(m-1) \times 1$ column of zeros, and every element of C is divisible by d_1. By induction, C can be converted to Smith normal form by elementary row operations. Suppose that its invariant factors are $d_2, \ldots, d_r$, where d_i divides d_{i+1} for $2 \le i \le r - 1$. The elementary operations applied to C do not change the property that all of its elements are divisible by d_1; so d_1 divides d_2, as required.

We prove that, if not every element of A is divisible by a_{11}, then we can apply elementary operations to find a matrix A' with $d(a'_{11}) < d(a_{11})$.

Case 1 Suppose that some element of the first row is not divisible by a_{11}, say a_{1j}, for $j > 1$. By the Euclidean property, we can write $a_{1j} = a_{11}q + r$, where $r \ne 0$ (since a_{11} does not divide a_{1j}) and hence $d(r) < d(a_{11})$. Now subtract q times the first column from the jth column. The new entry in the jth column is $a_{1j} - qa_{11} = r$. Now interchange the first and jth columns, to obtain a matrix with $(1,1)$ entry r.

Case 2 If some element of the first column is not divisible by a_{11}, the argument is similar, but using row operations instead of column operations.

Case 3 Finally, suppose that a_{11} divides every element in the first row and column, but does not divide the entry a_{ij}. Suppose that $a_{i1} = xa_{11}$. Subtracting $x - 1$ times the first row from the ith, we obtain a matrix with $(i, 1)$ entry equal to a_{11}. The new (i, j) entry is $a_{ij} - (x - 1)a_{1j} = b_{ij}$, say. Since a_{11} divides a_{1j} but not a_{ij}, it does not divide b_{ij}. Write $b_{ij} = qa_{11} + r$, where $d(r) < d(a_{11})$. Now, subtracting q times the first column from the jth, we obtain a matrix with (i, j) entry $b_{ij} - qa_{11} = r$. Finally, a row interchange and a column interchange bring r to the $(1, 1)$ position.

Now the value of $d(a_{11})$ can only decrease a finite number of times, since it is a non-negative integer. So, after a finite number of steps, we reach a matrix all of whose elements are divisible by a_{11}.

Finally, if $a_{1j} = q_j a_{11}$ for $j > 1$, then subtracting q_j times the first column from the jth produces a matrix with $(1, j)$ entry zero. Similarly, by row operations, we may assume that all entries a_{i1} are zero for $i > 1$.

This completes the existence proof. What about the uniqueness of the elements d_i (up to associates)?

In the first place, d_1 is determined: it is the greatest common divisor of all the elements in A. For it is easily checked that the elementary operations do not alter the greatest common divisor of all the entries up to associates (that is, at worst the g.c.d. is multiplied by a unit). For the final matrix, the greatest common divisor is d_1, since this is an entry and it divides all the others.

The other elementary divisors are determined by a somewhat more complicated rule:

> For $1 \le i \le r$, the greatest common divisor of all the determinants of $i \times i$ submatrices of A is equal to $d_1 d_2 \cdots d_i$; for $i > r$, this greatest common divisor is zero.

The proof is similar: one shows that the greatest common divisor of the $i \times i$ submatrices is multiplied by a unit (hence not changed, up to associates) by elementary operations, while the greatest common divisor for a matrix in the Smith normal form is $d_1 d_2 \cdots d_i$ if $i \le r$, or 0 if $i > r$.

So the theorem is proved.

Example Consider the matrix $\begin{pmatrix} 4 & 6 \\ 8 & 10 \end{pmatrix}$ over the integers. The greatest common divisor of the entries is 2, while the determinant is -8; so we expect the Smith normal form to be $\begin{pmatrix} 2 & 0 \\ 0 & 4 \end{pmatrix}$. Let us see what operations achieve this result.

First, 4 does not divide 6; indeed, $6 = 4 + 2$. So subtract the first column from the second, to obtain $\begin{pmatrix} 4 & 2 \\ 8 & 2 \end{pmatrix}$. Now interchange the first and second columns, obtaining $\begin{pmatrix} 2 & 4 \\ 2 & 8 \end{pmatrix}$.

Now 2 does divide all the other entries, so we subtract twice the first column from the second, giving $\begin{pmatrix} 2 & 0 \\ 2 & 4 \end{pmatrix}$, and then subtract the first row from the second, giving $\begin{pmatrix} 2 & 0 \\ 0 & 4 \end{pmatrix}$.

The final result is in Smith normal form.

Exercise 4.7 (a) Prove the First Isomorphism Theorem for vector spaces.

(b) Formulate and prove the Second and Third Isomorphism Theorems.

Exercise 4.8 Let $A = (a_{ij})$ be an $n \times n$ matrix over the field F. Define the (i, j) **cofactor** A_{ij} of A to be $(-1)^{i+j}$ times the determinant of the matrix obtained from A by deleting the ith row and jth column.

(a) Prove that $\sum_{j=1}^{n} a_{ij} A_{ij} = \det(A)$. [*Hint*: Show that the left-hand side satisfies the axioms for a determinant.]

(b) Prove that, for $i \ne k$, $\sum_{j=1}^{n} a_{ij} A_{kj} = 0$. [*Hint*: By (a), this is the determinant of a matrix with two equal rows.]

(c) Define the **adjoint** of A to be the matrix $\mathrm{Adj}(A)$ whose (i, j) entry is A_{ji}. Prove that $A \cdot \mathrm{Adj}(A) = \det(A)I$.

Exercise 4.9 Put the integer matrix

$$\begin{pmatrix} 6 & 10 & 0 \\ 6 & 0 & 15 \\ 0 & 10 & 15 \end{pmatrix}$$

into Smith normal form. Check your result by calculating determinants.

Exercise 4.10 Show that the reduced echelon form of a matrix over a field is unique.

5 Modules

A module bears the same relationship to a vector space as a ring does to a field. As this suggests, modules exist in much greater profusion than do vector spaces. We know everything about a vector space over a given field F once we know its dimension; but to specify a module, more detailed information is required. One of the themes of modern algebra is that the modules for a given ring capture a good deal of the structure of the ring. Our main goal in this chapter is the description of modules over Euclidean domains. The structure theorem tells us about finitely generated abelian groups, and also gives us canonical forms for matrices over fields.

Introduction

5.1 Definition of modules. A module is a 'vector space over a ring'. That is, it satisfies almost exactly the same axioms as a vector space, with scalars taken from a ring rather than a field. There are two small differences, resulting from the extra generality. First, since a ring does not necessarily have an identity element, we cannot impose the axiom involving the identity. Second, since ring multiplication may not be commutative, there are two forms of the axiom involving multiplication of scalars, leading to two different kinds of module. There is also a notational difference: sometimes we choose to write scalars on the right, instead of on the left as we did for vector spaces.

Formally, then, let R be a ring. We define a **right R-module** to be a set M with a binary operation of addition (written $+$) and, for each $r \in R$, a unary operation of scalar multiplication by r (where we write the result of multiplying m by r as mr), satisfying the following axioms:

(MA0) For all $m_1, m_2 \in M$, we have $m_1 + m_2 \in M$.

(MA1) For all $m_1, m_2, m_3 \in M$, we have $m_1 + (m_2 + m_3) = (m_1 + m_2) + m_3$.

(MA2) There exists $0 \in M$ such that $m + 0 = 0 + m = m$ for all $m \in M$.

(MA3) For all $m \in M$, there exists $-m \in M$ such that $m + (-m) = (-m) + m = 0$.

(MA4) For all $m_1, m_2 \in M$, $m_1 + m_2 = m_2 + m_1$.

(MM0) For all $r \in R$ and $m \in M$, we have $mr \in M$.

(MM1) For all $r \in R$, $m_1, m_2 \in M$, we have $(m_1 + m_2)r = m_1 r + m_2 r$.

(MM2) For all $r_1, r_2 \in R$, $m \in M$, we have $m(r_1 + r_2) = mr_1 + mr_2$.

(MM3) For all $r_1, r_2 \in R$, $m \in M$, we have $m(r_1 r_2) = (mr_1)r_2$.

If the ring R has an identity 1, then we call M a **unital** module if it also satisfies

(MM4) For all $m \in M$ we have $m1 = m$.

Axiom (MM3) says that scalar multiplication by r_1r_2 is the same as multiplication by first r_1 and then r_2. We could imagine structures in which this works the other way around, that is, first r_2, then r_1. These are called **left R-modules.** It would be possible simply to write an alternative version of (MM3) for left modules. But it is more natural to change the notation, so that scalar multiplication by r takes m to rm (with the scalar on the left). Formally, then, we define a **left R-module** to be a set M with a binary operation of addition and a unary operation of scalar multiplication by r for each $r \in R$ (written rm) such that axioms (MA0)–(MA4) above hold, and also

(MM0') For all $r \in R$ and $m \in M$, we have $rm \in M$.
(MM1') For all $r \in R$, $m_1, m_2 \in M$, we have $r(m_1 + m_2) = rm_1 + rm_2$.
(MM2') For all $r_1, r_2 \in R$, $m \in M$, we have $(r_1 + r_2)m = r_1m + r_2m$.
(MM3') For all $r_1, r_2 \in R$, $m \in M$, we have $(r_1r_2)m = r_1(r_2m)$.

Note that (MM0')–(MM2') are identical to (MM0)–(MM2), but written in different notation; only (MM3') is really different.

Again, if R has an identity 1, the left R-module is called **unital** if $1m = m$ for all $m \in M$.

Just as for a vector space, we can express the axioms more briefly in abstract language. Let M be a right R-module. Axioms (MA0)–(MA4) assert that M, with the operation of addition, is an abelian group. The map $\theta_r : M \to M$ given by $m\theta_r = mr$ is an endomorphism of the abelian group M (a homomorphism from M to itself). The set of all endomorphisms forms a ring $\mathrm{End}(M)$, and the map $\phi : R \to \mathrm{End}(M)$ given by $r\phi = \theta_r$ is a homomorphism. So we could say: *A right R-module is an abelian group M with a homomorphism from R to $\mathrm{End}(M)$. Moreover, it is a unital module if and only if the homomorphism ϕ maps the identity of R to the identity of $\mathrm{End}(M)$.*

For left modules, there is a complication. Our definition of the endomorphism ring of an abelian group takes multiplication of endomorphisms to be composition in the usual order. It is necessary to reverse this order. Accordingly, given any ring R, we define the **opposite ring** R° as follows: the elements of R° are the elements of R and addition is the same as in R; but multiplication (which we will denote by $\circ$) is given by the rule

$$r_1 \circ r_2 = r_2r_1.$$

It can be shown that R° is a ring, and shares most of the properties of R. Now we can say: *A left R-module consists of an abelian group M together with a homomorphism ψ from R° to $\mathrm{End}(M)$; it is unital if and only if ψ maps the identity of R° to the identity endomorphism.* In other words, a left R-module is a right R°-module.

Of course, all this confusing complication about left and right modules is unnecessary if the ring R is commutative. In this case, we simply speak of R-modules, without specifying left or right. This explains why we did not meet the left–right distinction when we were studying vector spaces.

5.2 Examples of modules.

Example 1 If F is a field, a vector space over F is an F-module (left or right), and conversely. Modules generalise vector spaces.

Example 2 Let R be any ring. Then we can make R into either a left or a right R-module, as follows. In either case, we take the module addition to be the ring addition. Also, in either case, we take the scalar multiplication to be the ring multiplication, but interpreted differently: for a right module, $r_1 r_2$ is the result of multiplying the module element r_1 by the ring element r_2; for a left module, r_2 is the module element, and r_1 the scalar.

If R has an identity, then these modules are automatically unital.

We call these modules the **free right and left R-modules of rank 1.**

Example 3 This generalises example 2. Let n be a positive integer, and let R^n denote the set of n-tuples of elements of R. We make R^n into a right R-module by defining

$$(x_1, x_2, \ldots, x_n) + (y_1, y_2, \ldots, y_n) = (x_1 + y_1, x_2 + y_2, \ldots, x_n + y_n),$$
$$(x_1, x_2, \ldots, x_n)r = (x_1 r, x_2 r, \ldots, x_n r),$$

or a left module by defining

$$(x_1, x_2, \ldots, x_n) + (y_1, y_2, \ldots, y_n) = (x_1 + y_1, x_2 + y_2, \ldots, x_n + y_n),$$
$$r(x_1, x_2, \ldots, x_n) = (r x_1, r x_2, \ldots, r x_n).$$

These are the **free right and left R-modules of rank n.**

Theorem 4.10 shows that any finite-dimensional vector space is a free module of rank n for a unique value of n. For general rings, things are much more complicated: most modules are not free, and free modules of different ranks may be isomorphic.

Example 4 Any module is an abelian group, if we forget the scalar multiplication and consider just the addition. Can we go back from abelian groups to modules? It turns out that any abelian group can be made into a $\mathbb{Z}$-module in a natural way.

Let M be an abelian group (whose operation is written as $+$). Now define a scalar multiplication by $\mathbb{Z}$ as follows:

if $n > 0$ and $x \in M$, then $nx = x + x + \cdots + x$ (n terms);
$0x = 0$, where the second zero is the identity element of M;
if $n = -m$ where $m > 0$, then $nx = -(mx)$.

So we have the general principle: *Abelian groups are the same thing as* $\mathbb{Z}$*-modules.*

Example 5 Let $M = F^n$, where F is a field. Then M is a vector space over F, and hence a (left) F-module. But it is also a right $M_n(F)$-module, where $M_n(F)$ is the ring of $n \times n$ matrices over F. ('Scalar multiplication' of a vector by a matrix is given by the usual formula for matrix multiplication, regarding the vector as a $1 \times n$ matrix.)

A structure M like this, which is a left module for R and a right module for S and satisfies the additional axiom

(MB) For all $r \in R$, $s \in S$ and $m \in M$, we have $(rm)s = r(ms)$,

is called an R–S **bimodule**.

Example 6 This is an important example for applications.

Let V be a vector space over F, and let S be a linear transformation from V to V. For any polynomial $f \in F[x]$, we can define a linear transformation $f(S)$ on V as follows: if

$$f(x) = a_n x^n + a_{n-1} x^{n-1} + \cdots + a_1 x + a_0,$$

then we put

$$f(S) = a_n S^n + a_{n-1} S^{n-1} + \cdots + a_1 S + a_0 I,$$

where S^n is the n-fold composition of S with itself $(vS^n = (\cdots (vS)S \cdots)S$, where S occurs n times), and I is the identity transformation.

Now we make V into a $F[x]$-module by the rule

$$vf(x) = vf(S).$$

We will see that the structure of this module reflects the properties of S in a very precise way.

Example 7 Here is a further generalisation of the free R-modules of rank 1 from Example 2. If I is an ideal in the ring R, then both I and R/I are (right or left) R-modules. For I, we take the addition and multiplication of R, but only add two elements of I, or multiply an element of I by an element of R. (Both of these operations yield results in I, by the definition of an ideal.)

For R/I, the construction is similar. We use the addition defined in the factor ring, and scalar multiplication $(I + x)r = I + xr$ (for a right module) or $r(I + x) = I + rx$ (for a left module). It is necessary to check that these operations are well defined, as well as proving the module axioms.

5.3 Submodules and homomorphisms. We formulate the notions of submodule and module homomorphism for right modules. The definitions for left modules are very similar.

Definition Let M be a right R-module. A **submodule** N of M is a subset of M which is an R-module (with respect to the addition and scalar multiplication of M).

As usual, in order to test whether N is a submodule, it suffices to check the closure laws, since all other laws hold automatically. Accordingly we have

Theorem 5.1 (Submodule Test) *The non-empty subset N of M is a submodule if and only if it is closed under subtraction and scalar multiplication.*

Proof The closure conditions are clearly necessary. Suppose that they hold. Closure under subtraction ensures that N is a subgroup of the abelian group $(M, +)$. So the result follows.   □

Let M and N be right R-modules. A (R-module) **homomorphism** from M to N is a map $\theta : M \to N$ which preserves addition and scalar multiplication:

$$(m_1 + m_2)\theta = m_1\theta + m_2\theta,$$

$$(mr)\theta = (m\theta)r,$$

for all $m, m_1, m_2 \in M$ and $r \in R$. (The second equation indicates that the homomorphism does not affect the scalars in R.)

An **isomorphism** is a homomorphism which is one-to-one and onto.

The **image** $\mathrm{Im}(\theta)$ of a homomorphism θ is $\{m\theta : m \in M\}$, and the **kernel** $\mathrm{Ker}(\theta)$ is $\{m \in M : m\theta = 0\}$.

If K is a submodule of M, then we can define a **factor module** M/K, whose elements are the cosets of K in M. (These cosets are defined because K is a subgroup of the abelian group $(M, +)$. For the same reason, addition on M/K is well defined.) Now scalar multiplication on M/K is given by the rule

$$(K + m)r = K + mr.$$

As usual, we can check that this is well defined, and that M/K is an R-module.

Theorem 5.2 *Let $\theta : M \to N$ be an R-module homomorphism. Then the image and kernel of θ are submodules of N and M respectively; and $M/\mathrm{Ker}(\theta) \cong \mathrm{Im}(\theta)$ (as R-modules).*

The proof is an exercise.

5.4   Annihilators, cyclic modules, direct sums. This section develops a few tools of module theory, which will be applied in the next. We assume from now on that our rings are commutative and have identities, and that our modules (for which the left–right distinction is not necessary) are unital.

Definition Let R be a commutative ring with identity, and M a unital R-module. The **annihilator** of M, written $\mathrm{Ann}(M)$, is the set

$$\{r \in R : mr = 0 \text{ for all } m \in M\}.$$

The submodule of M **generated by** $m_1, \ldots, m_n$ is the set

$$\langle m_1, \ldots, m_r \rangle = \{m_1 r_1 + \cdots + m_n r_n : r_1, \ldots, r_n \in R\}.$$

We say that M is **finitely generated** if there is a finite set of elements of M which generates M. We say that M is **cyclic** if it is generated by just one element.

Remark In the case where the module M is equal to R (as in Example 1), the submodule generated by a set of elements is exactly the same as the ideal generated by these elements; this is why we use the same notation.

Here is a general result about these concepts.

Theorem 5.3 *Let M be a unital module over a commutative ring R with identity.*

(a) $\operatorname{Ann}(M)$ is an ideal of R.
(b) For any $m_1, \ldots, m_n \in M$, the set $\langle m_1, \ldots, m_n \rangle$ is a submodule of M, and is the smallest submodule containing $m_1, \ldots, m_n$.
(c) If M is cyclic, then $M \cong R/\operatorname{Ann}(M)$ (as R-module).

Proof (a) We apply the Ideal Test.
If $r_1, r_2 \in \operatorname{Ann}(M)$, then $mr_1 = mr_2 = 0$, so $m(r_1 + r_2) = mr_1 + mr_2 = 0$, and $r_1 + r_2 \in \operatorname{Ann}(M)$.
If $r \in \operatorname{Ann}(M)$ and $s \in R$, then $mr = 0$, so $m(rs) = (mr)s = 0$, so that $rs \in \operatorname{Ann}(M)$.
Thus $\operatorname{Ann}(M)$ is an ideal of R.
(b) We apply the Submodule Test:
If $m_1 r_1 + \cdots + m_n r_n, m_1 s_1 + \cdots + m_n s_n \in \langle m_1, \ldots, m_n \rangle$, then the sum of these elements is

$$m_1 r_1 + \cdots + m_n r_n + m_1 s_1 + \cdots + m_n s_n$$
$$= m_1(r_1 + s_1) + \cdots + m_n(r_n + s_n) \in \langle m_1, \ldots, m_n \rangle.$$

If $m_1 r_1 + \cdots + m_n r_n \in \langle m_1, \ldots, m_n \rangle$ and $s \in R$, then

$$(m_1 r_1 + \cdots + m_n r_n)s = m_1(r_1 s) + \cdots + m_n(r_n s) \in \langle m_1, \ldots, m_n \rangle.$$

So $\langle m_1, \ldots, m_n \rangle$ is a submodule.
Moreover, this submodule contains

$$m_i = m_1 0 + \cdots + m_{i-1} 0 + m_i 1 + m_{i+1} 0 + \cdots + m_n 0$$

for $i = 1, \ldots, n$. (Here we used the unital property of M.) Any submodule N of M which contains $m_1, \ldots, m_n$ contains all multiples $m_i r_i$, and hence all linear combinations $m_1r_1 + \cdots + m_nr_n$, of these elements. So $\langle m_1, \ldots, m_n \rangle$, as defined, is the smallest submodule containing these elements.

(c) Let $M = \langle m \rangle$ be a cyclic R-module with annihilator I. We define a map $\theta : M \to R/I$ by the rule

$$(mr)\theta = I + r.$$

This will be the required isomorphism. First, we show that it is well defined and one-to-one. Note that $r \in I$ if and only if $mr = 0$. For, if $r \in I$, then $mr = 0$, since I is the annihilator of M and $m \in M$. Conversely, suppose that $mr = 0$. Take any element of M; by assumption, it has the form mx for some $x \in R$. Then $(mx)r = m(xr) = m(rx) = (mr)x = 0$. So $r \in \mathrm{Ann}(M) = I$.

Now

$$mr_1 = mr_2 \Leftrightarrow m(r_1 - r_2) = 0$$
$$\Leftrightarrow r_1 - r_2 \in I$$
$$\Leftrightarrow I + r_1 = I + r_2,$$

so θ is well defined and one-to-one.

Clearly, θ is onto. Finally,

$$(mr_1)\theta + (mr_2)\theta = (I + r_1) + (I + r_2) = I + (r_1 + r_2) = (m(r_1 + r_2))\theta,$$

and

$$((mr)\theta)s = (I + r)s = I + rs = (mrs)\theta,$$

so θ is a module isomorphism.

Remark   Conversely, if I is an ideal of R, then R/I is a cyclic R-module, generated by the coset $I + 1$: for any element of R/I has the form $I + r = (I + 1)r$ for some $r \in R$.

Definition   Let $M = \langle m_1, \ldots, m_n \rangle$ be a finitely generated R-module. We say that M is **freely generated by** $m_1, \ldots, m_n$ if

$$m_1r_1 + \cdots + m_nr_n = 0 \text{ implies } r_1 = \ldots = r_n = 0.$$

M is **free** if it is freely generated by some finite set.

Theorem 5.4   *Let M be a R-module, where R is a commutative ring with identity.*

(a) *M is freely generated by $m_1, \ldots, m_n$ if and only if every element of M can be uniquely expressed in the form $m_1r_1 + \cdots + m_nr_n$, for $r_1, \ldots, r_n \in R$.*

(b) *M is free if and only if it is isomorphic to R^n for some natural number n.*

Proof (a) The condition that every element of M has an expression of this form is just the statement that $M = \langle m_1, \ldots, m_n \rangle$. The uniqueness of the representation is equivalent to the definition of freeness, since $m_1 r_1 + \cdots + m_n r_n = m_1 s_1 + \cdots + m_n s_n$ if and only if $m_1(r_1 - s_1) + \cdots + m_n(r_n - s_n) = 0$.

(b) Suppose that M is freely generated by $m_1, \ldots, m_n$, so that every element of M can be written uniquely in the form $m_1 r_1 + \cdots + m_n r_n$. Then the map $\theta : M \to R^n$ given by

$$(m_1 r_1 + \cdots + m_n r_n)\theta = (r_1, r_2, \ldots, r_n)$$

is easily checked to be an R-module isomorphism.

Conversely, R^n is freely generated by $e_1, \ldots, e_n$, where e_i is the n-tuple with 1 in the ith position and 0 in all other positions, since

$$(r_1, \ldots, r_n) = e_1 r_1 + \cdots + e_n r_n.$$

$\square$

Definition Let M and N be R-modules. The **direct sum** $M \oplus N$ of M and N is the set of all ordered pairs (m, n), with $m \in M$ and $n \in N$, with addition given by

$$(m_1, n_1) + (m_2, n_2) = (m_1 + m_2, n_1 + n_2),$$

and scalar multiplication by

$$(m, n)r = (mr, nr).$$

Proposition 5.5 *If M and N are R-modules, then $M \oplus N$ is an R-module.*

The proof is an exercise.

The direct sum of modules can be extended to the sum of any finite number of terms, in an obvious way. The next result enables us to recognise a direct sum.

Theorem 5.6 *Let M be an R-module, where R is a commutative ring with identity. Suppose that M contains submodules $M_1, M_2, \ldots, M_n$ such that any element of M can be uniquely written as $m_1 + m_2 + \cdots + m_n$, with $m_i \in M_i$ for $i = 1, \ldots, n$. Then M is isomorphic to the direct sum of $M_1, \ldots, M_n$.*

Proof We define a map $\theta : M_1 \oplus \cdots \oplus M_n \to M$ by the rule

$$(m_1, m_2, \ldots, m_n)\theta = m_1 + m_2 + \cdots + m_n.$$

The hypothesis of the theorem guarantees that this mapping is one-to-one and onto, and it is easily checked that it is a homomorphism. $\square$

Remark The free module R^n is isomorphic to the direct sum of n copies of the free module R of rank 1.

Exercise 5.1 Show that the set of all $m \times n$ matrices over F is a $M_m(F)$–$M_n(F)$ bimodule (with the usual matrix addition and multiplication).

Exercise 5.2 Let M be an R-module, and let $I = \text{Ann}(M)$. Show that M can be regarded as an R/I-module, where scalar multiplication is given by the rule

$$m(I + r) = mr.$$

Exercise 5.3 Let R be a commutative ring with identity.

(a) Prove that, if M is an R-module generated by a single element, then $M \cong R/\text{Ann}(M)$ (where $R/\text{Ann}(M)$ is an R-module as in Example 7).

(b) Conversely, show that, if I is an ideal of R, then R/I (as R-module) is generated by a single element.

(c) Show that, if I and J are ideals of R, then there is an R-module homomorphism from R/I onto R/J if and only if $I \subseteq J$.

Remark Here we see the ideal structure of R reflected in the 1-generator R-modules and their epimorphisms.

Modules over a Euclidean domain

5.5   The structure theorem. In general, modules can exist in enormous profusion. In order to understand a ring, we should study its modules. In this section, we examine finitely generated modules over Euclidean rings, and prove a structure theorem.

Theorem 5.7 *A finitely generated module over a Euclidean domain is isomorphic to a direct sum of cyclic modules.*

This gives a very precise description of the structure of such modules. The direct sum is an explicit construction. Any cyclic R-module is isomorphic to R/I for some ideal I. If R is a Euclidean domain, then all its ideals are principal, so $I = (r)$ for some $r \in R$, unique up to associate.

The theorem also has important applications, as we will see.

The theorem will be deduced from another, seemingly unrelated theorem, which gives more detailed information.

Theorem 5.8 *Let R be a Euclidean domain, M a free module of rank n, and N a submodule of M. Then there exist elements $m_1, \ldots, m_n \in M$, a natural number $r \leq n$, and elements $d_1, \ldots, d_r \in R$ such that*

(a) M is freely generated by $m_1, \ldots, m_n$;
(b) N is freely generated by $m_1 d_1, \ldots, m_r d_r$;
(c) d_i divides d_{i+1} for $i = 1, \ldots, r - 1$.

Proof We will need to show that N is finitely generated. Then we will find that the rest of the work has already been done. So we begin with the assumption that N is finitely generated, and return later to justify this assumption.

We identify M with R^n, where n is its rank. Let $x_1, \ldots, x_m$ be generators of N. Each of these generators is an element of R^n, so we can summarise this information by taking them as the rows of a matrix A, of size $m \times n$, with elements in R. This matrix determines N; indeed, in the terminology of vector spaces, N is the 'row space' of A.

According to the Smith normal form Theorem 4.29, we can bring A to the form $\begin{pmatrix} D & O \\ O & O \end{pmatrix}$, where D is a diagonal matrix of size $r \times r$ whose diagonal entries satisfy $d_i \mid d_{i+1}$ for $i = 1, \ldots, r$, and O denotes a zero matrix of the appropriate size. What would the submodule corresponding to a matrix of this form look like? If e_i denotes the element of R^n with 1 in the ith position and zeros elsewhere, then M is the free module generated by $e_1, \ldots, e_n$, and N a submodule freely generated by $d_1 e_1, \ldots, d_r e_r$. In other words, the conclusions of the theorem would hold.

So we have to examine how the effect of elementary row and column operations on the matrix A translates into the module M and submodule N.

- *Elementary row operations merely change the generating set for N.* For suppose the rows of A are $x_1, \ldots, x_m$ as before. For each type of row operation, the submodule closure conditions imply that the new rows (such as $x_i + x_j c$ or $x_i c$ with c a unit) belong to N. So the submodule N' generated by the new rows is contained in N. However, in each case, we can undo the effect of the operation by another operation of the same kind; so a similar argument shows that N is contained in N', whence $N = N'$.

- *Elementary column operations change the free basis for M.* We will prove this for operations of type (Ec1); the argument for the other types is similar but easier. So, as above, let $e_1, \ldots, e_n$ be the free basis for M. Consider what happens when we replace e_j by $e_j - e_i c$, for some $c \in R$. It is still true that any element of M can be expressed uniquely in terms of these elements: for example,

$$e_1 r_1 + \cdots + e_n r_n = e_1 r_1 + \cdots + e_i (r_i + r_j c)$$
$$+ \cdots + (e_j - e_i c) r_j + \cdots + e_n r_n.$$

So we do have again a basis for M. In terms of coordinates, we have added c times the jth coordinate to the ith. If this process is done for every generator of N (every row of A), the result is to apply to A the column operation consisting of adding c times the jth column to the ith. In other words, the rows of the transformed matrix are the generators of N, expressed with respect to a different basis for M.

The conclusion is that elementary operations do not change the module M or the submodule N, but merely our representation of them. Hence we may indeed assume that N is the row space of a matrix in the Smith normal form, and the theorem is proved. □

It remains to show that N is indeed finitely generated, so that it can be represented as the row space of a matrix. The proof is by induction on n, there being nothing to prove when $n = 0$.

Let J be the set of all elements of R which occur as first component of an element of N:

$$J = \{r_1 \in R : \exists r_2, \ldots, r_n \in R \text{ with } (r_1, \ldots, r_n) \in N\}.$$

Since N is a submodule, it is closed under subtraction and multiplication by elements of R. It follows that the same is true for J, which is thus an ideal of R. Since R is a principal ideal domain, we have $J = (a)$ for some $a \in R$.

Choose $b_2, \ldots, b_n \in R$ such that $(a, b_2, \ldots, b_n) \in R$. Also, let

$$N_1 = \{(r_1, \ldots, r_n) \in N : r_1 = 0\}.$$

Now take any element $(r_1, \ldots, r_n) \in N$. We have $r_1 = as$ for some $s \in R$. Then

$$(r_1, \ldots, r_n) - (a, b_2, \ldots, b_n)s \in N_1.$$

Hence N is generated by $(a, b_2, \ldots, b_n)$ together with a generating set for N_1. But N_1 is a submodule of R^{n-1}, hence finitely generated (by induction); and so N is finitely generated.

This concludes the proof of the theorem.

Let M be an arbitrary finitely generated R-module. Suppose that $M = (m_1, \ldots, m_n)$. Define a map $\theta : R^n \to M$ by the rule

$$(r_1, r_2, \ldots, r_n)\theta = m_1 r_1 + m_2 r_2 + \cdots + m_n r_n.$$

It is straightforward to show that θ is an R-module homomorphism, and that $\mathrm{Im}(\theta) = M$. Let $N = \mathrm{Ker}(\theta)$.

By the submodule theorem, we can choose a basis $e_1, e_2, \ldots, e_n$ for R^n such that, for some $d_1, \ldots, d_r \in R$,

- $e_1 d_1, \ldots, e_r d_r$ is a basis for N;
- d_i divides d_{i+1} for $i = 1, \ldots, r-1$.

We claim that M is isomorphic to the direct sum of the cyclic submodules $(e_1\theta), \ldots, (e_n\theta)$, and that the annihilator of $(e_i\theta)$ is the ideal (d_i) of R (so that $(e_i\theta) \cong R/(d_i)$.

Using the characterisation of direct sums in Theorem 5.6, we have to show that every element of M is uniquely expressible in the form $x_1 + x_2 + \cdots + x_n$, with $x_i \in (e_i\theta)$ for all i. That any element can be so represented follows from the fact that $e_1, \ldots, e_n$ is a basis for R^n. Suppose that

$$x_1 + \cdots + x_n = y_1 + \cdots + y_n,$$

with $x_i, y_i \in (e_i\theta)$. Let $x_i = e_i\theta r_i$, and $y_i = e_i\theta s_i$. Then $\sum e_i(r_i - s_i) \in \mathrm{Ker}(\theta) = N$. By the structure of N, we conclude that d_i divides $(r_i - s_i)$ for all i, so that $e_i(r_i - s_i) \in \mathrm{Ker}(\theta)$, and so $e_i\theta r_i = e_i\theta s_i$ for all i, as required.

We have done better than originally promised. What we have actually proved is the following theorem:

Theorem 5.9 *Let M be a finitely generated module over a Euclidean domain. Then there exist elements $d_1, \ldots, d_n \in R$ such that*

(a) *none of the d_i are units;*
(b) *d_i divides d_{i+1} for $i = 1, \ldots, r-1$;*
(c) *$d_i = 0$ for $i > r$;*
(d) *$M \cong R/(d_1) \oplus \cdots \oplus R/(d_n)$.*

Note If d_i is a unit, then $(d_i) = R$, and so $R/(d_i) = \{0\}$. Taking the direct sum of a module with $\{0\}$ does not change the module. So we can delete all the units from among the d_i. (By the divisibility condition, they will occur only at the start.)

Note also that $R/\{0\} = R$, so the number of direct summands isomorphic to R is $n - r$.

Definition The number of indices i such that $d_i = 0$ is called the **torsion-free rank** of the module M, and the non-zero ring elements $d_1, \ldots, d_r$ are the **invariant factors**. They form a complete set of invariants for the module. Thus, two finitely generated R-modules are isomorphic if and only if they have the same torsion-free rank and the same invariant factors (up to associates). The meaning of the term 'invariant factors' should be fairly clear; that of 'torsion-free rank' somewhat more mysterious. Some light will be shed on this strange terminology in the section on abelian groups.

5.6 The primary decomposition. A module can often be written in many different ways as a direct sum of submodules. In this section, we discuss a particular decomposition for torsion modules over a principal ideal domain, which will lead to a simpler canonical form for matrices.

Proposition 5.10 *Let R be a principal ideal domain. Let M be an R-module for which $\mathrm{Ann}(M) = \langle r \rangle$ is non-zero. Suppose that $r = r_1 r_2$, where r_1 and r_2 are coprime. Then $M = M_1 \oplus M_2$, where M_1 and M_2 are submodules of M with $\mathrm{Ann}(M_1) = \langle r_1 \rangle$ and $\mathrm{Ann}(M_2) = \langle r_2 \rangle$.*

For example, the cyclic group C_6 of order 6 is a $\mathbb{Z}$-module with annihilator (6); and we saw that $C_6 \cong C_2 \oplus C_3$, where the subgroups C_2 and C_3 have annihilators (2) and (3) respectively.

Proof Let $M_1 = \{m \in M : mr_1 = 0\}$, and similarly $M_2 = \{m \in M : mr_2 = 0\}$. Then M_1 and M_2 are submodules:

$$m, n \in M_1 \Rightarrow mr_1 = nr_1 = 0 \Rightarrow (m+n)r_1 = 0 \Rightarrow m+n \in M_1,$$

$$m \in M_1, r \in R \Rightarrow mr_1 = 0 \Rightarrow (mr)r_1 = mr_1 r = 0 \Rightarrow mr \in M_1,$$

and similarly for M_2.

Now $\text{Ann}(M_1)$ is an ideal of R, and hence is of the form $\langle s_1 \rangle$ for some s_1. By definition, $r_1 \in \text{Ann}(M_1)$, so s_1 divides r_1, say $r_1 = s_1 x$. Now take any element $m \in M$. Then $0 = mr = (mr_2)r_1$; so, by definition, $mr_2 \in M_1$. Then $mr_2 s_1 = 0$. Since m was arbitrary, $s_1 r_2 \in \text{Ann}(M) = \langle r_1 r_2 \rangle$, so $r_1 r_2$ divides $s_1 r_2$, whence r_1 divides s_1. Since each of r_1 and s_1 divides the other, these elements are associates, and so $\text{Ann}(M_1) = \langle r_1 \rangle$. Similarly, $\text{Ann}(M_2) = \langle r_2 \rangle$.

It remains to show that $M = M_1 \oplus M_2$; equivalently, that any element $m \in M$ can be represented uniquely in the form $m = m_1 + m_2$ with $m_1 \in M_1$ and $m_2 \in M_2$.

First, we show that such a representation exists. Since R is a principal ideal domain, and $(r_1, r_2) = (1)$, there exist $x, y \in R$ such that $xr_1 + yr_2 = 1$. Then

$$m = m1 = myr_2 + mxr_1;$$

and $(myr_2)r_1 = (mx)r = 0$, so $myr_2 \in M_1$, and similarly $mxr_1 \in M_1$.

For the uniqueness, suppose that $m_1 + m_2 = m_1' + m_2'$, where $m_1, m_1' \in M_1$ and $m_2, m_2' \in M_2$. Then $m_1 - m_1' = m_2' - m_2$, so this element lies in both M_1 and M_2. So it is enough to prove that $M_1 \cap M_2 = \{0\}$. Take any element m which lies in this intersection. Then $mr_1 = mr_2 = 0$. So

$$m = m1 = m(xr_1 + yr_2) = (mr_1)x + (mr_2)y = 0,$$

as required. This completes the proof. $\square$

Reading the proof carefully, we see that there is an alternative description of the submodules: $M_1 = Mr_2 = \{mr_2 : m \in M\}$, and $M_2 = Mr_1 = \{mr_1 : m \in M\}$.

Theorem 5.11 (Primary decomposition) *Let M be a module over a principal ideal domain R. Let $\text{Ann}(M) = \langle r \rangle$, and suppose that $r = p_1^{n_1} p_2^{n_2} \cdots p_k^{n_k}$, where $p_1, \ldots, p_k$ are irreducible and $n_1, \ldots, n_k$ are positive integers. Then*

$$M = M_1 \oplus \cdots \oplus M_k,$$

where $M_1, \ldots, M_k$ are submodules of M and $\text{Ann}(M_i) = \langle p_i^{n_i} \rangle$.

This follows immediately by induction from the previous result. Note that there is no choice at all about the submodules M_i; for M_i consists of all elements $m \in M$ such that $mp_i^{n_i} = 0$. The M_i are called the **primary components** of M.

Proposition 5.12 *Let M be a finitely generated torsion module over a Euclidean domain. Then M is isomorphic to a direct sum of cyclic submodules whose annihilators are powers of irreducibles.*

Proof First we apply the primary decomposition to M, expressing it as a direct sum of submodules whose annihilators are powers of irreducibles. These submodules are finitely generated. (More generally, if $M = \langle m_1, \ldots, m_k \rangle$, then

$Mr = \langle m_1 r, \ldots, m_k r \rangle$.) So each of these submodules is a direct sum of cyclic submodules, the annihilator of each of which is a power of an irreducible.

We could alternatively first decompose M as a direct sum of cyclic modules, and then decompose each of these cyclic submodules according to the primary decomposition. (Taking $k = 1$ in the parenthetical remark above, we see that if M is cyclic, then so is Mr for any $r \in R$, and in particular the primary constituents of M are cyclic.) □

The generators of the annihilators of these cyclic submodules are called the **elementary divisors** of M. The elementary divisors are just the factors obtained when the invariant factors of M are factorised into powers of irreducibles. Like the invariant factors, they are unique:

Theorem 5.13 *Suppose that M is expressed in two different ways as a direct sum of cyclic submodules whose annihilators are powers of irreducibles. Then the annihilators are the same, up to associates.*

Proof It is enough to prove this when $\text{Ann}(M)$ is a prime power. (This follows because any cyclic submodule with prime power annihilator is contained in one of the primary components, and these are uniquely determined.) So assume that $\text{Ann}(M) = \langle p^n \rangle$, where p is irreducible.

Let $M = M_1 \oplus \cdots \oplus M_k$, where M_i is a cyclic module with annihilator $\langle p^{n_i} \rangle$, where $1 \le n_i \le n$.

Now $\langle p \rangle$ is a maximal ideal of R, and hence $R/\langle p \rangle$ is a field. Now consider the submodule Mp of M, and let N be the factor module M/Mp. We claim that M/Mp is a vector space over $R/\langle p \rangle$. The main point is that scalar multiplication is well defined: if $\langle p \rangle + r = \langle p \rangle + s$, then $s = r + px$ and so mr and ms differ by an element of Mp. Now the cosets containing the generators $m_1, \ldots, m_k$ of the cyclic summands form a basis for this vector space; so $\dim(M/Mp) = k$ (as $R/\langle p \rangle$-vector space).

Now suppose that $n_i > 1$ for $i \le k_1$, while $n_{k_1+1} = \ldots = n_k = 1$. Then Mp is the direct sum of cyclic modules generated by $m_1 p, \ldots, m_{k_1} p$. (We have $m_{k_1+1} p = \ldots = m_k p = 0$, so these elements generate trivial submodules which can be ignored.) So $\dim(Mp/Mp^2) = k_1$ (as $R/\langle p \rangle$-vector space).

Continuing in this way, we find that the dimension of Mp^j/Mp^{j+1} is equal to the number of $n_1, \ldots, n_k$ which are greater than j.

But, given a set of positive integers $n_1, \ldots, n_k$, if we are told how many of them are greater than j for each $j \ge 0$, then we can recover the numbers n_i. □

For example, if the dimensions of the spaces M/Mp, Mp/Mp^2, Mp^2/Mp^3, Mp^3/Mp^4, Mp^4/Mp^5 are respectively $7, 3, 3, 1, 0$, then there are seven numbers n_i, and they are $1, 1, 1, 1, 3, 3, 4$.

Remark The elementary divisors are obtained from the invariant factors by factorising each of them into prime powers and taking all the prime powers obtained. Conversely, suppose that we are given the elementary divisors. Take the largest power of each irreducible which occurs, and multiply them all together; this is the largest invariant factor. Now remove these elementary divisors and repeat the procedure.

For example, if the elementary divisors are $2, 4, 4, 3, 9, 5$, then the invariant factors (in the reverse of the usual order) are $4 \cdot 9 \cdot 5 = 180$, $4 \cdot 3 = 12$, and 2.

Applications

5.7   Finitely generated abelian groups.   We have seen that an abelian group is exactly the same thing as a $\mathbb{Z}$-module. Since $\mathbb{Z}$ is our prototype of a Euclidean domain, we can immediately apply our structure theorem to obtain the structure of finitely generated abelian groups:

Theorem 5.14   *Let A be a finitely generated abelian group. Then*

$$A \cong C_{d_1} \oplus \cdots \oplus C_{d_r} \oplus C_\infty \oplus \cdots \oplus C_\infty,$$

where $d_1, \ldots, d_r$ are positive integers with $d_i \mid d_{i+1}$ for $i = 1, \ldots, r - 1$, C_d is a cyclic group of finite order d, and C_∞ is an infinite cyclic group.

The uniqueness part of the module structure theorem gives us extra information.

Theorem 5.15   *Suppose that*

$$C_{d_1} \oplus \cdots \oplus C_{d_r} \oplus C_\infty \oplus \cdots \oplus C_\infty \cong C_{e_1} \oplus \cdots \oplus C_{e_s} \oplus C_\infty \oplus \cdots \oplus C_\infty,$$

where $d_i, e_j > 1$ for all i, j, $d_i \mid d_{i+1}$ for $i = 1, \ldots, r - 1$ and $e_j \mid e_{j+1}$ for $j = 1, \ldots, s - 1$. Let the numbers of C_∞ summands of the two groups be u and v respectively. Then $u = v$, $r = s$, and $d_i = e_i$ for $i = 1, \ldots, r$.

Note that the theorem is false without the divisibility condition. For example, $C_2 \oplus C_3 \cong C_6$.

In the abelian group $C_{d_1} \oplus \cdots \oplus C_{d_r} \oplus C_\infty \oplus \cdots \oplus C_\infty$, we call the sum $C_{d_1} \oplus \cdots \oplus C_{d_r}$ of the finite cyclic groups the **torsion part**, and the sum of the infinite cyclic groups the **torsion-free part**; the number of infinite summands is the **torsion-free rank** of the group. Note that the torsion part consists of all the elements of finite order.

Where does this terminology come from? The answer lies in the field of 'algebraic topology', and it would take another book to explain it in detail; what follows is only a rough sketch. One of the central problems of topology is how to distinguish between different topological spaces (surfaces, etc.), or to decide whether two quite different recipes give the same or different spaces. (The kind of recipe that we are thinking of here can be described by example. If we take a rectangular strip of paper, bend it round, and join the ends, we obtain a cylinder. If we give the end a 180° twist before joining the ends, we obtain instead a Möbius band.)

Topologists discovered that it is possible to associate a collection of abelian groups with a space, so that if two spaces are the same, then the groups associated with them are isomorphic. These groups can be calculated from a description of the space (as above). So, if we calculate the groups from the descriptions of two spaces, and the groups turn out not to be isomorphic, then the spaces are really different. (The converse is false; different spaces may give the same groups.)

Now the presence of elements of finite order (other than the identity) in the group indicates some 'twisting' of the space (as in the Möbius band). Hence the name **torsion elements** was used for elements of finite order in an abelian group, and the group was called **torsion-free** if it has no elements of finite order except the identity.

If we regard an abelian group as a $\mathbb{Z}$-module, then a torsion element is one whose annihilator is not the zero ideal. Hence, as in the last section, we can generalise to modules over an arbitrary commutative ring: a **torsion element** of such a module is one whose annihilator is not the zero ideal, and a module is **torsion-free** if 0 is its only torsion element.

Theorems 5.14 and 5.15 enable us to count abelian groups. The result is given in terms of a famous number-theoretic function.

Definition The **partition function** $p(n)$ is the function whose value on the positive integer n is the number of ways of writing n as the sum of positive integers, where the order of the summands is unimportant.

For example, $p(4) = 5$, because

$$4 = 3 + 1 = 2 + 2 = 2 + 1 + 1 = 1 + 1 + 1 + 1.$$

(We do not count $3 + 1$ and $1 + 3$ separately.)

By convention, $p(0) = 1$.

Proposition 5.16 *Let $f_A(n)$ be the number of abelian groups of order n.*

(a) If $m = p_1^{m_1} \cdots p_k^{m_k}$, where $p_1, \ldots, p_k$ are distinct primes, then

$$f_A(n) = f_A(p_1^{m_1}) \cdots f_A(p_k^{m_k}).$$

(b) If p is prime, then $f_A(p^m) = p(m)$ (the partition function of m).

Proof (a) By the primary decomposition, an abelian group of order n is the direct sum of abelian groups of orders $p_1^{m_1}, \ldots, p_k^{m_k}$; it determines, and is determined by, the choices of these groups.

(b) To each expression $n = a_1 + a_2 + \cdots + a_r$ corresponds a group, the direct sum of cyclic groups of orders $p^{a_1}, p^{a_2}, \ldots, p^{a_r}$. These groups are all different, and every abelian group of order p^n is isomorphic to one of them. $\square$

For example, the number of abelian groups of order $108 = 2^2 3^3$ is $p(2)p(3) = 2 \cdot 3 = 6$. The groups are given in the following table, which lists the forms given by the invariant factors as well as the elementary divisors:

$$C_4 \oplus C_{27} \cong C_{108},$$
$$C_2 \oplus C_2 \oplus C_{27} \cong C_2 \oplus C_{54},$$
$$C_4 \oplus C_3 \oplus C_9 \cong C_3 \oplus C_{36},$$
$$C_2 \oplus C_2 \oplus C_3 \oplus C_9 \cong C_6 \oplus C_{18},$$
$$C_4 \oplus C_3 \oplus C_3 \oplus C_3 \cong C_3 \oplus C_3 \oplus C_{12},$$
$$C_2 \oplus C_2 \oplus C_3 \oplus C_3 \oplus C_3 \cong C_3 \oplus C_6 \oplus C_6.$$

It can be shown, from the above result, that the number of abelian groups of order n is not greater than n (Exercise 5.4).

5.8   Normal forms of matrices.   In this section, we tackle a problem which appears similar to the canonical form under equivalence (see Section 4.9). There, we began with a linear transformation $\theta : U \to V$, and remarked that any choice of bases in U and V gives rise to a matrix representing θ, and matrices A, B representing the same transformation relative to different bases are related by $B = PAQ^{-1}$. Moreover, there is a choice of bases such that the matrix representing θ has the simple form $\begin{pmatrix} I & O \\ O & O \end{pmatrix}$. The submatrix I is $r \times r$, where r is the rank of θ (the dimension of its image), so there is unique matrix of this form which represents θ.

The situation we consider here is that θ is a linear transformation from a vector space V to itself. We represent θ by choosing a basis $\{v_1, \ldots, v_n\}$ for V, and letting $v_i\theta = \sum a_{ij} v_j$; then θ is represented by the matrix $A = (a_{ij})$. The difference is that, instead of choosing two different bases in the source and target spaces, we have only the freedom to choose one basis. If we use a different basis, with transition matrix P, then the new matrix B representing θ is given by $B = PAP^{-1}$. (Because there is only one basis to change, we must have $P = Q$ in the earlier formalism.) Can we find a set of 'simple' matrices so that each linear transformation from V to itself can be represented by one of them? Since our freedom to transform is less, there will be more matrices in such a set. (We are seeking canonical forms for a 'finer' equivalence relation than before.)

We solve this problem by using θ to make V into a module for the polynomial ring $F[x]$, such that the module captures the structure of θ. As in Example 6 of Section 5.2, we make V an $F[x]$-module by setting

$$vf(x) = vf(\theta),$$

where powers of θ are calculated by composition. This module is finitely generated, since a basis for it as F-vector space certainly generates it as a module.

We will apply the structure theorems for modules over Euclidean domains (since $F[x]$ is certainly a Euclidean domain). First, we have to see what cyclic modules look like.

Definition Let $f(x)$ be a monic polynomial in $F[x]$.

$$f(x) = x^n + a_{n-1}x^{n-1} + \cdots + a_1x + a_0.$$

The **companion matrix** $C(f)$ is the $n \times n$ matrix given by

$$C(f) = \begin{pmatrix} 0 & 1 & 0 & \cdots & 0 & 0 \\ 0 & 0 & 1 & \cdots & 0 & 0 \\ \ddots & \ddots & \ddots & \ddots & \ddots & \ddots \\ 0 & 0 & 0 & \cdots & 0 & 1 \\ -a_0 & -a_1 & -a_2 & \cdots & -a_{n-2} & -a_{n-1} \end{pmatrix}.$$

(In other words, in the first $n - 1$ rows, the entries immediately to the right of the diagonal are 1, and all others zero; the last row consists of the coefficients of f (excluding the coefficient of x^n), in reverse order and with the sign changed.) Furthermore, if m is a positive integer, let the **extended companion matrix** $C(m, f)$ be the matrix with $m \times m$ blocks, each $n \times n$, given by

$$C(m, f) = \begin{pmatrix} C & J & O & \cdots & O & O \\ O & C & J & \cdots & O & O \\ \ddots & \ddots & \ddots & \ddots & \ddots & \ddots \\ O & O & O & \cdots & C & J \\ O & O & O & \cdots & O & C \end{pmatrix}$$

with C on the diagonal, J immediately to the right, and zeros elsewhere, where $C = C(f)$ is the companion matrix of f, and J is a matrix with 1 in the southwest corner (row n and column 1) and zero elsewhere. Note that $C(f) = C(1, f)$.

Proposition 5.17 *Let θ be a linear transformation on V, and suppose that the corresponding $F[x]$-module is cyclic, with annihilator (g^m). Then there is a basis for V relative to which the matrix of θ is $C(m, g)$.*

Proof Suppose that $\theta : V \to V$ is a linear transformation, and V is a cyclic $F[x]$-module, say $V = vF[x]$. We claim that V has a basis $v_1, \ldots, v_n$, where $v_i = v\theta^{i-1}$ for $i = 1, \ldots, n$. It is clear that the vectors $v, v\theta, v\theta^2, \ldots$ span V. Choose n minimal such that $v, v\theta, \ldots, v\theta^n$ are linearly dependent. Then it must be the case that the coefficient of $v\theta^n$ in such a linear combination is non-zero; so $v\theta^n$ can be expressed as a linear combination of $v, v\theta, \ldots, v\theta^{n-1}$. Hence the span of these vectors is mapped to itself by θ, and hence is the whole of V (since $V = vF[\theta]$). By minimality of n, the vectors $v, v\theta, \ldots, v\theta^{n-1}$ are linearly independent; so they form a basis for V.

Set $v_i = v\theta^{i-1}$ for $i = 1, \ldots, n$. Then $v_i\theta = v_{i+1}$ for $i = 1, \ldots, n-1$. Suppose that

$$v_n\theta = -a_0v_1 - \cdots - a_{n-1}v_n$$

for some $a_0, \ldots, a_{n-1} \in F$. Then the matrix representing θ relative to this basis is the companion matrix of

$$f(x) = x^n + a_{n-1}x^{n-1} + \cdots + a_0.$$

In addition, $vf(\theta) = 0$. It follows that, for any polynomial h, we have

$$vh(\theta)f(\theta) = vf(\theta)h(\theta) = 0.$$

Since $V = vF[x]$, we see that $\mathrm{Ann}(V) = (f)$. So we have now proved the proposition for $m = 1$.

We showed above that $\{v\theta^{i-1} : 1 \le i \le n\}$ is a basis for the cyclic module V, where $n = \dim(V)$. It is easy to see that, if $f_{i-1}(x)$ is any polynomial of degree $i - 1$, for $1 \le i \le n$, then the set $\{vf_{i-1}(\theta) : 1 \le i \le n\}$ is also a basis.

Suppose that $\mathrm{Ann}(V) = (f)$ where $f = g^m$, and g is a polynomial of degree k (with $km = n$). Then $x^{i-1}g(x)^{j-1}$ is a polynomial of degree $(i-1) + k(j-1)$. Taking $1 \le i \le k$ and $1 \le j \le m$, these degrees take all values from 0 to $km - 1$, so we obtain an alternative basis for V, namely the vectors $w_{ij} = v\theta^{i-1}g(\theta)^{j-1}$.

What is the matrix representing θ relative to this basis? We have $w_{ij}\theta = w_{i+1\,j}$ for $i < k$, and, if

$$g(x) = x^k + b_{k-1}x^{k-1} + \cdots + b_0,$$

then

$$w_{kj}\theta = w_{1j}\theta^k = -b_0 w_{1j} - \cdots - b_{k-1}w_{kj} + w_{1\,j+1},$$

with the convention that $w_{1\,m+1} = 0$. Thus the matrix of θ is exactly $C(m, g)$, as claimed. $\qquad\qquad\square$

Theorem 5.18 (Normal forms of matrices) *Let $\theta : V \to V$ be a linear transformation, and regard V as an $F[x]$-module in the usual way.*

(a) If the invariant factors of the module are $d_1(x), \ldots, d_r(x)$, then there is a basis for V relative to which θ is represented by a block diagonal matrix with $C(d_1), \ldots, C(d_r)$ on the diagonal and zeros elsewhere.

(b) If the elementary divisors of the module are $e_1(x)^{m_1}, \ldots, e_k(x)^{m_k}$, then there is a basis for V relative to which θ is represented by a block diagonal matrix with $C(m_1, e_1), \ldots, C(m_k, e_k)$ on the diagonal and zeros elsewhere.

Matrices of the shape described in (a) or (b) of this theorem are said to be in **rational canonical form** or **primary rational canonical form** respectively. (Note that, in the rational canonical form we require that d_i divides d_{i+1} for $i = 1, \ldots, r - 1$, while in the primary rational canonical form we require that each e_i is irreducible.) So we can restate the theorem in matrix form as follows:

Theorem 5.19 *Let A be a $n \times n$ matrix over F.*

(a) There is an invertible $n \times n$ matrix P such that PAP^{-1} is in rational canonical form.

(b) There is an invertible $n \times n$ matrix Q such that QAQ^{-1} is in primary rational canonical form.

Moreover, the rational and primary rational canonical forms of a given matrix are unique (up to the order of the diagonal blocks in the primary rational case).

There is one particularly important case of this theorem. Suppose that the field F is algebraically closed. (Traditionally, this analysis is given for the field $\mathbb{C}$.) Then any non-constant polynomial has a root in F, and hence a linear factor. So the only irreducible polynomials are those of the form $x - a$ for $a \in F$. Now $C(x - a)$ is the 1×1 matrix (a). Hence $C(m, x - a)$ has the form

$$\begin{pmatrix} a & 1 & 0 & \ldots & 0 & 0 \\ 0 & a & 1 & \ldots & 0 & 0 \\ \ddots & \ddots & \ddots & \ddots & \ddots & \ddots \\ 0 & 0 & 0 & \ldots & a & 1 \\ 0 & 0 & 0 & \ldots & 0 & a \end{pmatrix}.$$

Such a matrix is called a **Jordan block**. So the primary rational form immediately gives the following result.

Theorem 5.20 (Jordan form) *Let A be a $n \times n$ matrix over an algebraically closed field F. Then there is an invertible $n \times n$ matrix Q over F such that QAQ^{-1} is a block diagonal matrix with Jordan blocks on the diagonal and zeros elsewhere.*

5.9   The Cayley–Hamilton Theorem.   Let A be an $n \times n$ matrix over a field F. Then the $n^2 + 1$ matrices $A^0 = I, A^1 = A, A^2, \ldots, A^{n^2}$ lie in the n^2-dimensional vector space $M_n(F)$, and so they are linearly dependent. Hence A satisfies a polynomial equation of degree at most n^2.

The Cayley–Hamilton Theorem shows that there is a specific polynomial equation of degree n which is satisfied by A, the so-called characteristic equation of A.

The **minimal polynomial** of A is the monic polynomial of least degree which is satisfied by A. It is unique. Indeed,

$$J = \{f \in F[x] : f(A) = O\}$$

is an ideal of $F[x]$, and the minimal polynomial of A is the unique monic generator of this ideal. Hence any polynomial f such that $f(A) = O$ is divisible by the minimal polynomial.

Definition The **characteristic polynomial** of the matrix A is the polynomial $\det(xI - A)$.

Theorem 5.21 (The Cayley–Hamilton Theorem) *Let $c(x)$ and $m(x)$ be the characteristic and minimal polynomials of the matrix A. Then:*

(a) $c(A) = O$;
(b) $m(x)$ divides $c(x)$.

Remarks The two parts of the theorem are clearly equivalent. The theorem can be proved by a direct calculation, which does not require all the background of the rational canonical form. But the proof given here may provide more insight.

Proof The strategy is in three parts. First, we show that it suffices to deal with matrices in rational canonical form. Next, we show that it suffices to deal with companion matrices of polynomials. Finally, we prove the theorem directly for these matrices. The proof is an illustration of the way in which a canonical form theorem can be used to simplify calculations.

Step 1 We show that, if P is invertible, then A and PAP^{-1} have the same characteristic polynomials and the same minimal polynomials. Since, for every A, there is an invertible P such that PAP^{-1} is in rational canonical form, it suffices to prove the theorem for these.

For the characteristic polynomial, we have

$$\det(xI - PAP^{-1}) = \det(P(xI - A)P^{-1})$$
$$= \det(P)\det(xI - A)\det(P)^{-1}$$
$$= \det(xI - A),$$

using the multiplicative property of determinants.

For the minimal polynomial, we observe that, if f is any polynomial, then

$$f(PAP^{-1}) = Pf(A)P^{-1},$$

so $f(A) = O$ if and only if $f(PAP^{-1}) = O$.

Step 2 Let A be in rational canonical form. Thus, A has diagonal blocks $C(f_1), C(f_2), \ldots, C(f_r)$, and O elsewhere, where $C(f)$ is the companion matrix of f, and $f_1, \ldots, f_r$ are the invariant factors of A (so that f_i divides f_{i+1} for $1 \le i \le r - 1$). We *claim* that a companion matrix $C(f)$ has characteristic and minimal polynomial both equal to f. (The proof of this is given in Step 3.) Now the determinant of a block diagonal matrix is the product of the determinants of the diagonal blocks. Hence the characteristic polynomial of A is the product of the characteristic polynomials of $C(f_1), \ldots, C(f_r)$; that is, it is $f_1 \cdots f_r$.

Also, f_i is the minimal polynomial of $C(f_i)$. Since f_i divides f_r for all i, we see that $f_r(C(f_i)) = O$ for all i, and hence that $f_r(A) = O$. So the minimal polynomial divides f_r, and hence divides the characteristic polynomial.

In fact, f_r is the minimal polynomial, since a polynomial of smaller degree would not be satisfied by the block $C(f_r)$.

Step 3 Let $A = C(f)$, where

$$f(x) = x^n + a_{n-1}x^{n-1} + \cdots + a_1 x + a_0.$$

Thus,

$$xI - A = \begin{pmatrix} x & -1 & 0 & \cdots & & 0 \\ 0 & x & -1 & \cdots & & 0 \\ \cdots & \cdots & \cdots & \cdots & & \cdots \\ 0 & \cdots & 0 & x & & -1 \\ a_0 & a_1 & \cdots & a_{n-2} & & x + a_{n-1} \end{pmatrix}.$$

We prove that $\det(xI - A) = f(x)$ by induction on n.

For $n = 1$, we have $f(x) = x - a_0$ and $A = (a_0)$, so the assertion is true.

Suppose that it holds for $n - 1$. Consider the formula for $\det(xI - A)$ as a sum over permutations. Since there are only two non-zero elements in the first row, the only permutations which contribute to the sum are those with $1g = 1$ or $1g = 2$.

If $1g = 1$, the $(1, 1g)$ entry of $xI - A$ is equal to x. Apart from this factor, the terms are just those in the determinant of the matrix with the first row and column deleted. This matrix is $xI - B$, where B is the companion matrix of the polynomial $x^{n-1} + a_{n-1}x^{n-2} + \cdots + a_1$. By the inductive hypothesis, the contribution is $x(x^{n-1} + \cdots + a_1)$.

If $1g = 2$, then $2g \neq 2$. The only other non-zero element in the second row is in the third column, so we can assume that $2g = 3$. Similarly $3g = 4, \ldots$, $(n-1)g = n$, $ng = 1$. Thus, g is a cyclic permutation, and its sign is $(-1)^{n-1}$. The term that we obtain is $(-1)^{n-1}a_0$, since the $(i, i+1)$ entry of $xI - A$ is -1 for $1 \leq i \leq n - 1$, while the $(n, 1)$ entry is a_0. So there is a single term a_0.

Thus

$$\det(xI - A) = x(x^{n-1} + \cdots + a_1) + a_0 = f(x),$$

as required.

For the minimal polynomial, let V be a vector space with basis $v_1, \ldots, v_n$, and let θ be a linear transformation of V with matrix A relative to this basis. Thus $v_i\theta = v_{i+1}$ for $i + 1, \ldots, n - 1$, while

$$v_n\theta = -a_0 v_1 - \cdots - a_{n-1}v_n.$$

Thus, $v_i = v_1\theta^{i-1}$ for $i = 1, \ldots, n$, while $v_1 f(\theta) = 0$. Then

$$v_i f(\theta) = v_1\theta^{i-1}f(\theta) = v_1 f(\theta)\theta^{i-1} = 0,$$

so $f(\theta)$ is represented by the zero matrix. Thus, $f(A) = 0$. Clearly no polynomial equation of smaller degree can be satisfied by A; so $f(x)$ is the minimal polynomial.

The proof is complete. □

Note that the minimal polynomial of a matrix is equal to the last elementary divisor, while the characteristic polynomial is the product of all the elementary divisors.

This observation adds a little extra to the Cayley–Hamilton Theorem:

Proposition 5.22 *An irreducible polynomial divides the characteristic polynomial if and only if it divides the minimal polynomial.*

Proof Since the minimal polynomial divides the characteristic polynomial, the reverse implication is clear. Conversely, any irreducible which divides the characteristic polynomial must divide one of the invariant factors, and hence must divide the last invariant factor, which is the minimal polynomial. □

We conclude with a very brief discussion of one of the most important topics in linear algebra, namely, eigenvalues and eigenvectors. Again, this approach is not the most direct, but does show the usefulness of the rational canonical form.

Definition An **eigenvector** of the $n \times n$ matrix A over F is a non-zero vector $v \in F^n$ such that $vA = \lambda v$ for some scalar λ. The corresponding **eigenvalue** of A is λ.

Theorem 5.23 *Let $A \in M_n(F)$. The following conditions for the scalar λ are equivalent:*

(a) λ is an eigenvalue of A;
(b) λ is a root of the characteristic polynomial of A;
(c) λ is a root of the minimal polynomial of A.

Proof (a) implies (c): Let $f(x)$ be the minimal polynomial of A. If $vA = \lambda v$ with $v \neq 0$, then $0 = vf(A) = f(\lambda)v$; so $f(\lambda) = 0$.

(c) implies (b) by the Cayley–Hamilton Theorem 5.21(b).

(b) implies (a): If $\det(\lambda I - A) = 0$, then $\lambda I - A$ is not invertible. A non-zero vector $v \in \text{Ker}(\lambda I - A)$ is an eigenvector of A with eigenvalue λ. □

The concepts of eigenvalue and eigenvector can be applied to linear transformations also. We omit the details.

5.10 An application: league tables. In many league competitions, teams are awarded a fixed number of points for a win or a draw. It may happen that two teams win the same number of matches and so are equal on points, but the opponents beaten by one team are clearly 'better' than those beaten by the other. How can we take this into account?

You might think of giving each team a 'score' to indicate how strong it is, and then adding the scores of all the teams beaten by team T to see how well T has performed. Of course this is self-referential, since the score of T depends on the scores of the teams that T beats. So suppose we ask simply that the score of T should be proportional to the sum of the scores of all the teams beaten by T.

Now we can translate the problem into linear algebra. Let $T_1, \ldots, T_n$ be the teams in the league. Let A be the $n \times n$ matrix whose (i, j) entry is equal to 1 if T_j beats T_i, and 0 otherwise. Now for any vector $(x_1, x_2, \ldots, x_n)$ of scores, the jth entry of xA is equal to the sum of the scores x_i for all teams T_i beaten by T_j. So our requirement is simply that

x should be an eigenvector of A with all entries positive.

Here is an example. There are six teams A, B, C, D, E, and F. Suppose that

A beats B, C, D, E;
B beats C, D, E, F;
C beats D, E, F;
D beats E, F;
E beats F;
F beats A.

The matrix A is

$$\begin{pmatrix} 0 & 0 & 0 & 0 & 0 & 1 \\ 1 & 0 & 0 & 0 & 0 & 0 \\ 1 & 1 & 0 & 0 & 0 & 0 \\ 1 & 1 & 1 & 0 & 0 & 0 \\ 1 & 1 & 1 & 1 & 0 & 0 \\ 0 & 1 & 1 & 1 & 1 & 0 \end{pmatrix}.$$

We see that A and B each have four wins, but that A has generally beaten the stronger teams; there was one upset when F beat A. Also, E and F have the fewest wins, but F took A's scalp and should clearly be better.

Calculation with MAPLE shows that the vector

$$(0.7744, 0.6452, 0.4307, 0.2875, 0.1920, 0.3856)$$

is an eigenvector of A with eigenvalue 2.0085. This confirms our view that A is top of the league and that F is ahead of E; it even puts F ahead of D.

But perhaps there is a different eigenvalue and/or eigenvector which would give us a different result?

In fact, there is a general theorem called the **Perron–Frobenius theorem** which gives us conditions for this method to give a unique answer. Before we state it, we need a definition.

Let A be an $n \times n$ real matrix with all its entries non-negative. We say that A is **indecomposable** if, for any i, j with $1 \leq i, j \leq n$, there is a number m such that the (i, j) entry of A^m is strictly positive.

This odd-looking condition means, in our football league situation, that for any two teams T_i and T_j, there is a chain $T_{k_0}, \ldots, T_{k_m}$ with $T_{k_0} = T_i$ and $T_{k_m} = T_j$, such that each team in the chain beats the next one. Now it can be shown that the only way that this can fail is if there is a collection C of teams such that each team in C beats each team not in C. In this case, obviously the teams in C occupy the top places in the league, and we have reduced the problem to ordering these teams. So we can assume that the matrix of results is indecomposable.

In our example, we see that B beats F beats A, so the $(2, 1)$ entry in A^2 is non-zero. Similarly for all other pairs. So A is indecomposable in this case.

Theorem 5.24 (Perron–Frobenius Theorem) *Let A be a $n \times n$ real matrix with all its entries non-negative, and suppose that A is indecomposable. Then, up to scalar multiplication, there is a unique eigenvector $v = \begin{pmatrix} x_1 & \cdots & x_n \end{pmatrix}^{\top}$ for A with the property that $x_i > 0$ for all i. The corresponding eigenvalue is the largest eigenvalue of A.*

So the Perron–Frobenius eigenvector solves the problem of ordering the teams in the league.

Remarks 1. Further refinements are clearly possible. For example, instead of just putting the (i, j) entry equal to 1 if T_j beats T_i, we could take it to be the number of goals by which T_j won the game.

2. This procedure has wider application. How does an Internet search engine find the most important web pages that match a given query? An important web page is one to which a lot of other web pages link; this can be described by a matrix, and we can use the Perron–Frobenius eigenvector to do the ranking.

Exercise 5.4 (a) Prove that $p(n) \leq 2^n$ for all n. [*Hint*: The number of partitions of n containing at least one part i for $i < n$ is $p(n - i)$; so

$$p(n) \leq \sum_{i=1}^{n} p(n - i),$$

where we have $\leq$ since some partitions are counted more than once. Now use induction.]
   (b) Hence show that $f_A(p^m) \leq 2^m$ for any prime p.
   (c) Hence show that $f_A(n) \leq n$ for any n.

Exercise 5.5 Prove the Cayley–Hamilton Theorem for 2×2 and 3×3 matrices by direct calculation. [This was done by Cayley, and the 4×4 case by Hamilton; it was Frobenius who produced the first general proof.]

Exercise 5.6 Prove that two 3×3 complex matrices are similar if and only if they have the same characteristic and minimal polynomials.

Is the same true for 4×4 matrices?

Exercise 5.7 Show that the invariant factors of a matrix A over a field F are the non-constant diagonal elements in the Smith normal form of the matrix $xI - A$ over $F[x]$.

Exercise 5.8 The name **rational canonical form** comes from the fact that if a matrix A has rational entries, then so does its rational canonical form, even if we work over a larger field. More generally, if A is an $n \times n$ matrix over F, and K is a field containing F, then the rational canonical forms of A over F and over K are identical. Prove this.

[This is not true for the primary rational canonical form, since enlarging the field may cause irreducible polynomials to become reducible. For example, the real matrix $C(x^2 + 1) = \begin{pmatrix} 0 & 1 \\ -1 & 0 \end{pmatrix}$ is in (primary) rational canonical form over $\mathbb{R}$, but over $\mathbb{C}$, its Jordan form is $\begin{pmatrix} i & 0 \\ 0 & -i \end{pmatrix}$.]

Exercise 5.9 (a) Prove that, for any eigenvalue λ of the matrix A, the set of eigenvectors with eigenvalue λ, together with the zero vector, is a subspace of F^n.

(b) Prove that, if $v_1, \ldots, v_k$ are eigenvectors associated with distinct eigenvalues $\lambda_1, \ldots, \lambda_k$ respectively, then $v_1, \ldots, v_k$ are linearly independent.

Exercise 5.10 Find the eigenvalues and eigenvectors of the real matrix

$$\begin{pmatrix} 11 & -1 & -4 \\ -1 & 11 & -4 \\ -4 & -4 & 14 \end{pmatrix}.$$

Exercise 5.11 Give a proof of Theorem 5.23 in the order (a) implies (b) implies (c) implies (a).

Exercise 5.12 In Section 3.2, we made the easy observation that any abelian group is the additive group of a ring. Prove that any finitely generated abelian group is the additive group of a ring with identity.

Exercise 5.13 An abelian group G is generated by elements $x_1, \ldots, x_n$ satisfying the relations

$$a_{i1}x_1 + \cdots + a_{in}x_n = 0$$

for $i = 1, \ldots, n$, where $a_{ij} \in \mathbb{Z}$. (We assume that all relations satisfied by $x_1, \ldots, x_n$ are consequences of these.) Let A be the matrix (a_{ij}). Prove that

(a) if $\det(A) = 0$, then G is infinite;
(b) if $\det(A) \neq 0$, then $|G| = |\det(A)|$.

Exercise 5.14 An abelian group G is generated by elements x, y, z satisfying the relations

$$6x + 10y = 0,$$
$$6x + 15z = 0,$$
$$10y + 15z = 0$$

(with the same convention as in the preceding exercise). Write G as a direct sum of cyclic groups.

6   The number systems

The nineteenth-century German mathematician Leopold Kronecker said, 'God made the integers; the rest is the work of man'. Many others, including non-mathematicians, have felt similarly

> What could be more general than 2, which can represent two galaxies or two pickles, or one galaxy plus one pickle (the mind doth boggle), or just 2 gently bobbing—where? It, like God, is an "I am" and many have thought that it must be a precipitate of ultimate reality.
>
> Alfred W. Crosby (1997).

We may take it that by the integers, Kronecker meant the counting numbers (the positive integers). Any civilisation that has left records knew how to count. The other number systems (zero and negative integers, rational numbers, real numbers, and complex numbers) are very much more recent. Historians of mathematics can trace for us the origin and development of these systems.

If we start with the positive integers, we have a system in which addition and multiplication can be performed, but subtraction and division cannot. More precisely, subtraction and division are not operations on natural numbers; they do not satisfy the closure law. In Section 2.14, we saw how to embed an integral domain into a field (its field of fractions); in other words, such a ring can be embedded in a larger ring in which division is possible. The prototype for this process is the enlargement of the integers to the rational numbers. We will see that a very similar process can be used to build the integers from the natural numbers.

It is more difficult to construct the real numbers from the rationals. We want to enlarge the rationals to include not only the roots of polynomials (such as $\sqrt{2}$), but also other useful numbers such as π and e. In effect, we have to plug the gaps in the rationals, and this is not entirely an algebraic process; some ideas from analysis, such as Cauchy sequences, are needed in this procedure. Once we have the real numbers, we obtain the complex numbers by a process that we have already seen in Section 2.16: adjoining a root of the polynomial equation $x^2 + 1 = 0$.

However, science marches on, and various tasks that were once God's preserve are now carried out by white-coated technicians. So it is that mathematical logicians have created the natural numbers, and indeed have created them out of nothing (more precisely, starting from the empty set).

In the first section of this chapter, we will look at these constructions in some detail. In the second section, we examine the distinction between algebraic numbers (those which satisfy some polynomial equation with integer coefficients) and transcendental numbers (those which do not): we prove in three different ways that transcendental numbers exist, and show that the classical problems of squaring the circle, duplicating the cube and trisecting the angle are insoluble with ruler and compass. In the final section, we treat in a little more detail some aspects of set theory: cardinality and the Axiom of Choice.

To the complex numbers

6.1   The natural numbers.   The natural numbers were invented for the purpose of counting. It is easy to believe that the earliest pastoral societies relied on counting for keeping tallies of their flocks. There is some evidence that this was done by establishing a bijection between the animals in a flock and a collection of pebbles or marks on a stick. A more sophisticated method is to have names for the natural numbers, independent of any physical representation of them. From that, it is a short step to algorithmic manipulation of numbers, so that subtraction can be used to establish how many animals are missing when the herd returns.

It does not really matter what names are used for the numbers, any more than it matters what material is used to construct the standard metre. However, it is very convenient if, for example, the standard number 248 is a set with 248 elements. If we adopt this principle, then zero should be a set with no elements. As we saw, there is a unique empty set, and this we take as zero. Then the number 1 should be a set with one element; using what we have to hand, we take it to be $\{0\}$. Then we take $2 = \{0, 1\}$, $3 = \{0, 1, 2\}$, and so on. We see that each natural number is the set consisting of all smaller natural numbers (including zero). This leads us to the formal construction.

Definition

- The empty set is a natural number (called **zero**, and written 0).
- If n is a natural number, then so is $n \cup \{n\}$.
- Every natural number is generated by these two rules.

The natural number $n \cup \{n\}$ is called the **successor** of n. We temporarily write it as $s(n)$; later we will call it $n + 1$.

The most important property of the natural numbers is the **principle of induction**.

Theorem 6.1   *Let $P(n)$ be a proposition about the natural number n. Suppose that*

(a) $P(0)$ is true; and
(b) $P(n)$ implies $P(s(n))$ for any natural number n.

Then $P(n)$ holds for all natural numbers n.

Proof The definition of the natural numbers makes clear that the set of natural numbers satisfying P is the same as the set of all natural numbers. ☐

The Principle of Induction can be used for definitions as well as proofs. In order to define a function f on the set of natural numbers, it is enough to define $f(0)$ and to define $f(s(n))$ in terms of $f(n)$. For example, we define addition by

$$m + 0 = m,$$
$$m + s(n) = s(m + n),$$

for all natural numbers m and n, and then define multiplication by

$$m \cdot 0 = 0,$$
$$m \cdot s(n) = m \cdot n + m,$$

for all m and n. From these definitions, it is possible to prove all the 'elementary' properties of addition and multiplication. The proofs are quite complicated to follow, since they use the idea of 'double induction'. We are trying to prove a proposition such as the commutative law $m + n = n + m$ which depends on two variables. We prove it by induction on n, so we have to show that $m + 0 = 0 + m$ and also that $(m + n = n + m) \Rightarrow (m + s(n) = s(n) + m)$. Each of these sub-propositions is proved by induction on m (since at the beginning, induction is the only tool we have!) We can formalise 'double induction' as follows:

Theorem 6.2 *Let $P(m, n)$ be a proposition about pairs of natural numbers. Assume that:*

(a) $P(0, 0)$ is true.
(b) $P(0, n)$ implies $P(0, s(n))$ for all n.
(c) $P(m, 0)$ implies $P(s(m), 0)$ for all m.
(d) For a given value of m, from the truth of $P(m, x)$ for all x, and also that of $P(s(m), n)$ for some n, we can infer the truth of $P(s(m), s(n))$.

Then $P(m, n)$ is true for all m and n.

Proof Hypotheses (a) and (b) are the base case and inductive step for the proof of $P(0, n)$ for all n. Similarly, hypotheses (c) and (d) are the base case and inductive hypothesis for a proof (by induction on n) of the statement that, if $P(m, n)$ holds for all n, then $P(s(m), n)$ holds for all n. Together these show by induction on m that $P(m, n)$ holds for all m. ☐

Using this principle, we show the commutative law for addition.

Proposition 6.3 $m + n = n + m$ *for all natural numbers m, n.*

Proof For double induction, we have to show:

(a) $0 + 0 = 0 + 0$;
(b) $0 + n = n + 0$ implies $0 + s(n) = s(n) + 0$;

(c) $m + 0 = 0 + m$ implies $s(m) + 0 = 0 + s(m)$;

(d) $m + x = x + m$ for all x, and $s(m) + n = n + s(m)$, imply $s(m) + s(n) = s(n) + s(m)$.

Here (a) is a triviality. To prove (b), assume that $0 + n = n + 0$; that is, $0 + n = n$, since $n + 0 = n$ by definition of addition. Now, by definition, $0 + s(n) = s(0 + n)$, which is equal to $s(n)$ by the assumption, and is equal to $s(n) + 0$ by the definition of addition again. Now (c) is just (b) in disguise, so it is also true.

To prove (d), we assume that $m + x = x + m$ for all x (and the fixed value of m), and that $s(m) + n = n + s(m)$ (where n has some fixed value also). We have

$$s(m) + s(n) = s(s(m) + n) = s(n + s(m)) = s(s(n + m)),$$

where the first and third equalities follow from the definition of addition, and the second from our assumptions. Similarly,

$$s(n) + s(m) = s(s(n) + m) = s(m + s(n)) = s(s(m + n)).$$

By assumption, $m + n = n + m$; it follows that $s(m) + s(n) = s(n) + s(m)$, as required.   □

It follows from the definition of addition that $n + 1 = n + s(0) = s(n)$. So we can replace the notation $s(n)$ by the more familiar $n + 1$.

Various further properties can be proved in a similar way. I will not give the proofs. Here is a list of what is needed; if you have great powers of perseverance, and work through them all, you will have put the natural numbers on a firm logical basis.

- *Closure laws*: For all $a, b \in \mathbb{N}$, we have $a + b, ab \in \mathbb{N}$.
- *Associative laws*: For all $a, b, c \in \mathbb{N}$, we have $a + (b + c) = (a + b) + c$ and $a(bc) = (ab)c$.
- *Commutative laws*: For all $a, b \in \mathbb{N}$, we have $a + b = b + a$ and $ab = ba$.
- *Distributive law*: For all $a, b, c \in \mathbb{N}$, we have $a(b + c) = ab + ac$.
- *Zero law*: For all $a \in \mathbb{N}$, we have $a + 0 = a$.
- *Identity law*: For all $a \in \mathbb{N}$, we have $a1 = a$.
- *Cancellation laws*: If $a + b = a + c$, then $b = c$; if $a \neq 0$ and $ab = ac$, then $b = c$.

We can also define the usual ordering on the natural numbers: $a \leq b$ holds if and only if there is a natural number c such that $a + c = b$.

6.2 The integers.

The motivation for extending the natural numbers to include negative numbers is to allow subtraction; more precisely, to produce a number system which is closed under subtraction. Accordingly, we want to construct numbers called $a - b$ for all $a, b \in \mathbb{N}$, where $a - b$ is a solution x to the equation $x + b = a$. There are a couple of problems. First, if $a \geq b$, then

N already contains an element $a - b$. Second, the same integer will have many different names, since (for example) $2 - 3 = 5 - 6$.

Accordingly, we represent the integer $a - b$ by the ordered pair (a, b), and we want to ensure that, if $a - b = c - d$ (in other words, if $a + d = b + c$), then the pairs (a, b) and (c, d) should represent the same integer. This is a job for an equivalence relation!

So here is the formal definition:

Define a relation $\sim$ on the set of ordered pairs of natural numbers by the rule that $(a, b) \sim (c, d)$ if and only if $a + d = b + c$. Then $\sim$ is an equivalence relation:

- Since $a + b = b + a$, we have $(a, b) \sim (a, b)$; so $\sim$ is reflexive.
- If $a + d = b + c$, then $c + b = d + a$; so $(a, b) \sim (c, d)$ implies $(c, d) \sim (a, b)$, and $\sim$ is symmetric.
- Suppose that $(a, b) \sim (c, d)$ and $(c, d) \sim (e, f)$. Then $a + d = b + c$ and $c + f = d + e$. Thus

$$a + c + f = a + d + e = b + c + e,$$

and the cancellation law implies that $a + f = b + e$; that is, $(a, b) \sim (e, f)$, and $\sim$ is transitive.

Definition An **integer** is an equivalence class of the relation $\sim$. We denote the equivalence class containing (a, b) temporarily by $[a, b]$. We define addition and multiplication of equivalence classes by the rules

$$[a, b] + [c, d] = [a + c, b + d],$$
$$[a, b] \cdot [c, d] = [ac + bd, ad + bc].$$

Let $\mathbb{Z}$ denote the set of integers with these operations.

Where do these definitions come from? The symbol $[a, b]$ is going to be the integer $a - b$; and we have

$$a - b = c - d \Leftrightarrow a + d = b + c,$$
$$(a - b) + (c - d) = (a + c) - (b + d),$$
$$(a - b) \cdot (c - d) = (ac + bd) - (ad + bc).$$

Theorem 6.4 *The set $\mathbb{Z}$, with the above-defined operations, is a commutative ring with identity, and is an integral domain.*

Proof Before we begin the verification of the axioms, we must first prove that the operations are well defined. That is, we must prove that, if $(a, b) \sim (a', b')$ and $(c, d) \sim (c', d')$, then $(a + c, b + d) \sim (a' + c', b' + d')$ and $(ac + bd, ad + bc) \sim (a'c' + b'd', a'd' + b'c')$. In yet other words, we must show that, if $a + b' = b + a'$ and $c + d' = d + c'$, then $(a + c) + (b' + d') = (b + d) + (a' + c')$ and $(ac + bd) +$

$(a'd' + b'c') = (ad + bc) + (a'c' + b'd')$. These facts follow from the properties of $\mathbb{N}$ by tedious but elementary algebraic manipulations.

Now we verify the eight ring axioms, the commutative law, and the existence of an identity. Relatively straightforward calculations are needed. Take the left distributive law as an example. For any natural numbers a, b, c, d, e, f,

$$[a, b]([c, d] + [e, f]) = [a, b][c + e, d + f]$$
$$= [a(c + e) + b(d + f), a(d + f) + b(c + e)],$$
$$[a, b][c, d] + [a, b][e, f] = [ac + bd, ad + bc] + [ae + bf, af + be]$$
$$= [ac + bd + ae + bf, ad + bc + af + be]$$

and the equality of the right-hand sides follows from properties of $\mathbb{N}$.

The zero element is the class $[a, a]$; the negative of $[a, b]$ is $[b, a]$; and the identity is the class $[1, 0]$.

To see that $\mathbb{Z}$ is an integral domain, it is convenient to choose representatives for the equivalence classes as follows: If $a > b$, then $[a, b] = [a - b, 0]$; if $a < b$, then $[a, b] = [0, b - a]$. So we may assume that either a or b is zero. Now take two non-zero integers, each represented as either $[a, 0]$ or $[0, b]$ with a or b non-zero. For their product there are several cases. For example,

$$[a, 0] \cdot [0, b] = [0, ab] \neq 0,$$

since if $a, b \neq 0$ then $ab \neq 0$ by the Cancellation Law for $\mathbb{N}$. □

The set $\mathbb{N}$ is embedded isomorphically into $\mathbb{Z}$ by the map taking a to $[a, 0]$. Since $[a, b] + [b, 0] = [a + b, b] = [a, 0]$, we see that $x = [a, b]$ is the solution to the equation $x + b = a$, where we identify a and b with the corresponding integers $[a, 0]$ and $[b, 0]$ respectively. So we can denote $[a, b]$ by the more usual notation $a - b$. Moreover,

$$[a, b] - [c, d] = [a + d, b + c],$$

so subtraction is everywhere defined on $\mathbb{Z}$.

The ordering on $\mathbb{Z}$ can be defined by the rule that $a \leq b$ if and only if $b - a$ is a natural number. Equivalently, $[a, b] \leq [c, d]$ if and only if the inequality $a + d \leq b + c$ holds in the natural numbers.

6.3 The rational numbers. The construction of the rational numbers from the integers is quite similar to the construction of the integers from the natural numbers. In $\mathbb{Z}$, we can add, subtract, and multiply, but not always divide; we want to add elements a/b, that is, solutions of $bx = a$, whenever $b \neq 0$. Similar comments about the non-uniqueness of the representation a/b apply. If you studied Section 2.14, you will have seen the process: $\mathbb{Q}$ is the field of fractions of $\mathbb{Z}$. I will run through the construction more briefly.

Let X be the set of all pairs (a, b) of integers with $b \neq 0$. Define a relation $\sim$ on X by the rule that $(a, b) \sim (c, d)$ if and only if $ad = bc$. This relation is:

- *reflexive*, since $ab = ba$;
- *symmetric*, since $ad = bc$ implies $cb = da$;
- *transitive*, since $ad = bc$ and $cf = de$ imply $adf = bcf = bde$, and hence that $af = be$ (since $d \neq 0$ and $\mathbb{Z}$ is an integral domain).

Hence $\sim$ is an equivalence relation. We let $[a, b]$ denote the equivalence class containing (a, b), and let $\mathbb{Q}$ be the set of equivalence classes (its elements are called **rational numbers**).

Now we define addition and multiplication of rational numbers by the rules

$$[a, b] + [c, d] = [ad + bc, bd],$$
$$[a, b] \cdot [c, d] = [ac, bd].$$

(These are motivated by the rules for adding and multiplying fractions, since $[a, b]$ is to represent the rational number a/b.) It can now be shown that these operations are well defined (independent of the choice of representatives), and also that the following holds:

Theorem 6.5 $\mathbb{Q}$, *with the above operations, is a field.*

Moreover, $\mathbb{Z}$ is embedded isomorphically into $\mathbb{Q}$ by the map taking a to $[a, 1]$. If (as usual) we denote this element by the same symbol a, then if $b \neq 0$ we have $b^{-1} = [1, b]$, and hence $[a, b] = ab^{-1} = a/b$, as we intended.

Of course, any equation $bx = a$, with $a, b \in \mathbb{Q}$ and $b \neq 0$, can now be solved. If $a = [a_1, a_2]$ and $b = [b_1, b_2]$ with $b_1 \neq 0$, then $a/b = [a_1 b_2, a_2 b_1]$.

The ordering of the rational numbers can be defined. The rational number $[a, b]$ is positive if a and b have the same sign (both positive or both negative); then $q \leq r$ if and only if $r - q$ is zero or positive.

6.4 The real numbers. At each stage after the natural numbers so far, we have been enlarging the number system so as to make an operation (either subtraction or division) defined everywhere it should be; in other words, to ensure that equations ($x + b = a$ or $bx = a$ respectively) have solutions.

The next stage is different. There are still many equations which do not have solutions (for example, the equation $x^2 = 2$ which gave the Pythagoreans so much trouble). There are also various non-algebraic equations that we would like to solve, such as $\sin x = 1$ or $\log x = 1$. These fail to have solutions because, although we can approximate the solutions as closely as we like, we cannot express them exactly by rational numbers. Although the rationals are dense (in the sense that between any two we can always find another), there are still many gaps between them that we have to fill.

The technique follows the general outline that we have used before. First, we find a method to represent one of the 'missing' numbers. We calculate what it means for two such representations to define the same number, and hence define an equivalence relation on them. Then we define a 'real number' to be an equivalence class of this relation, and give rules for addition and multiplication of real numbers. However, unlike the previous case, we can now use the machinery of rings and ideals to help.

There are two methods commonly used for the construction of the real numbers, namely **Cauchy sequences** and **Dedekind cuts**. The fields constructed by these two approaches turn out to be isomorphic. I will use Cauchy sequences, which permit a more algebraic approach. The motivation is the representation of a real number by an infinite decimal expansion. Such a decimal has the form $n.a_1a_2a_3\ldots$, where n is an integer and $a_1, a_2, a_3, \ldots$ are decimal digits. This number is the limit of the sequence

$$n, n.a_1 = n + a_1/10, n.a_1a_2 = n + a_1/10 + a_2/100, \ldots$$

Each term in the sequence is rational, and the sequence is a Cauchy sequence according to the following definition. We use the notation $|q|$ for the **modulus** of the rational number q; that is,

$$|q| = \begin{cases} q & \text{if } q \geq 0, \\ -q & \text{if } q \leq 0. \end{cases}$$

Definition

- A **Cauchy sequence** of rational numbers is defined to be a sequence $q_0, q_1, q_2, \ldots$ with $q_n \in \mathbb{Q}$ for all n, satisfying the following condition: for any positive rational ϵ, there exists a positive integer N such that $|q_m - q_n| < \epsilon$ whenever $m, n > N$.
- A **null sequence** of rational numbers is a sequence $q_0, q_1, q_2, \ldots$ with $q_n \in \mathbb{Q}$ for all n, satisfying the following condition: for any positive rational ϵ, there exists a positive integer N such that $|q_n| < \epsilon$ whenever $n > N$.

We define **addition** and **multiplication** of sequences componentwise: that is, if (q_n) and (r_n) denote the sequences with nth terms q_n and r_n respectively, their sum has nth term $q_n + r_n$, and their product has nth term $q_n r_n$.

Let $\mathcal{C}$ and $\mathcal{N}$ denote the sets of Cauchy sequences and null sequences respectively.

Theorem 6.6 *$\mathcal{C}$ is a commutative ring with identity, and $\mathcal{N}$ is a maximal ideal in $\mathcal{C}$.*

Proof The proof that $\mathcal{C}$ is a commutative ring involves some fairly standard verification of axioms; a kind of mixture of algebra and analysis. Its zero and identity are the constant sequences with values 0 and 1 respectively. Two examples will illustrate.
Closure under multiplication: We use the fact that any Cauchy sequence is bounded. For let (q_n) be a Cauchy sequence. Choosing $\epsilon = 1$ in the definition, let N have the property that $|q_m - q_n| < 1$ for $m, n > N$. It follows

that $|q_n| < |q_{N+1}| + 1$ for $n > N$. So, if B is the greatest of the numbers $|q_1|, |q_2|, \ldots, |q_N|, |q_{N+1}| + 1$, then we have $|q_n| \le B$ for all n.

Now let (q_n) and (r_n) be Cauchy sequences, bounded by B and C respectively. Given $\epsilon > 0$, choose N so that $|q_m - q_n| < \epsilon/2C$ for $m, n > N$, and choose M so that $|r_m - r_n| < \epsilon/2B$ for $m, n > M$. Let P be the greater of M and N. Then, for $m, n > P$, we have

$$|q_m r_m - q_n r_n| = |q_m r_m - q_m r_n + q_m r_n - q_n r_n|$$
$$\le |q_m| \cdot |r_m - r_n| + |r_n| \cdot |q_m - q_n|$$
$$< B(\epsilon/2B) + C(\epsilon/2C)$$
$$= \epsilon.$$

So $(q_n r_n)$ is a Cauchy sequence.

Commutative law for multiplication: The nth terms of $(q_n)(r_n)$ and $(r_n)(q_n)$ are, respectively, $q_n r_n$ and $r_n q_n$, which are equal, by the commutative law for multiplication in $\mathbb{Q}$.

It is not obvious that $\mathcal{N}$ is a subset of $\mathcal{C}$. Let (q_i) be a null sequence, and let $\epsilon > 0$ be given. Choose N so that $|q_n| < \epsilon/2$ for $n > N$ (this is done by applying the definition of a null sequence with $\epsilon/2$ replacing ϵ). Then, if m and n are both greater than N, we have $|q_m| < \epsilon/2$ and $|q_n| < \epsilon/2$; so $|q_m - q_n| < \epsilon$. Thus (q_n) is a Cauchy sequence.

Now further standard verification shows that $\mathcal{N}$ is an ideal of $\mathcal{C}$. (We again require the fact that Cauchy sequences are bounded).

To show that $\mathcal{N}$ is a maximal ideal, let J be any ideal of $\mathcal{C}$ which properly contains it; we must show that $J = \mathcal{C}$. Take any sequence (q_n) which lies in J but not in $\mathcal{N}$. Then (q_n) is not a null sequence. Negating the definition of a null sequence yields the following: *there exists some $\epsilon > 0$ such that, for any N, there are terms q_n of the sequence with $n > N$ such that $|q_n| \ge \epsilon$.* But (q_n) is a Cauchy sequence; so, taking $\epsilon/2$ in the definition, we find a number M such that $|q_m - q_n| < \epsilon/2$ for all $m, n > M$. We know that there exists some $m > M$ with $|q_m| \ge \epsilon$; it follows that $|q_n| > \epsilon/2$ for **all** $n > M$. In other words, apart from finitely many terms at the start, the sequence (q_n) is bounded away from zero.

Now let (x_n) be any Cauchy sequence. Define a sequence (r_n) by

$$r_n = \begin{cases} 0 & \text{if } q_n = 0, \\ x_n/q_n & \text{if } q_n \ne 0. \end{cases}$$

The case $q_n = 0$ can only occur finitely often. Using this and the fact that q_n is bounded away from zero for $n > M$, it can be shown that (r_n) is a Cauchy sequence. Hence $(r_n)(q_n) \in J$.

But $r_n q_n = x_n$ whenever $q_n \ne 0$; that is, for all but finitely many values. So the sequence $(x_n - r_n q_n)$ is zero from some point onwards, and hence certainly a null sequence, and thus in J. Then $(x_n) = (x_n - r_n q_n) + (r_n)(q_n) \in J$ also. Since (x_n) was an arbitrary Cauchy sequence, we have $J = \mathcal{C}$, as required. $\square$

Definition A **real number** is an element of the ring $\mathbb{R} = \mathcal{C}/\mathcal{N}$.

Theorem 6.7 $\mathbb{R}$ *is a field.*

This follows from the preceding theorem and Theorem 2.27.

There is an isomorphic embedding of $\mathbb{Q}$ into $\mathbb{R}$: we map any rational number q to the coset of $\mathcal{N}$ containing the constant sequence (q). (This coset consists of all sequences of rational numbers which have the limit q.) Furthermore, we can now define Cauchy sequences of real numbers, and it is possible to show that every Cauchy sequence of real numbers converges to a real number; that is, the field of real numbers is **complete**. (To define Cauchy sequences of real numbers, we need first to define the modulus of a real number, which itself depends on the ordering, defined below.)

In accordance with our original motivation, we note that the decimal expansion of a real number does indeed give a Cauchy sequence representing that number. Also, it can be shown that every real number has a decimal expansion (that is, the coset of $\mathcal{N}$ representing that number contains a particular Cauchy sequence of the type arising from a decimal expansion).

The ordering of the real numbers can be defined as follows: We say that a Cauchy sequence (q_n) is positive if it is not a null sequence and $q_n > 0$ for all but finitely many values of n. Then we say that $(q_n) < (r_n)$ if $(r_n - q_n)$ is positive. Now it can be shown that, if $(q_n) < (r_n)$, then $(q'_n) < (r'_n)$ for any Cauchy sequences (q'_n) and (r'_n) which differ from (q_n) and (r_n) respectively by null sequences. This means that we have a well defined ordering on the cosets of $\mathcal{N}$ in $\mathcal{C}$, that is, on the real numbers.

6.5   The complex numbers.   We have already met the construction of the complex numbers from the reals. The aim is to enlarge the real numbers to a field containing a square root of -1 (a root of the polynomial equation $x^2 + 1 = 0$). We will see that we get much else too.

As an instance of the construction of a field extension in which a given irreducible polynomial has a root (described in Section 2.16), we define the field $\mathbb{C}$ of complex numbers as the factor ring $\mathbb{R}[x]/(x^2 + 1)$. (The polynomial $x^2 + 1$ is irreducible in $\mathbb{R}[x]$, because $a^2 + 1 \geq 1 > 0$ for all $a \in \mathbb{R}$.) If i denotes the root of the polynomial $x^2 + 1$ (that is, the coset $(x^2 + 1) + x$), then every element of $\mathbb{C}$ can be written in the form $a + bi$, where a and b are real, and the expression for a given complex number in this form is unique. Now the addition and multiplication are given by the usual 'rules of arithmetic', putting $i^2 = -1$:

$$(a + bi) + (c + di) = (a + c) + (b + d)i,$$
$$(a + bi) \cdot (c + di) = (ac - bd) + (ad + bc)i.$$

One of the most important properties of $\mathbb{C}$ is the 'Fundamental Theorem of Algebra'. This shows that it is not necessary to construct still larger fields to include roots of more complicated polynomials.

Theorem 6.8 (Fundamental Theorem of Algebra) *Any non-constant polynomial in* $\mathbb{C}[x]$ *has a root in* $\mathbb{C}$.

Despite its name, the 'Fundamental Theorem of Algebra' is not a theorem of algebra at all. All known proofs of it (and there are many) require some arguments from analysis. The best-known proof uses Liouville's Theorem, a result which comes at the end of a first course on complex analysis. Liouville's Theorem states that a complex analytic (that is, everywhere differentiable) function which is bounded must be constant. If f were a non-constant polynomial with no roots in $\mathbb{C}$, then it can be shown that $1/f(z) \to 0$ as $z \to \infty$, and hence that $1/f$ is bounded; Liouville's Theorem would then imply that $1/f$ (and hence also f) is constant, a contradiction.

In Chapter 8, there is a completely different proof. It replaces most of the analysis by algebra, using only facts about $\mathbb{R}$ which are consequences of the Intermediate Value Theorem. By contrast, it uses some fairly sophisticated group theory.

Definition We say that a field F is **algebraically closed** if it has the property that any non-constant polynomial in $F[x]$ has a root in F. Thus we can express the conclusion of the Fundamental Theorem of Algebra more simply: $\mathbb{C}$ *is algebraically closed*.

Exercise 6.1 Prove that $2 + 2 = 4$.

Remark Bertrand Russell, in his *History of Western Philosophy*, says:

> '3' means '2+1', and '4' means '3+1'. Hence it follows (though the proof is long) that '4' means the same as '2+2'. Thus mathematical knowledge ceases to be mysterious.

Exercise 6.2 Prove that, if a and b are natural numbers (regarded as sets, as in the construction), then the following are equivalent:

(a) $a \leq b$; (b) $a \subseteq b$; (c) $a \in b$.

Exercise 6.3 In the construction of natural numbers, we take

$$0 = \emptyset, \quad 1 = \{\emptyset\}, \quad 2 = \{\emptyset, \{\emptyset\}\}, \quad 3 = \{\emptyset, \{\emptyset\}, \{\emptyset, \{\emptyset\}\}\}, \quad \ldots$$

So each natural number is represented by a string of symbols from the alphabet with four symbols: $\emptyset$, opening and closing braces, and comma. Calculate the number of occurrences of each symbol in the string representing n.

Exercise 6.4 Prove carefully that $\mathbb{Z}$ is a ring.

Exercise 6.5 (∗) Prove the division algorithm for $\mathbb{Z}$: if $a, b \in \mathbb{Z}$ with $b > 0$, then there exist $q, r \in \mathbb{Z}$ with $a = bq + r$ and $0 \leq r < b$.

Exercise 6.6 (∗) Prove the **Principle of the Supremum** for $\mathbb{R}$: *if S is a non-empty subset of $\mathbb{R}$ which has an upper bound, then S has a least upper bound.*

Algebraic and transcendental numbers

Among the real (or complex) numbers, there are some (such as $\sqrt{2}$ or i) which satisfy polynomial equations with integer coefficients, and others (like π and e) which do not. In this section, we examine the distinction between the two classes of numbers, and give an application to ruler-and-compass constructions.

6.6   Algebraic numbers. We followed the traditional approach to the construction of the number systems. Another way to proceed, having constructed the rationals, would be to add next the roots of polynomials, and then put in all the other useful numbers that we require.

Instead of doing that, we now look back and examine the algebraic numbers (the roots of polynomials over the rationals), and establish that they do form a field.

Definition Let F be a field, E a subfield of F, and $a \in F$. We say that a is **algebraic over** E if there is a non-zero polynomial $f \in E[x]$ such that $f(a) = 0$ (evaluated in F).

We will prove that the set of all elements of F which are algebraic over E is a field. In order to do this, we require some results about field extensions.

Definition Let F be a field, E a subfield of F.

(a) For $a \in F$, the **field generated by** a **over** E is defined to be the smallest subfield of F which contains both E and a, denoted by $E(a)$.
(b) The **degree of** F **over** E, denoted by $[F : E]$, is the dimension of F as a vector space over E (when we allow multiplication only of elements of F by elements of E, as in Example 3 in Section 4.2.)

Proposition 6.9 *Let E be a subfield of F, and $a \in F$. Then a is algebraic over E if and only if $[E(a) : E]$ is finite.*

Proof Suppose first that $[E(a) : E] = n$ is finite. Then the $n + 1$ elements $1, a, a^2, \ldots, a^n$ of the vector space $E(a)$ over E must be linearly dependent. So there exist scalars $c_0, c_1, c_2, \ldots, c_n \in E$, not all zero, such that

$$c_0 + c_1 a + c_2 a^2 + \cdots + c_n a^n = 0.$$

This equation says that the non-zero polynomial $f(x) = c_0 + c_1 x + \cdots + c_n x^n$ has a as a root; so a is algebraic over E.

Conversely, suppose that a is algebraic over E. Let $f \in E[x]$ be a monic polynomial of least degree satisfied by a. (This is called the **minimal polynomial** of a over E.) Now f is irreducible in $E[x]$. For, if $f = gh$, where g and h have smaller degree than f, then we have $g(a)h(a) = 0$ in F; so either $g(a) = 0$ or $h(a) = 0$, contrary to the choice of f as the polynomial of smallest degree that has a as a root.

Let g be any polynomial in $E[x]$. Obviously, if f divides g, then $g(a) = 0$. Conversely, suppose that $g(a) = 0$. Write $g = fq + r$, where $r = 0$ or r has degree less than $\deg(f)$. But $r(a) = g(a) - f(a)q(a) = 0$; so, by choice of f, we have $r = 0$, and f divides g.

Now let θ be the 'evaluation' homomorphism from $E[x]$ to F defined by $g\theta = g(a)$. We have $\mathrm{Ker}(\theta) = \langle f \rangle$ (by the preceding paragraph: for this equation says that $g(a) = 0$ if and only if f divides g). Hence $\mathrm{Im}(\theta) \cong E[x]/\langle f \rangle$, by the First Isomorphism Theorem. Since f is irreducible, $E[x]/\langle f \rangle$ is a field. Thus, $\mathrm{Im}(\theta)$ is a field containing E and a. Also, we know from Section 4.2 that $E[x]/\langle f \rangle$, and hence $\mathrm{Im}(\theta)$, is a finite-dimensional vector space over E. So $[E(a) : E]$ is finite, as required. □

Theorem 6.10 *Suppose that E, F, G are fields with $E \subseteq F \subseteq G$. Then $[G : E]$ is finite if and only if both $[G : F]$ and $[F : E]$ are finite. If this holds, then*

$$[G : E] = [G : F] \cdot [F : E].$$

Proof If $[G : E]$ is finite, then so is $[F : E]$ (since F is a subspace of G, as E-vector spaces); and so also is $[G : F]$ (since a basis for G as E-vector space certainly spans G as F-vector space).

Conversely, suppose that $[F : E] = m$ and $[G : F] = n$ are finite. Let $f_1, \ldots, f_m$ be a basis for F as E-vector space, and let $g_1, \ldots, g_n$ be a basis for G as F-vector space. We claim that the mn elements $f_i g_j$, for $i = 1, \ldots, m$ and $j = 1, \ldots, n$, form a basis for G as E-vector space. Proof of this claim will show that $[G : E]$ is finite, and also prove the product formula for the degree.
Spanning: Take $a \in G$. Express it in terms of the basis over F; say

$$a = b_1 g_1 + \cdots + b_n g_n.$$

Now each b_j is an element of F, so can be written in terms of the basis over E:

$$b_j = c_{1j} f_1 + \cdots + c_{mj} f_m.$$

Substitution gives

$$a = \sum_{i=1}^{m} \sum_{j=1}^{n} c_{ij} f_i g_j.$$

So the elements $f_i g_j$ form a spanning set.

Linearly independent: Suppose that

$$\sum_{i=1}^{m}\sum_{j=1}^{n} c_{ij} f_i g_j = 0.$$

Each term in the sum over j, namely $\sum_{i=1}^{m} c_{ij} f_i$, is an element of F. Since $g_1, \ldots, g_n$ are linearly dependent over F, we must have $\sum_{i=1}^{m} c_{ij} f_i = 0$ for each j. Now the linear independence of $f_1, \ldots, f_m$ over E shows that all the coefficients c_{ij} are zero.

This completes the proof. $\square$

Theorem 6.11 *Let E and F be fields with $E \subseteq F$. Then the set of all elements of F which are algebraic over E is a field containing E.*

Proof   In Chapter 2, we did not specifically develop a subfield test. But we will be done if we prove that the set A of elements algebraic over E is a subring containing the inverses of all its non-zero elements. For it is certainly commutative (since it is contained in the field F) and has an identity (since it contains the field E). So we have to show that, for any $a, b \in A$, we have $a - b \in A$, $ab \in A$, and $a^{-1} \in A$ if $a \neq 0$.

So choose any $a, b \in A$. Then a is algebraic over E, so $[E(a) : E]$ is finite. And b is algebraic over E, hence certainly algebraic over $E(a)$; so $[E(a, b) : E(a)]$ is finite (where we have used $E(a, b)$ as an abbreviation for $E(a)(b)$, the field generated by a and b over E). By the above proposition, $[E(a, b) : E]$ is finite. But $E(a, b)$ contains $a - b$, ab, and a^{-1} (if $a \neq 0$); so all these elements are algebraic over E, and lie in A, as required. $\square$

Now we let $\mathbb{A}$ be the set of all complex numbers which are algebraic over $\mathbb{Q}$. Then $\mathbb{A}$ is a field. Its elements are called **algebraic numbers**. Using the Fundamental Theorem of Algebra, we can show:

Theorem 6.12 $\mathbb{A}$ *is an algebraically closed field.*

The proof depends on the following result. If E is a subfield of F, we say that F is **algebraic over** E if every element of F is algebraic over E.

Proposition 6.13   *Let E, F, G be fields with $E \subseteq F \subseteq G$. If F is algebraic over E, and G is algebraic over F, then G is algebraic over E.*

Proof   Take any element $c \in G$. Since c is algebraic over F, there is a polynomial $f(x) = x^n + a_{n-1}x^{n-1} + \cdots + a_0$ in $F[x]$ such that $f(c) = 0$. Now each of $a_0, a_1, \ldots, a_{n-1}$ is algebraic over E. So each of $[E(a_0) : E]$, $[E(a_0, a_1) : E(a_0)]$, and so on, is finite. So $[E(a_0, a_1, \ldots, a_{n-1}) : E]$ is finite. If F_0 is the field $E(a_0, a_1, \ldots, a_{n-1})$, then c satisfies the polynomial f with coefficients in F_0; so c is algebraic over F_0, and $[F_0(c) : F_0]$ is finite. Thus $[F_0(c) : E]$ is finite, and c lies in $F_0(c)$; so $[E(c) : E]$ is finite, and c is algebraic over E. So, by definition, G is algebraic over E. $\square$

Proof of the Theorem Take any non-constant polynomial $f \in \mathbb{A}[x]$. By the Fundamental Theorem of Algebra, f has a root $c \in \mathbb{C}$. Then $\mathbb{A}[c]$ is algebraic over $\mathbb{A}$, which is itself algebraic over $\mathbb{Q}$ by definition; so $\mathbb{A}[c]$ is algebraic over $\mathbb{Q}$. But every complex number algebraic over $\mathbb{Q}$ is in $\mathbb{A}$; so $c \in \mathbb{A}$. Thus f has a root in $\mathbb{A}$, and we conclude that $\mathbb{A}$ is algebraically closed. $\qquad\square$

6.7 Transcendental numbers. Nothing we have said so far allows us to conclude that the field of complex numbers is really different from the field $\mathbb{A}$ of algebraic numbers. If these two fields were the same, there would be no need to go through the construction of the real numbers by Cauchy sequences; everything could be obtained from $\mathbb{Q}$ by adjoining roots of polynomials.

To lend an air of mysticism to the proceedings, we define a **transcendental number** to be a complex number which is not algebraic, that is, an element of $\mathbb{C} \setminus \mathbb{A}$. The question is: *Do transcendental numbers exist?*

The answer is that they do; but there are three entirely different ways of reaching this conclusion. The first is to show that some very familiar number such as e (the base of natural logarithms) or π (the ratio of the circumference of a circle to its diameter) is transcendental. This was achieved for e by Hermite in 1873, and for π by Lindemann in 1882. (A modification of Hermite's proof is given below, involving various simplifications.) The second approach, taken by Liouville in 1844, was to write down a particular number which is easy to prove transcendental. The third, most revolutionary, approach is that of Cantor in 1874. He gave an argument which shows that 'almost all' numbers are transcendental, but without exhibiting even a single example!

First proof: The transcendence of e

Proposition 6.14 e *is transcendental.*

Proof We assume, to the contrary, that e is algebraic, and let

$$a_n e^n + \cdots + a_1 e + a_0 = 0,$$

where the coefficients a_i are rational. Multiplying this equation by the least common multiple of the denominators, we may assume that the a_i are integers. Furthermore, assuming we took the minimal polynomial of e, we have $a_0 \neq 0$.

We let p be any prime number, and define the polynomial

$$f(x) = \frac{x^{p-1}(x-1)^p(x-2)^p \cdots (x-n)^p}{(p-1)!}.$$

Now f is a polynomial of degree $np + p - 1$ with rational coefficients. Let $f^{(i)}(x)$ be the polynomial obtained by differentiating $f(x)$ i times. Note that $f^{(i)}(x) = 0$ for $i \geq np + p$. We also require the following property:

For all i, and for $j = 0, 1, \ldots, n$, $f^{(i)}(j)$ is an integer, and is divisible by p unless $j = 0$ and $i = p - 1$.

The proof uses Leibniz's rule for differentiating a product. Suppose that $j \neq 0$. To evaluate $f^{(i)}(j)$, we add all of the terms obtained by writing i as a sum of $n+1$ non-negative integers $m_0, \ldots, m_n$, differentiating the kth factor $(x-k)^p$ (or x^{p-1} if $k = 0$) m_k times, multiplying the product of the results by a suitable integer coefficient (a so-called multinomial coefficient), and dividing by the denominator $(p-1)!$ of f. If $(x-j)^p$ is differentiated fewer than p times, there is a factor $(x-j)$, which vanishes when we substitute $x = j$; if it is differentiated more than p times, then it is zero; and if it is differentiated exactly p times, then the result is $p!$, which leaves p on division by $(p-1)!$. The contribution from all of the other factors multiplies this by an integer. So the result is a multiple of p.

Now suppose that $j = 0$. Again, we only obtain a non-zero contribution if the first term x^{p-1} is differentiated $p-1$ times; the resulting term $(p-1)!$, divided by the denominator $(p-1)!$, leaves 1. Again, the effect of the other factors is to multiply the result by an integer. If any other factor is differentiated even once, there will be a factor of p in the product, and so the result is divisible by p. So only in the case when $j = 0$ and $i = p-1$ is the result not divisible by p (and it is definitely not divisible by p in that case).

Let $F(x)$ be the sum of $f(x)$ and its derivatives of all orders. (As we saw, this is a finite sum.) Now $F'(x)$ is the sum of all the derivatives; so $F'(x) - F(x) = -f(x)$. Now

$$\frac{\mathrm{d}}{\mathrm{d}x}\left(\mathrm{e}^{-x}F(x)\right) = \mathrm{e}^{-x}(F'(x) - F(x)) = -\mathrm{e}^{-x}f(x).$$

Hence, for $j = 0, \ldots, n$,

$$a_j \int_0^j \mathrm{e}^{-x} f(x)\, \mathrm{d}x = a_j F(0) - a_j \mathrm{e}^{-j} F(j).$$

Multiply this equation by e^j and sum over $j = 0, 1, \ldots, n$,

$$\sum_{j=0}^n \left(a_j \mathrm{e}^j \int_0^j \mathrm{e}^{-x} f(x)\, \mathrm{d}x \right) = F(0) \sum_{j=0}^n a_j \mathrm{e}^j - \sum_{j=0}^n a_j F(j)$$

$$= -\sum_{j=0}^n \sum_{i=0}^{np+p-1} a_j f^{(i)}(j).$$

Here we have used the supposed equation $\sum_{j=0}^n a_j \mathrm{e}^j = 0$ satisfied by e, together with the definition of F as the sum of f and its derivatives.

This last equation is the key to the proof. We show that the left-hand side can be made arbitrarily close to zero by choosing p to be sufficiently large. We also show that the right-hand side is an integer not divisible by p, and hence has modulus at least 1. This is a contradiction, which completes the proof. So it remains to establish the two assertions.

First, consider the left-hand side. For $0 \leq x \leq n$, we have

$$|f(x)| \leq n^{np+p-1}/(p-1)!,$$

so that

$$\left| \sum_{j=0}^{n} \left(a_j e^j \int_0^j e^{-x} f(x) \, dx \right) \right| \leq \sum_{j=0}^{n} |a_j e^j| \int_0^j \frac{n^{np+p-1}}{(p-1)!} \, dx$$

$$= \sum_{j=0}^{n} |a_j e^j| j \cdot \frac{n^{np+p-1}}{(p-1)!},$$

and the last expression tends to zero as $p \to \infty$.

Now consider the right-hand side; call it R. Our observation about f shows that R is indeed an integer. Moreover, all terms except $-a_0 f^{(p-1)}(0)$ are integers divisible by p. So $R \equiv_p -a_0 f^{(p-1)}(0)$. If we choose p greater than $|a_0|$ then, using the fact that $a_0 \neq 0$ (see the start of the proof) and $f^{(p-1)}(0)$ is not divisible by p, we see that R is not divisible by p, as claimed. This completes the proof. □

This proof is taken (with thanks) from Ian Stewart's book *Galois Theory*. In the same book, you will find an account of the proof that π is transcendental. This is similar but more complicated. (In the exercises after this chapter, Stewart gives a number of 'true or false?' questions, one of which reads: 'Sometimes the only way to prove a theorem is to pull rabbits out of hats.' You may agree!)

Second proof: Transcendence of Liouville's number   Liouville devised a simpler proof that transcendental numbers exist. He first showed that it is impossible to find very good rational approximations to irrational algebraic numbers (in a suitable sense). Then he wrote down a number which has very good rational approximations, and deduced that this number must be transcendental.

Given any real number α, we can find rational numbers as close to it as we choose. The rational numbers with denominator q cover the real line with a gap of $1/q$ between two consecutive ones, so we can find one of these within $1/q$ of any real number. What makes a good rational approximation to α is that the difference between α and the approximation p/q is much smaller than $1/q$.

This is the result about approximating algebraic numbers.

Theorem 6.15   *Let α be an algebraic number whose minimal polynomial has degree n. Then there is a constant $c > 0$ such that there are only finitely many rational numbers p/q satisfying $|\alpha - p/q| < 1/cq^n$.*

Proof   Let $f(x)$ be the minimal polynomial of α, the polynomial of least degree having α as a root. Multiplying by the least common multiple of the denominators of the coefficients of f, we may assume that all the coefficients are integers.

Consider the derivative f'. This is a continuous function, and so it is bounded on any closed interval; say, $|f'(x)| \leq c$ for $x \in [\alpha - 1, \alpha + 1]$.

Now there are only finitely many rational numbers p/q such that $1 < |\alpha - p/q| < 1/cq^n$ (this can only hold if $q < (1/c)^{1/n}$). So any other rational approximation p/q for which $|\alpha - p/q| < 1/cq^n$ must lie in the interval $[\alpha - 1, \alpha + 1]$. We show that there are no such rationals.

The Mean Value Theorem tells us that

$$f(\alpha) - f(p/q) = (\alpha - p/q)f'(\xi),$$

where ξ is some real number between α and p/q. Now $f(\alpha) = 0$, by assumption. Also $f(p/q) \neq 0$: for f is irreducible in $\mathbb{Q}[x]$ and has degree greater than 1, and so has no rational root. Since the coefficients of f are integers, $f(p/q)$ is a rational number with denominator q^n, so $|f(p/q)| \geq 1/q^n$. Also, since ξ is in the interval $[\alpha - 1, \alpha + 1]$, we have $|f'(\xi)| \leq c$.

Putting all this together, we get $|\alpha - p/q| \geq 1/cq^n$. $\square$

Liouville's number is the real number with decimal expansion

$$\alpha = 0.11000100000000000000000001\ldots$$

(the ones occur in positions 1!, 2!, 3!, ...). In other words,

$$\alpha = \sum_{n=1}^{\infty} 10^{-n!}.$$

Now α is irrational, since any rational number has a terminating or periodic decimal expansion. Let a_n be the nth partial sum of the series for α; that is, $a_n = \sum_{m=1}^{n} 10^{-m!}$. Then a_n is a rational number p/q, where $q = 10^{n!}$. Also, $\alpha - a_n < 2 \cdot 10^{-(n+1)!} = 2/q^{n+1}$. So, given n, and given any positive constant c, we have $|\alpha - a_m| < 1/cq_m^n$ for all sufficiently large m, where q_m is the denominator of the rational number a_m. So it is impossible that α satisfies a polynomial of degree n. Since this is true for all n, we see that α *is transcendental*.

Third proof: Transcendence of almost all numbers   In the late nineteenth century, Cantor was developing the concept of the **cardinal number** of elements in an infinite set, allowing him to compare the sizes of infinite sets. One of the triumphs of his theory is the following proof of the existence of transcendental numbers, which is technically much simpler than either of the other proofs.

Definition   An infinite set X is **countable** if there is is a bijection between X and the set $\mathbb{N}$ of natural numbers; that is, if the elements of X can be labelled as $x_0, x_1, x_2, \ldots$ (indexed by the natural numbers).

Among Cantor's discoveries was the following result:

Theorem 6.16   *(a) The set $\mathbb{A}$ is countable.*
(b) The sets $\mathbb{R}$ and $\mathbb{C}$ are not countable.

Proof (a) For $\mathbb{A}$, we have to list all the algebraic numbers in a sequence. Each algebraic number is the root of a polynomial f, and by multiplying by the least common multiple of the denominators we can assume that f has integer coefficients. Now each polynomial has only finitely many roots, so if we can list all the polynomials then we can list the algebraic numbers.

To list the integer polynomials, we define the **height** of the non-constant polynomial $a_n x^n + a_{n-1} x^{n-1} + \cdots + a_0$ to be the positive integer $n + |a_n| + |a_{n-1}| + \cdots + |a_0|$. There are only a finite number of polynomials with any given height, and so we can list the roots of all the polynomials of height 2 (in fact, there are just two such polynomials, namely x and $-x$), then height 3, and so on.

(We are using here an instance of a general principle, according to which the union of a countable number of finite sets is countable.)

(b) We will show that the unit interval $(0,1) \subseteq \mathbb{R}$ is uncountable; the same assertion then follows for the larger set $\mathbb{C}$. Here are two proofs:

First proof This is Cantor's famous 'diagonal argument'. We represent the real numbers in the unit interval as decimals, which may be finite (terminating), or recurring, or neither. We assume that all the decimals are infinite, by appending zeros to the finite ones (so, for example, $1/2 = 0.50000\ldots$).

Suppose that the unit interval is countable; that is, we can list all its members as $r_1, r_2, r_3, \ldots$. Let r_i have the decimal expansion $0.x_{i1} x_{i2} x_{i3} \ldots$. We show that the assumption is wrong, by constructing a number s which is not in the list. We define s to be the number whose decimal expansion is $0.b_1 b_2 b_3 \ldots$, where

$$b_i = \begin{cases} 7, & \text{if } a_{ii} = 5, \\ 5 & \text{if } a_{ii} \neq 5. \end{cases}$$

Now, by construction, $s \neq r_i$, since the ith decimal digit of s is different from that of r_i. So s is not equal to any number in the list, and the assumption that we have listed all the real numbers in $(0,1)$ must be wrong. $\qquad\square$

Second proof Again suppose that $r_1, r_2, r_3, \ldots$ is a list of all the real numbers in $(0,1)$. Take any positive real number ϵ, and for $n = 1, 2, \ldots$, let I_n be the interval of length $\epsilon/2^n$ with centre at the point r_n. (Possibly this interval is not entirely contained in $(0,1)$, but this does not matter.) Since $r_n \in I_n$, our assumption that all the numbers in $(0,1)$ have been listed implies that the interval $(0,1)$ is contained in the union of the intervals I_n. But $(0,1)$ has length 1, while the union of the intervals I_n is at most $\sum \epsilon/2^n = \epsilon$. (The length may be smaller since the intervals may overlap.) $\qquad\square$

Since $\mathbb{C}$ is uncountable but $\mathbb{A}$ is countable, there is at least one transcendental number. But the proof gives no recipe to find one. (The first proof appears constructive: it shows that, given a list of all the algebraic numbers, if we knew the ith digit in the decimal expansion of the ith algebraic number in the list, we could construct a transcendental number. But this is only a theoretical possibility.)

However, the proof gives more. We can see that the set of transcendental numbers, like the set of complex numbers, is uncountable. (If it were countable, and was enumerated as $b_1, b_2, b_3, \ldots$, then we could take a list of all the algebraic numbers as in part (a) of the theorem, say $a_1, a_2, a_3, \ldots$, and produce a list of all the complex numbers as $a_1, b_1, a_2, b_2, a_3, \ldots$, contrary to part (b).)

The second proof shows a little more. The algebraic numbers in the unit interval, or indeed any countable set, can be covered by a union of intervals with total length less than any preassigned positive number ϵ. So, if we choose a number at random from the unit interval, with probability 1 it will be transcendental. (In the language of measure theory, we say that the algebraic numbers form a **null set**.)

Remark The similar fact that the rational numbers form a null set seems to have been known to the remarkable fourteenth-century French mathematician Nicole Oresme. In his work *De proportionibus proportionum*, he states the proposition

It is probable that two proposed unknown ratios are incommensurable.

On the basis of this and related ideas he argued that the future is unpredictable and hence that astrology is futile. See Crosby's book cited earlier, or Karl Petersen's *Ergodic Theory*, for more information about Oresme.

6.8   Ruler-and-compass constructions. The topic of ruler-and-compass constructions has fascinated mathematicians since the days of the ancient Greeks. The three famous problems (trisecting a general angle, duplicating a cube, and squaring a circle), are all now known to be impossible; but this has not halted the steady stream of 'solutions' which arrive at mathematics departments all over the world!

The general set-up can be described as follows: We are given a finite set S of points in the Euclidean plane, and wish to construct another finite set T. To perform the construction, we are allowed two tools:

- a **ruler** or **straightedge**, which can be used to draw a line of arbitrary length through any two points already constructed;
- a **compass**, which can be used to draw a circle whose centre is any constructed point, and whose radius is equal to the distance between any two constructed points.

New points are constructed as the intersections of two lines, of a line and a circle, or of two circles. The construction process is required to take only finitely many steps.

The set S should contain at least two points, since otherwise nothing can be constructed. It is conventional to assume that two of the points in S are the origin $(0,0)$ and the point $(1,0)$. (This convention simply sets the position and scale of the coordinate axes.)

The works of Euclid, and the versions of them which have been fed to many generations of schoolchildren in the past, contain the details of many constructions: bisecting a line segment or an angle, constructing an equilateral triangle, a square, or regular pentagon, dividing a line segment in the golden ratio, and so on. But we show that certain things cannot be constructed by giving a necessary condition.

If S is a finite set of points in $\mathbb{R}^2$, we define $\mathbb{Q}(S)$ to be the field generated by the coordinates of all points in S; that is, the smallest field containing all these coordinates.

Theorem 6.17 *If T is constructible from S, then the coordinates of the points in T lie in a field F containing $\mathbb{Q}(S)$ such that $[F : \mathbb{Q}(S)]$ is a power of 2.*

Proof We look at one step in the construction, where one new point is created, and show that its coordinates lie in an extension of the preceding field with degree at most 2. Then, by the multiplicative formula for degrees of extensions (Theorem 6.10), the final field has degree a power of 2 over the initial one.

Suppose that we have constructed a set U of points. Their coordinates lie in the field $E = \mathbb{Q}(U)$. Now we may draw a line joining two of these points, or a circle with centre in U and radius equal to the distance between two points in U. Coordinate geometry tells us how to find the equation of such a line or circle; we claim that all coordinates in such an equation belong to E. For:

(a) The line joining (a,b) to (c,d) has equation $(x-a)(d-b) = (y-b)(c-a)$.
(b) The circle with centre (a,b) and radius r has equation $(x-a)^2+(y-b)^2 = r^2$.
    If r is the distance between (c,d) and (e,f), then $r^2 = (c-e)^2 + (d-f)^2$.

Now suppose that we construct a new point as the intersection of two such curves.

(a) If both curves are lines, then we have to solve two linear equations with coefficients in E; the solution lies in E.
(b) If one is a line and the other a circle, we can use the equation of the line to find y in terms of x (or vice versa), and substitute into the equation of the circle; we obtain a quadratic equation for x (or y). If the quadratic is reducible in $E[x]$, the solution lies in E; otherwise, it lies in an extension E' of E with $[E' : E] = 2$. Once x has been found, the equation of the line yields y without further enlarging the field.
(c) Suppose that both curves are circles, with equations $x^2+y^2+ax+by+c = 0$ and $x^2 + y^2 + dx + ey + f = 0$. Any solution must also satisfy the difference of these equations, namely $(a-d)x + (b-e)y + (c-f) = 0$. This is the equation of a line; so we are back in the situation of case (b).

The theorem is proved. $\qquad\qquad\qquad\qquad\qquad\qquad\qquad\qquad\qquad\qquad\qquad\square$

We use this result to show that the classical problems are insoluble.

Duplicating the cube   This problem gives the side of a cube of volume 1, and asks for the side of a cube of volume 2 to be constructed. We may take $S = \{(0,0), (1,0)\}$, so that $\mathbb{Q}(S) = \mathbb{Q}$. If the required length can be constructed, then we can construct the point $(2^{1/3}, 0)$, and so the final field F contains $\mathbb{Q}(2^{1/3})$.

However, the polynomial $x^3 - 2$ is irreducible over $\mathbb{Q}$, since $2^{1/3}$ is irrational. So $[\mathbb{Q}(2^{1/3}) : \mathbb{Q}] = 3$. Since $[F : \mathbb{Q}] = 2^n$ for some n, we conclude that $[F : \mathbb{Q}(2^{1/3})] = 2^n/3$, an obvious impossibility.

Trisecting the angle   What is required in this problem is a *general* procedure to trisect any angle. This can be refuted by showing just one angle that cannot be trisected. (Some angles can be trisected, for example, a right angle: we can construct an angle of 30° by bisecting the angle of an equilateral triangle.)

We show that the angle 60° cannot be trisected.

Note that, if an angle θ can be constructed, then the length $\cos\theta$ is constructible. Let $c = \cos 20°$. Then

$$4c^3 - 3c = \cos 60° = \tfrac{1}{2},$$

using the formula $\cos 3\theta = 4\cos^3\theta - 3\cos\theta$. So c is a root of the polynomial $8x^3 - 6x - 1 = 0$. This polynomial is easily seen to be irreducible. Then the proof continues exactly as for duplicating the cube.

Squaring the circle   The problem is to construct a square whose area is equal to that of a circle of unit radius; in other words, to construct a length of $\sqrt{\pi}$. Since π is transcendental, by the result of Lindemann, it cannot lie in any extension of $\mathbb{Q}$ of finite degree, whether a power of 2 or not. So the construction is impossible.

It may be the fact that this argument uses the relatively complicated proof of the transcendence of π that leads to the existence of an army of 'circle-squarers' who do not accept it.

Exercise 6.7   Prove that, if a is transcendental over E, then $E(a)$ is isomorphic to the field of fractions of the polynomial ring $E[x]$.

Exercise 6.8   Prove that it is not possible to construct a regular 7-gon with ruler and compass. [*Hint:* The real number $\alpha = 2\cos(2\pi/7)$ satisfies the cubic equation $x^3 + x^2 - 2x - 1 = 0$.]

More about sets

We used some arguments about the cardinalities of infinte sets in the preceding section. This chapter finishes will a general account of Cantor's theory of cardinality, and a look at the Axiom of Choice.

6.9 Cardinality. We used some basic notions of Cantor's theory of cardinal number to prove that almost all numbers are transcendental. Here is a more detailed account, mostly without proofs.

We constructed each natural number as a set: the number n is the set $\{0, 1, \ldots, n-1\}$. We say that a set A has **cardinality** n if there is a bijection between A and n. (This is well defined; there cannot be a bijection between two different natural numbers.) We say that A is **finite** if it has cardinality n for some natural number n, and **infinite** otherwise. Furthermore, A is **countable** if there is a bijection between A and $\mathbb{N}$, and is **uncountable** if it is infinite but not countable.

One of the foundations of the theory is the **Schröder–Bernstein theorem**:

Theorem 6.18 (Schröder–Bernstein Theorem) *If there are injective maps from A to B and from B to A, then there is a bijection between A and B.*

Using this, we can write $|A| = |B|$ if there is a bijection from A to B; $|A| \leq |B|$ if there is an injective map from A to B; and $|A| < |B|$ if $|A| \leq |B|$ but there is no bijection between them. [We have not defined the cardinality of a set, simply these three relations between pairs of sets.] Clearly, if $A \subseteq B$, then $|A| \leq |B|$. Now the Schröder–Bernstein theorem states that, if $|A| \leq |B|$ and $|B| \leq |A|$, then $|A| = |B|$.

An example of Cantor's diagonal argument is the following result, where $\mathcal{P}(A)$ is the **power set** of A, the set of all subsets of A.

Theorem 6.19 $|A| < |\mathcal{P}(A)|$.

Proof We can define an injection from A to $\mathcal{P}(A)$ by mapping the element $a \in A$ to the set $\{a\}$.

Suppose that F is a bijection between A and $\mathcal{P}(A)$. Let

$$B = \{a \in A : a \notin F(a)\}.$$

Then $B \in \mathcal{P}(A)$, so by assumption $B = F(b)$ for some $b \in A$. Now we ask: is $b \in B$? If so, then by definition of B we have $b \notin f(b) = B$; while if not, then $b \in f(b) = B$. So either assumption leads to a contradiction. Hence no such bijection can exist. $\qquad\square$

So the cardinal numbers of sets go on for ever; there is no largest set!

Some of Cantor's discoveries are summarised in the following theorem:

Theorem 6.20   (a) *The union and Cartesian product of countable sets are countable.*

(b) *For any natural number n, $|\mathbb{N}^n| = |\mathbb{N}|$.*

(c) *A subset of a countable set is finite or countable.*

(d) *$|\mathbb{Q}| = |\mathbb{Z}| = |\mathbb{N}|$, so $\mathbb{Z}$ and $\mathbb{Q}$ are countable.*

(e) $|\mathbb{R}| = |\mathcal{P}(\mathbb{N})| > |\mathbb{N}|$, *so* $\mathbb{R}$ *is uncountable.*
(f) $|\mathbb{C}| = |\mathbb{R}|$.

In general, there may be sets A and B for which neither $|A| \le |B|$ nor $|B| \le |A|$ holds; that is, there may be no injective map in either direction between A and B. A set-theoretical principle called the **Axiom of Choice** ensures that any two sets are comparable. We discuss this principle further in the next section. It is independent of the other axioms for set theory; that is, it can be neither proved nor disproved from them.

Cantor posed the famous **Continuum Hypothesis**, according to which there does not exist a set A satisfying $|\mathbb{N}| < |A| < |\mathbb{R}|$; in other words, $|\mathbb{R}|$ is the next cardinal number after $|\mathbb{N}|$. This also turns out to be independent of the other axioms (even including the Axiom of Choice).

6.10   The Axiom of Choice.   This section contains a self-contained account of the Axiom of Choice and its best-known application in algebra. We will use this in the next chapter.

A **family of sets** is a collection $(X_i : i \in I)$ of sets, where I is an index set. Formally, we can regard it as a function F whose domain is the index set I, with $F(i) = X_i$ for all $i \in I$. A **choice function** for the family is a function f whose domain is the index set I, satisfying $f(i) \in X_i$ for all $i \in I$. Informally, f chooses a member of each set X_i.

Of course, for a choice function to exist, it is necessary that each set X_i should be non-empty. The **Axiom of Choice** asserts that this condition is also sufficient:

Any family of non-empty sets has a choice function.

The Axiom of Choice cannot be proved or disproved from the other axioms in a standard list of axioms for set theory, such as those of Zermelo and Fraenkel. Note, however, that we do not need to invoke it to choose an element from a single non-empty set, or even to choose elements from finitely many non-empty sets; the other axioms justify doing this.

Bertrand Russell's explanation shows what is going on here. Suppose that you have a wardrobe containing infinitely many pairs of shoes. Can you choose one shoe from each pair? Yes, just choose the left shoes. But if, instead, you have infinitely many pairs of socks, then can you choose one from each pair? The Axiom of Choice asserts that such a selection exists, even if (as here) there is no rule for doing it.

It is this non-constructive nature of the Axiom of Choice which makes its use somewhat controversial. Most mathematicians accept it, but perhaps more because they cannot do without its remarkable consequences than because of any philosophical reason.

We abbreviate the Axiom of Choice to AC.

We now discuss two equivalent principles. First, some definitions:

Definitions A **partial order** on a set A is a reflexive, antisymmetric, and transitive relation on A, usually written as $\leq$. In other words,

- if $a \leq b$ and $b \leq c$ then $a \leq c$;
- $a \leq b$ and $b \leq a$ if and only if $a = b$.

We write $a < b$ to mean $a \leq b$ and $a \neq b$; and $b > a$ means the same as $a < b$. The relation is a **total order** if in addition the **trichotomy law** holds:

- for any a and b, exactly one of $a < b$, $a = b$, $a > b$ holds.

A **chain** in a partially ordered set is a subset which is totally ordered by the relation. An **upper bound** for a subset B is an element a satisfying $b \leq a$ for all $b \in B$. A **maximal element** of A is an element $a \in A$ satisfying $a \not< b$ for all $b \in A$. The terms **lower bound** and **minimal element** are defined similarly. A maximal or minimal element of a chain is usually called a **greatest element** or **least element** respectively.

A **well-order** of A is a total order with the property that every non-empty subset has a least element. (Apologies for the ugly grammar, which is a back-formation from 'well-ordered set'.)

Theorem 6.21 *The following statements are equivalent:*

(Axiom of Choice, AC) Any family of non-empty sets has a choice function.
(Zorn's Lemma, ZL) If a partial order has the property that every chain has an upper bound, then there exists a maximal element.
(Well-ordering Principle, WO) Every set has a well-order.

Proof (AC) implies (ZL): The idea of the proof is that, in order to find a maximal element in a partial order, we start anywhere, and move upwards until we come to one. This obviously works in a finite set; for an infinite set, more care is required.

Assume AC, and let $(A, \leq)$ be a partially ordered set in which every chain has an upper bound. Consider the family of upper bounds of chains, and let f be a choice function for it; that is, for any chain C, $f(C)$ is an upper bound for C. Assume, for a contradiction, that A has no maximal element. Then, for any $a \in A$, the set of elements greater than a is non-empty; again by AC, we may let $g(a)$ be an element greater than a.

Construct a chain B as follows. Start by including any element $a \in A$ in B. At any stage, if b is the greatest element of B so far, add $g(b)$ to B; if B has no greatest element, add $f(B)$ to B. Each move retains the property that B is a chain.

Now, by assumption, the chain B resulting from this construction has an upper bound. But this is a contradiction, since if B has an upper bound b, we can add either $g(b)$ (if $b \in B$) or $f(B)$ to it, so we had not finished.

(ZL) implies (WO): Take a set A which we wish to well-order. Let X be the set X of ordered pairs $B, \leq_B$), for which B is a subset of A and $\leq_B$ is a well-order on B. Now define a relation $\preceq$ on X by the rule that $(B, \leq_B) \preceq (C, \leq_C)$ if $B \subseteq C$ and $\leq_B$ is the restriction of $\leq_C$ to B. Check that $\preceq$ is a partial order on X. Check also that every chain in $(X, \prec)$ has an upper bound (take the union of all the sets and orderings in the chain). Assuming (ZL), X has a greatest element, say $(B, \leq_B)$. Now we must have $B = A$; for if not, choose an element $a \notin B$ and define it to be greater than all elements of B to obtain a larger member of X. So $\leq_A$ is a well-order on A.

(WO) implies (AC): We are given a family $(X_i : i \in I)$ of non-empty sets. By (WO), there is a well-order of their union $\bigcup_{i \in I} X_i$. Now define $f(i)$ to be the least element of X_i with respect to this well-order; then f is a choice function. $\qquad \square$

Remark Convince yourself of the non-constructive nature of AC by trying to construct a well-order of $\mathbb{R}$.

Here is a typical (and very important) application of AC to algebra.

Theorem 6.22 (Krull's Theorem) *Assume AC. Then every ring with identity has a maximal ideal.*

Proof Let R be a ring with identity. Let A be the set of proper ideals of R. The relation $\leq$ on A defined by $I \leq J$ if $I \subseteq J$ is a partial order.

Suppose that B is a chain in A, and let K be the union of the ideals in B. We claim that $K \in A$; that is, K is a proper ideal of R.

- If $a, b \in K$, then $a \in I$ and $b \in J$ for some $I, J \in B$. Since B is a chain, then $I \subseteq J$ or $J \subseteq I$ holds; suppose the former. Then $a, b \in J$, so $a - b \in J$. Hence $a - b \in K$.
- If $a \in K$ and $r \in R$, then $a \in I$ for some $i \in B$. Then $ar, ra \in I$, so $ar, ra \in K$. Thus K passes the Ideal Test.
- To show that K is a proper ideal, suppose for a contradiction that $K = R$. Then $1 \in K$, so $1 \in I$ for some $I \in B$. But this is impossible, since members of B are proper ideals by definition. (This is the only point where we use the fact that R has an identity.)

By Zorn's Lemma, A has a maximal element; that is, R has a maximal (proper) ideal. $\qquad \square$

Remark Wilfrid Hodges has shown that the converse holds; the conclusion of Krull's Theorem is equivalent to the Axiom of Choice.

The Well-ordering Principle gives us a new proof technique: **transfinite induction**. Let $\leq$ be a well-ordering of a non-empty set A. Then A has a least element, which we will call 0. Moreover, if a is an element of A which is not the greatest element, then the set of elements greater than A has a least element $s(a)$, called the **successor** of a.

Now suppose that B is a subset of A. Assume that

- $0 \in B$;
- if $a \in B$, then $s(a) \in B$;
- if $b \neq 0$, b is not a successor, and every element smaller than b is in B, then $b \in B$.

Then we can conclude that $B = A$. For if not, let c be the smallest element of $A \setminus B$. Then c cannot be 0; it cannot be a successor; and then the final clause shows that it cannot exist.

Here is an application, promised in the preceding subsection.

Proposition 6.23 *Assume AC. Then for any two sets A and B, either there is an injective map from A to B, or there is an injective map from B to A. In other words, either $|A| \leq |B|$, or $|B| \leq |A|$.*

Proof We may assume that both A and B are well-ordered. We attempt to define a map $f : A \to B$ as follows:

- $f(0_A) = 0_B$;
- if $f(a) = b$, then $f(s(a)) = s(b)$;
- if $a \neq 0$ and a is not a successor, then $f(a)$ is the least element of B not of the form $f(a')$ for any $a' < a$.

If we succeed in defining f, then it is an injective map from A to B. If we fail, it is because at a certain point we have used up all the elements of B; then we have an injective map from a subset of A onto B, whose inverse is an injective map from B to A. $\qquad \square$

Exercise 6.9 Prove Theorem 6.20.

Exercise 6.10 Assuming AC, show that any infinite set contains a countable subset.
   [*Hint*: Take a choice function f for the family of non-empty subsets of A. Now define by induction $a_0 = f(A)$ and

$$a_n = f(A \setminus \{a_0, \ldots, a_{n-1}\})$$

for all positive integers n.]

Exercise 6.11 Assume AC. Show that, if R is a ring with identity, and I is a proper ideal of R, then I is contained in a maximal ideal of R.

Exercise 6.12 Assume AC. Show that any Boolean ring R is isomorphic to a subring of the ring of subsets of some set X.
   [*Hint*: Take X to be the set of all maximal ideals of R.]

Exercise 6.13 A subset B of an arbitrary vector space V is a **basis** if every finite subset of B is linearly independent, and every vector of V is in the span of some finite subset of B. Assuming AC, show that every vector space has a basis.

Exercise 6.14 Assume AC. Prove that there exists a discontinuous function $f : \mathbb{R} \to \mathbb{R}$ satisfying

$$f(x + y) = f(x) + f(y)$$

for all $x, y \in \mathbb{R}$. *Hint*: Show first that every continuous solution to the displayed equation has the form $f(x) = cx$ for some real number c. Now regard $\mathbb{R}$ as a vector space over the field $\mathbb{Q}$. Show that the displayed equation is equivalent to the linearity of f, and use a basis for the vector space to construct a discontinuous solution.

7 Further topics

In this chapter, we delve a little further into groups, rings, and fields, and examine some other algebraic systems which have been studied.

Further group theory

The emphasis in this section is on finite groups. We prove Sylow's theorems on the existence of subgroups of prime power order, and the Jordan–Hölder Theorem, which reduces the problem of describing all finite groups to describing the finite simple groups and fitting them together. There is also some discussion of these two sub-problems.

7.1   Permutation groups and group actions.   As we already know, a permutation group is a group whose elements are permutations; that is, a subgroup of a symmetric group. It is more in keeping with the spirit of abstract algebra that we should not tie down the elements of a group to being permutations. Accordingly, we define an **action** of a group G on a set Ω. This will associate to every group element a permutation, so that the permutations arising will form a permutation group. But we do not require that the correspondence between group elements and permutations is one-to-one. The formal definition is as follows:

An **action** of a group G on a set Ω is a function $\mu : \Omega \times G \to \Omega$ with the following two properties:

(GA1) $\mu(\mu(x,g),h) = \mu(x,gh)$ for all $x \in \Omega$, $g, h \in G$.
(GA2) $\mu(x,1) = x$ for all $x \in \Omega$, where 1 is the identity of G.

These axioms are obviously related to the closure and identity laws for the group G. You might have expected an axiom corresponding to the inverse law,

(GA3) $\mu(\mu(x,g),g^{-1}) = \mu(\mu(x,g^{-1}),g) = x$ for all $x \in \Omega$, $g \in G$;

but in fact this follows from (GA1) and (GA2) (Exercise 7.1).

You should think of $\mu(x,g)$ as the image of x under the permutation of Ω corresponding to g. The next result guarantees that it does indeed work like that.

Proposition 7.1   (a) *For any $g \in G$, the map $\pi_g : \Omega \to \Omega$ defined by $x\pi_g = \mu(x,g)$ is a permutation.*

(b) *The map $\theta : G \to \mathrm{Sym}(\Omega)$ defined by $g\theta = \pi_g$ is a homomorphism.*

(c) *Conversely, given any homomorphism $\theta : G \to \mathrm{Sym}(\Omega)$, there is an action μ of G on Ω given by $\mu(x,g) = x(g\theta)$.*

Proof (a) The 'derived axiom' (GA3) says that the functions π_g and $\pi_{g^{-1}}$ are inverses of one another. A function that has an inverse is a permutation.

(b) This says that $\pi_{gh} = \pi_g \pi_h$, which is the content of (GA1).

(c) Straightforward checking shows this. □

Examples We will use three examples in which the action is derived from the abstract group structure. In each case the axioms (GA1) and (GA2) are easily checked.

(a) Let H be a subgroup of G. Let Ω be the set of all right cosets of H in G. Define an action by $\mu(Ha, g) = H(ag)$. This is the action of **right multiplication**.

(b) Define an action of G on itself (that is, $\Omega = G$) by the rule $\mu(x, g) = g^{-1}xg$. This is the action of **conjugation**.

(c) Let Ω be the set of all subgroups of G. Then G acts on Ω by **conjugation**: $\mu(H, g) = g^{-1}Hg$.

Now we develop a little of the theory of group actions.

Let μ be an action of G on Ω. Define a relation $\sim$ on Ω by the rule that $x \sim y$ if there exists $g \in G$ with $\mu(x, g) = y$. The reflexive, symmetric, and transitive laws for $\sim$ follow almost immediately from the axioms (GA2), (GA3), and (GA1) for an action (in other words, from the properties of identity, inverses, and closure). So $\sim$ is an equivalence relation. Its equivalence classes are called **orbits**. So x and y lie in the same orbit if the permutation corresponding to some element of G carries x to y. The set Ω decomposes into a disjoint union of orbits.

We say that the action is **transitive** if there is just one orbit, and **intransitive** otherwise. In our examples, the action of G by right multiplication on the right coset space is transitive, whereas (if $G \neq \{1\}$) the action of G on itself by conjugation is not; the orbits for the latter action are the **conjugacy classes** of G.

The **stabiliser** of an element $x \in \Omega$ is the set

$$\{g \in G : \mu(x, g) = x\}$$

of elements of G for which the corresponding permutation fixes x. It is denoted G_x.

Theorem 7.2 (Orbit–Stabiliser Theorem) *Given an action of G on Ω, and $x \in \Omega$:*

(a) the stabiliser G_x is a subgroup of G;

(b) there is a bijection between the orbit of x and the set of right cosets of G_x in G.

Proof (a) Apply the subgroup test: the composition or inverse of permutations fixing x fixes x. (Argue formally with the action μ if you prefer.)

(b) We show that, for any y belonging to the orbit of x, the set

$$X(x, y) = \{g \in G : \mu(x, g) = y\}$$

is a right coset of $H = G_x$, and every right coset arises in this way. First, since y lies in the orbit of x, there is an element of g such that $\mu(x, g) = y$. Then every element of the right coset Hg maps x to y. Conversely, if g' maps x to y, then $g'g^{-1}$ fixes x, so lies in $G_x = H$; then $Hg' = Hg$. So the set $X(x, y)$ is a right coset of G_x. Conversely, every right coset $G_x g$ is contained in (and hence equal to) $X(x, \mu(x, g))$. $\qquad \square$

Remark If G is finite, the size of the orbit of x is equal to $|G : G_x| = |G|/|G_x|$.

In our examples,

(a) In the action of G on the right cosets of H by right multiplication, the stabiliser of the coset H is the subgroup H. Show that the stabiliser of the coset Hx is $x^{-1}Hx$. The proposition is clear in this case.

(b) In the action of G by conjugation, the stabiliser of x is its **centraliser** $C_G(x)$, and we recover the formula $|G : C_G(x)|$ for the size of the conjugacy class.

(c) In the action of G by conjugation on its subgroups, the stabiliser of a subgroup H is its **normaliser**

$$N_G(H) = \{g \in G : g^{-1}Hg = H\}.$$

This subgroup contains H, and indeed it is the largest subgroup of G in which H is contained as a normal subgroup. If H is a normal subgroup of G, then $N_G(H) = G$.

Remark It can be shown that, with a suitably defined notion of isomorphism of actions,

(a) every transitive action is isomorphic to the action by right multiplication on the right cosets of a subgroup;

(b) the actions on the right cosets of two subgroups H and K are isomorphic if and only if H and K are conjugate.

This gives us a complete classification of the transitive actions, and hence of arbitrary actions, of a given group G; we just have to classify the subgroups up to conjugacy.

Using the Orbit–Stabiliser Theorem, we can give a formula for the number of orbits of a finite group acting on a finite set. If G acts on Ω, let fix(g) denote the number of elements $\omega \in \Omega$ which satisfy $\mu(\omega, g) = \omega$ (that is, the number of points fixed by G).

Theorem 7.3 (Orbit–Counting Lemma) *The number of orbits of G on Ω is given by the formula*

$$\frac{1}{|G|} \sum_{g \in G} \mathrm{fix}(g).$$

Proof We count in two ways the number of pairs (ω, g) for which $\omega \in \Omega$, $g \in G$, and g fixes ω.

On one hand, g fixes $\mathrm{fix}(g)$ points, so the number of pairs is

$$\sum_{g \in G} \mathrm{fix}(g).$$

On the other hand, take $\omega \in \Omega$. The number of group elements fixing ω is the order $|G_\omega|$ of its stabiliser. Let O be the orbit containing Ω. Now every point of O has stabiliser of the same order as ω, namely $|G_\omega|$; so the number of pairs with $\omega \in O$ is $|O| \cdot |G_\omega| = |G|$, by the Orbit–Stabiliser Theorem. That is, every orbit contributes $|G|$ pairs to the sum; so the total number of pairs is $|G|$ multiplied by the number of orbits.

Equating the two values and dividing by $|G|$ gives the result. $\square$

Example How many ways are there of painting the faces of a cube with three colours (say red, green, and blue), if two colourings differing by a rotation are identified?

A colouring is a function from the six faces of the cube to the set $\{\text{red}, \text{green}, \text{blue}\}$ of colours; so there are 3^6 colourings. Let Ω be the set of these colourings. We are asked to count the number of orbits on Ω of the group G of rotations of the cube.

The group G has order 24 and consists of the following elements:

(a) The identity;
(b) three rotations through $180°$ about axes through face centres;
(c) six rotations through $\pm 90°$ about axes through face centres;
(d) eight rotations through $\pm 120°$ about axes through vertices;
(e) six rotations through $180°$ about axes through midpoints of edges.

Type (a) fixes all 3^6 colourings. For any other type, a colouring is fixed if and only if faces in the same cycle get the same colour, so the number of fixed colourings will be 3^c, where c is the number of cycles of the rotation acting on the faces of the cube. These numbers are 4, 3, 2, and 3 in cases (b)–(e) respectively.

So the number of orbits is

$$\frac{1}{24} \left(3^6 + 3 \cdot 3^4 + 6 \cdot 3^3 + 8 \cdot 3^2 + 6 \cdot 3^3 \right) = 57.$$

We deduce from the Orbit-Counting Lemma a simple but useful result of Jordan.

Corollary 7.4 (Jordan's Theorem)   *(a) Let G act transitively on the finite set Ω, where $|\Omega| > 1$. Then there is an element of G which fixes no point of Ω.*

(b) Let H be a proper subgroup of a finite group G. Then

$$\bigcup_{g \in G} g^{-1}Hg \neq G.$$

Proof   (a) By the Orbit-Counting Lemma, the average number of fixed points of elements of G is equal to 1 (the number of orbits). The identity fixes more than one point; so some element must fix less than one point.

(b) Let G act on the set of right cosets of H by right multiplication. The action is transitive. The point stabilisers are the conjugates $g^{-1}Hg$. So the element guaranteed by (a) lies in none of these conjugates. $\square$

7.2   Sylow's Theorem.   We next prove what is arguably the most important theorem about finite groups, **Sylow's Theorem**. It is motivated by the question: Does Lagrange's Theorem have a converse? Lagrange's Theorem asserts that the order of any subgroup of G divides the order of G. But not every divisor occurs. The alternating group A_4 has order 12 but has no subgroup of order 6 (Exercise 3.36).

Theorem 7.5 (Sylow's Theorem)   *Let G be a group of order $n = p^a m$, where p is prime and p does not divide m.*

(a) G contains a subgroup of order p^a.
(b) The number of subgroups of order p^a is congruent to 1 mod p, and all these subgroups are conjugate.
(c) Any subgroup of G of order a power of p is contained in a subgroup of order p^a.

Proof   The proof involves the ideas of group actions.

For (a), let Ω be the set of all subsets of G of cardinality p^a. This is a very large set, of cardinality $\binom{p^a m}{p^a}$. We define an action μ of G on Ω by 'right multiplication':

$$\mu(X, g) = Xg = \{xg : x \in X\}.$$

Now Ω splits into orbits for this action. We note that the sets making up any orbit must cover the whole of G: for, if $x \in X$, then $g \in \mu(X, x^{-1}g)$. So the size of the orbit is at least $n/p^a = m$, with equality if and only if the stabiliser is a subgroup of order p^a. Conversely, if X is a subgroup of order p^a, then the orbit of X consists of the right cosets of X, of which there are just m. If the size of an orbit is larger than m, then (as it divides $p^a m$) it must be a multiple of p.

So there are two kinds of orbits:

(a) orbits of size m, whose stabilisers have order p^a;
(b) orbits of size divisible by p.

If we can show that the size of Ω is not divisible by p, then we can conclude that orbits of the first type exist, and hence that G has subgroups of order p^a.

Now the size of Ω is $\binom{p^a m}{p^a}$. It is possible to show, using some number theory, that this number is not divisible by p. But this can be done with a trick. The size of Ω is completely independent of which group of order n we have chosen. So consider the cyclic group of order n. We know that it has a unique subgroup of order p^a, hence has just one orbit of type (a). It follows that

$$\binom{p^a m}{p^a} \equiv_p m,$$

so this number is not divisible by p. Thus part (a) of the Theorem is proved.

To prove parts (b) and (c), we take a different action. Let Ω be the set of all subgroups of G of order p^a (we know by (a) that this set is non-empty), and let G act on Ω by conjugation: $\mu(X, g) = g^{-1}Xg$.

Let Q be any non-trivial subgroup of G whose order is a power of p. We consider the action of Q on Ω obtained by restricting the action of G.

Suppose first that $|Q| = p^a$, so that Q is one of the members of Ω. Clearly Q fixes itself, and so lies in an orbit of size 1. We claim that Q fixes no other member of Ω. If Q fixes a different subgroup X in this action, then by Exercise 7.3, we have that QX is a subgroup, and

$$|QX| = |Q| \cdot |X|/|Q \cap X| = p^a \cdot p^a/p^b = p^{2a-b},$$

where $|Q \cap X| = p^b$. But since the subgroups Q and X are different, their intersection is a proper subgroup of each, and so $b < a$, whence $2a - b > a$. But this is impossible, since no higher power of p than p^a divides $|G| = p^a m$.

So all the other orbits of Q have sizes which are greater than 1 but divide $|Q| = p^a$, and hence are divisible by p. It follows that $|\Omega| \equiv_p 1$.

What about the orbits of G? These are obtained by glueing together orbits of Q in some way. So the orbit containing Q has size congruent to 1 mod p, and all the others have size congruent to 0 mod p. Could there be more than one orbit? If P lies in a different orbit to Q, then the same argument would show that the orbit of P has size congruent to 1 mod p and the others 0 mod p, which is clearly impossible. So there is only one orbit for the action of G by conjugation; in other words, all subgroups of order p^a are conjugate. Thus, (b) is proved.

Finally, consider (c). Let Q have order p^b, where $0 < b \leq a$. Every orbit of Q on Ω has size dividing $|Q|$, and hence a power of p. Since $|\Omega| \equiv_p 1$, at least one orbit must have size 1. If this orbit is $\{P\}$, then as above we find that $P \cap Q = Q$, whence $Q \subseteq P$. $\qquad\square$

7.3 p-groups. If a group has prime power order, it has a number of special properties not shared by arbitrary groups. These will be described later in this section. I have chosen to give a composite result, which proves the basic property of such groups at the same time as proving the first part of Sylow's Theorem. First, we prove **Cauchy's Theorem**.

Theorem 7.6 (Cauchy's Theorem) *If a prime number p divides the order of a group G, then G contains an element of order p.*

Proof Let

$$\Omega = \{(g_1, \ldots, g_p) \in G^p : g_1 \cdots g_p = 1\}.$$

We define an action of the cyclic group of order p on Ω by

$$(g_1, \ldots, g_p)\pi = (g_p, g_1, \ldots, g_{p-1})$$

where π generates C_p (that is, π shifts the coordinates cyclically). We have to check that π maps Ω to itself. If $(g_1, \ldots, g_p) \in \Omega$, then $g_p = (g_1 \cdots g_{p-1})^{-1}$, so $g_p g_1 \cdots g_{p-1} = 1$ as required.

Now C_p has orbits of size 1 and p on Ω. An orbit of size 1 contains an element $(g, g, \ldots, g)$ of Ω, where $g^p = 1$; any other element of Ω lies in an orbit of size p. Since $|\Omega| = |G|^{p-1}$ is divisible by p, the number of orbits of size 1 (and hence the number of solutions of $g^p = 1$) is also a multiple of p. One of these solutions is $g = 1$, so there must be at least $p - 1$ more; these are elements of order p. $\square$

The principle used here states:

> A group of p-power order, acting on a set of size divisible by p, has the property that the number of fixed points is divisible by p; hence, if there is at least one fixed point, then there are at least p.

We use this in the next proof.

Theorem 7.7 *Let G be a group of order $p^a m$, where p is a prime not dividing m. Then, for $0 \le i \le a$,*

A_i: *G contains a subgroup of order p^i;*
B_i: *if $i < m$, then any subgroup of order p^i is contained normally in a subgroup of order p^{i+1}.*

Proof The argument is an induction: we show that

$$A_0 \Rightarrow B_0 \Rightarrow A_1 \Rightarrow \cdots \Rightarrow B_{a-1} \Rightarrow A_a$$

(the last statement B_a is vacuous). In other words, we have to start the induction by showing A_0, and the inductive step has two parts: $A_i \Rightarrow B_i$ and $B_i \Rightarrow A_{i+1}$.

Statement A_0 is easy: the identity subgroup will do.

Suppose that A_i is true, with $i < a$. Let P be a subgroup of order p^i, and consider the action of P by right multiplication on its own right cosets. The number of cosets is $p^{a-i}m$, a multiple of p; since P has p-power order and fixes itself, it must fix another coset, say Px. Now the statement $Pxg = Px$ for all $g \in P$ shows that x belongs to the normaliser of P. So $[N_g(P) : P]$ is equal to the number of fixed points of P in this action, hence divisible by p.

By Cauchy's Theorem, $N_G(P)/P$ contains a subgroup $\overline{Q}$ of order p. This subgroup corresponds to a subgroup Q of $N_G(P)$ of order $p|P| = p^{i+1}$ in which P is normal.

Finally, it is trivial that B_i implies A_{i+1} for $i < a$. So our inductive proof is complete. $\square$

Corollary 7.8   (a) *(First part of Sylow's Theorem) A group of order $p^a m$, where p is a prime not dividing m, contains subgroups of order p^a.*
(b) *A group P of prime power order p^a has the property that any proper subgroup is properly contained in its normaliser; so there is a chain*

$$P_0 < P_1 < \cdots < P_a = P$$

of subgroups, where $|P_i| = p^i$ and each is a normal subgroup of the next.

The last part can be strengthened. We use the term p-**group** for a group whose order is a power of a prime p.

Theorem 7.9   (a) *The centre of a non-trivial p-group is non-trivial.*
(b) *If $|P| = p^a$, then P has a chain*

$$P_0 < P_1 < \cdots < P_a = P$$

of subgroups, where $|P_i| = p^i$ and each is a normal subgroup of P. Moreover, $P_{i+1}/P_i \cong C_p$.

Proof   Let P act on itself by conjugation. By our general principle stated earlier, the number of fixed points (which is the order of the centre $Z(P)$) is greater than 1. This proves (a). For (b), we proceed by induction. We take $P_0 = \{1\}$, and P_1 the subgroup generated by an element of order p in $Z(P)$. (Every subgroup of $Z(P)$ is normal in P.) If we have constructed P_i, then we take $\overline{Q}$ to be a normal subgroup of order p in P/P_i, and let P_{i+1} be the corresponding subgroup of P. $\square$

7.4   The Jordan–Hölder Theorem.   In the remainder of this section, we examine the structure of finite groups. It is not possible to give the kind of description we gave for finite fields, where there is a unique field of each prime power order; groups are much more complicated. First, we prove the **Jordan–Hölder Theorem**, according to which any finite group is built from a unique collection of simple groups. (A group G is **simple** if it is not the identity group but its only normal subgroups are the trivial ones, the whole group and the identity.) This breaks the problem of describing groups into two parts: describing the simple groups, and describing how they can be fitted together. The first part has been completed, but the length and complexity of the proof have meant that no self-contained account has yet appeared. The second problem is fairly well understood, but we do not have a complete 'solution' to it.

Let G be a finite group. It is possible to choose a normal subgroup G_1 of G which is **maximal**; that is, which is contained in no normal subgroup of G except for itself and G. By the Second Isomorphism Theorem, the normal subgroups of G/G_1 correspond to subgroups of G containing G_1; hence there are just two of them, namely G/G_1 and the identity. In other words, G/G_1 is simple. Now we repeat the procedure with G_1. We end up with a sequence

$$G = G_0 \geq G_1 \geq G_2 \geq \ldots \geq G_r = \{1\},$$

with the properties that, for $i = 1, 2, \ldots, r$, we have $G_i \trianglelefteq G_{i-1}$ and G_{i-1}/G_i is simple. The series displayed is called a **composition series** for G, and the simple groups G_{i-1}/G_i are called the **composition factors** of G.

Note that we obtain a list of composition factors (so that the same simple group may occur more than once), and also that the list of composition factors is associated with a particular composition series. Note also that the product of the orders of the composition factors of G is equal to the order of G. Furthermore, given any descending series of subgroups, each normal in the preceding one, we can refine the series by adding more terms to obtain a composition series, which is just such a series in which no more terms can be inserted (since each term is a maximal normal subgroup of its predecessor).

For example, S_4 has a composition series

$$S_4 \geq A_4 \geq V_4 = C_2 \times C_2 \geq C_2 \geq \{1\},$$

with composition factors $S_4/A_4 \cong C_2$, $A_4/V_4 \cong C_3$, $V_4/C_2 \cong C_2$, and $C_2/\{1\} \cong C_2$; in other words, C_2 three times and C_3 once. We have $|S_4| = 4! = 2^3 3$.

The **Jordan–Hölder Theorem** asserts that, no matter what composition series we choose for G, we will obtain the same composition factors (each repeated the same number of times in both series).

Theorem 7.10 (Jordan–Hölder Theorem) *Let*

$$G = G_0 \geq G_1 \geq G_2 \geq \ldots \geq G_r = \{1\}$$

and

$$G = H_0 \geq H_1 \geq H_2 \geq \ldots \geq H_s = \{1\}$$

be two composition series for the finite group G. Then the lists of composition factors obtained from the two series are the same. In particular, the series have the same length (that is, $r = s$).

Proof Our proof is by induction on the order of G. The induction begins with the trivial group, which has the empty list of composition factors. So we assume, inductively, that the theorem holds for all groups smaller than G.

If $G_1 = H_1$, then deleting G from the given series gives two composition series for G_1. By the induction hypothesis, they have the same lists of composition factors (and the same length): adding G/G_1 gives the list of composition factors of G, and we are done. So we may suppose that $G_1 \neq H_1$.

Let $N = G_1 \cap H_1$. Then N is the intersection of two normal subgroups of G, and so is a normal subgroup. Also, $G_1 H_1$ is a normal subgroup containing G_1, so $G_1 H_1 = G_1$ or $G_1 H_1 = G$ (by the maximality of G_1 as normal subgroup). The first alternative is impossible, since it implies $H_1 \leq G_1$, whence $H_1 = G_1$ (since H_1 is also maximal normal), contrary to assumption. So $G_1 H_1 = G$.

From this it follows that

$$G/G_1 = G_1 H_1/G_1 \cong H_1/(G_1 \cap H_1) = H_1/N,$$

and similarly $G/H_1 \cong G_1/N$.

Now let $N = N_0 \geq N_1 \geq \ldots \geq N_t = \{1\}$ be a composition series for N. Let $\mathcal{L}$ be the list of composition factors derived from this series. Adding G_1 at the start of the series gives a composition series for G_1. By the inductive hypothesis, $\mathcal{L} \cup \{G_1/N\}$ is the list of composition factors for G_1, and is the same as obtained from the composition series $G_1 \geq G_2 \geq \ldots \geq \{1\}$. Hence the list of composition factors of G obtained from the first composition series is $\mathcal{L} \cup \{G_1/N, G/G_1\}$.

In the same way, the list obtained from the second series is $\mathcal{L} \cup \{H_1/N, G/H_1\}$.

But we have already showed that $G/G_1 \cong H_1/N$ and $G/H_1 \cong G_1/N$. So the two lists are the same. $\qquad\square$

7.5 Soluble groups. The Jordan–Hölder Theorem suggests that finite simple groups are the 'building blocks' of finite groups; any finite group is built from a unique collection of simple groups (its composition factors). There are two fundamentally different kinds of simple groups, and groups built entirely from the first type have very different properties from those with some factors of the second type.

The first type of simple group consists of the cyclic groups of prime order. By Lagrange's Theorem, these groups have no non-trivial subgroups at all, and so certainly they are simple. A simple abelian group is necessarily cyclic of prime order.

The second type consists of the non-abelian simple groups, which we will discuss further in the next section.

A finite group G is **soluble** if all its composition factors are cyclic of prime order. The strange name for this class of groups will remain mysterious until we discuss the work of Galois in Chapter 8, connecting these groups with polynomial equations which are 'soluble by radicals'. In this section, we provide the groundwork for that theory, by giving some alternative characterisations of soluble groups.

First, some more definitions. If x and y are elements of the group G, their **commutator**, written $[x,y]$, is the element $x^{-1}y^{-1}xy$. (Note that x and y commute—that is, $xy = yx$—if and only if $[x,y] = 1$.) The **commutator subgroup**, or **derived group**, of G, is the subgroup generated by all commutators in G; it is written G' or $[G,G]$. Finally, the **derived series** of G is the series

$$G^{(0)} \geq G^{(1)} \geq G^{(2)} \geq \ldots$$

defined by $G^{(0)} = G$, $G^{(i+1)} = [G^{(i)}, G^{(i)}]$ for $i \geq 0$.

Lemma 7.11 *For any group G, the subgroup $[G,G]$ is normal in G and $G/[G,G]$ is abelian. Moreover, if $N \lhd G$ and G/N is abelian, then $N \geq [G,G]$.*

Proof Calculation shows that

$$g^{-1}[x,y]g = [g^{-1}xg, g^{-1}yg].$$

Hence conjugation by g merely permutes the commutators, and fixes the subgroup they generate. Hence $[G,G]$ is normal in G. Let $H = [G,G]$. Then, for any $x,y \in G$, we have

$$[xH, yH] = [x,y]H = H,$$

since $[x,y] \in H$; and thus G/H is abelian. This argument shows further that G/N is abelian if N contains H. Conversely, suppose that G/N is abelian. Then, for all x,y,

$$N = [xN, yN] = [x,y]N,$$

so $[x,y] \in N$. Since this holds for all x and y, we have $[G,G] \leq N$. □

Theorem 7.12 *The following conditions for the finite group G are equivalent:*

(a) there is a series

$$G = G_0 \geq G_1 \geq G_2 \geq \ldots \geq G_r = \{1\}$$

of subgroups of G with $G_i \unlhd G_{i-1}$ and G_{i-1}/G_i cyclic of prime order for $i = 1, \ldots, r$;
(b) there is a series

$$G = H_0 \geq H_1 \geq H_2 \geq \ldots \geq H_s = \{1\}$$

of subgroups of G with $H_i \unlhd H_{i-1}$ and H_{i-1}/H_i abelian for $i = 1, \ldots, s$;
(c) $G^{(d)} = \{1\}$ for some d.

Proof   Clearly (a) implies (b). Moreover, if (b) holds, then we can refine the given series to a composition series; the composition factors are abelian simple groups, and hence are cyclic of prime order, so (a) holds.

Suppose that (b) holds. Since H_0/H_1 is abelian, the lemma implies that $G^{(1)} = [G, G] \leq H_1$. This is the first stage of a proof by induction that $G^{(i)} \leq H_i$ for all i. Suppose that this holds for $i = k$. Then all commutators $[g, h]$, for $g, h \in G^{(k)}$, lie in H_k'; so $G^{(k+1)} \leq H_k'$. Also, $H_k' \leq H_{k+1}$, since H_k/H_{k+1} is abelian. So the equation $G^{(i)} \leq H_i$ holds also for $i = k + 1$. By induction, it holds for all i; and so $G^{(s)} \leq H_s = \{1\}$.

Conversely, if $G^{(d)} = 1$, then the normal series

$$G = G^{(0)} \geq G^{(1)} \geq \ldots \geq G^{(d)} = \{1\}$$

has abelian factors, so (b) holds.   □

Corollary 7.13   *A group of prime power order is soluble.*

Proof   Corollary 7.9 and Theorem 7.12.   □

Condition (a) is our definition of solubility. So the theorem gives three equivalent characterisations of finite soluble groups.

For infinite groups, conditions (b) and (c) are equivalent (check that the proof given above is valid), but not equivalent to (a) (which only holds for finite groups!). So, in general, we take (b) or (c) as the definition of solubility.

7.6   Simple groups.   If all finite simple groups were cyclic of prime order, then all finite groups would be soluble. Unfortunately, this is not so. The alternating group A_5 is simple and non-abelian.

As mentioned earlier, the finite simple groups have been determined. This theorem is probably the most complex ever proved, and is well beyond the scope of this text. Even the description of the groups in the classification is more than I can attempt here, except in broad outline.

According to the classification, the finite non-abelian simple groups are of three types. First, there are the alternating groups A_n for $n \geq 5$. (We will prove their simplicity below.)

Then there are the so-called 'groups of Lie type', which are defined as certain groups of matrices over finite fields. The easiest type to describe are the **projective special linear groups**. The **special linear group** $\mathrm{SL}(n, q)$ consists of all $n \times n$ matrices over the finite field $\mathrm{GF}(q)$ which have determinant 1; the group operation is matrix multiplication. The subgroup Z of this group consisting of all scalar matrices with determinant 1 (that is, all cI_n, where $c \in \mathrm{GF}(q)$ and $c^n = 1$), is normal; we set $\mathrm{PSL}(n, q) = \mathrm{SL}(n, q)/Z$. Now it can be shown that $\mathrm{PSL}(n, q)$ is simple for all $n \geq 2$ and all prime powers q, with the exception of $\mathrm{PSL}(2, 2)$ and $\mathrm{PSL}(2, 3)$.

Finally, there are just 26 so-called 'sporadic groups', which have no uniform definition and have to be constructed individually. The smallest has order 7920; the largest, approximately 10^{54}.

Theorem 7.14 *The alternating group A_n is simple for all $n \geq 5$.*

Proof We use a similar method to the one we employed to find the normal subgroups of S_5. We recall from the proof of Proposition 3.30 that the conjugacy classes in S_5 have sizes 1, 24, 15, 20, 10, 30, and 20, where the first four consist of even permutations and make up A_5. Obviously, a conjugacy class in S_5 will be a union of conjugacy classes in A_5: we have to see how they split up.

Lemma 7.15 *Let C be a conjugacy class in S_n, which is contained in A_n. Then either C is a conjugacy class in A_n, or C is the union of two conjugacy classes C' and C'' of the same size in A_n. The first alternative holds if and only if some member of C commutes with an odd permutation.*

Proof Since S_n acts transitively on C by conjugation, there is a bijection between elements of C and cosets of the stabiliser of an element $c \in C$. Now, in this case, the stabiliser of c is

$$\{g \in S_n : g^{-1}cg = c\} = \{g \in S_n : cg = gc\} = C_{S_n}(c),$$

the centraliser of c.

If no odd permutation commutes with c, then $H = C_{S_n}(c) \leq A_n$, and so half of the cosets of H lie in A_n and the other half in $S_n \setminus A_n$. So the orbit of A_n containing c contains just half of the conjugacy class C, and C splits into two classes of equal size.

On the other hand, suppose that H contains an odd permutation. Then $C_{A_n}(c) = C_{S_n}(c) \cap A_n$ is a subgroup of index 2 in H. Now the size of the A_n-conjugacy class is the index of the stabiliser, which is

$$|A_n : C_{A_n}(c)| = |S_n : C_{A_n}(c)|/2 = |S_n : C_{S_n}(c)| = |C|.$$

So C is a conjugacy class in A_n. $\qquad\square$

Now the conjugacy class of size 15 consists of elements with cycle structure $(2, 2, 1)$ (products of two transpositions). Each of these commutes with a transposition, so the class does not split in A_n. The class of size 20 consists of 3-cycles; the 3-cycle (1 2 3) commutes with the transposition (4 5), and again the class does not split. The class of size 24 consists of 5-cycles. It can be shown that a 5-cycle commutes only with its own powers, all of which are even permutations. So this class splits into two classes of size 12 in A_5.

We conclude that the conjugacy classes in A_5 have sizes 1, 15, 20, 12, and 12. No sum of any proper sublist of these, including 1, is a divisor of 60. So A_5 has no normal subgroups except itself and the identity, and therefore it is simple.

Now we show by induction that A_n is simple for $n > 5$. The inductive hypothesis is that A_{n-1} is simple. So let N be a non-trivial normal subgroup of A_n.

Then NA_{n-1} is a subgroup of A_n containing A_{n-1}. If $NA_{n-1} = A_{n-1}$, then $N \leq A_{n-1}$; since N is normal, it is contained in all of the point stabilisers (the conjugates of A_{n-1}), and so $N = 1$, contrary to assumption. Now A_{n-1} is a maximal subgroup of A_n, so it follows that $NA_{n-1} = A_n$.

The subgroup $A_{n-1} \cap N$ is normal in A_{n-1}. By the inductive hypothesis, $A_{n-1} \cap N = A_{n-1}$ or $A_{n-1} \cap N = \{1\}$. In the first case, $N \geq A_{n-1}$, and the equation $NA_{n-1} = A_n$ implies $N = A_n$. In the second case, we have $|N| = n$, and so A_n has a conjugacy class of size at most $n - 1$. By the lemma, S_n has a conjugacy class of size at most $2(n - 1)$, which can be seen to be impossible for $n \geq 6$ (Exercise 7.13). So this case cannot occur, and we are done. $\qquad\square$

7.7 Extensions. The cyclic group C_4 and the Klein group V_4 both have composition factors C_2 (twice). So the composition factors alone are not enough to determine a group. If simple groups are the building bricks of finite groups, then extension theory is the mortar used to stick them together.

The general problem of extension theory is as follows: given groups A and B, describe all possible groups G which have a normal subgroup N such that N is isomorphic to A while G/N is isomorphic to B. (Such groups are called **extensions** of A by B.) It is clear that a complete solution to this problem, together with the list of finite simple groups, would allow us to describe finite groups completely. Such a complete solution does not exist.

In this section, we examine an important special case of the extension problem, when the group A is abelian. We will see that two pieces of information are needed to define an extension of A by B: first, an action of B on A; and then a **factor set**, a certain function from $B \times B$ to A. Unfortunately, different actions and factor sets may define isomorphic groups, so we do not get information as precise as we would like; but we will prove some results based on this approach.

Let A be an abelian normal subgroup of the group G, with $G/A \cong B$. For any $g \in G$, the map $\sigma_g : A \to A$ defined by $a\sigma_g = g^{-1}ag$ is well defined (by the normality of A), and is an automorphism of A. Now the map $\theta : G \to \operatorname{Aut}(A)$ defined by $g\theta = \sigma_g$ is a homomorphism, since it is easily checked that $\sigma_g \sigma_h = \sigma_{gh}$. Now A is abelian, so $\sigma_a = 1$ for $a \in A$. Thus A lies in the kernel of the homomorphism θ. It follows that the value of $g\theta$ is the same for all elements of the coset Ag. So θ induces a homomorphism $\phi : G/A = B \to \operatorname{Aut}(A)$. This homomorphism is the **action** of B on A.

There is a special kind of extension of A by B known as a **semidirect product** or **split extension**. This is an extension G containing a complement to A, a subgroup H which satisfies $AH = G$, $A \cap H = \{1\}$. Now a semidirect product is determined up to isomorphism by the action. For we have $H \cong G/A = B$, and every element of G is uniquely expressible in the form ha for $h \in H$, $a \in A$. We have

$$(h_1 a_1)(h_2 a_2) = (h_1 h_2)(h_2^{-1} a_1 h_2) a_2 = (h_1 h_2)(a_1 (h_2 \phi) a_2).$$

So, if we identify H with B (to which it is isomorphic), and define an operation $\circ$ on $B \times A$ by the rule

$$(b_1, a_1) \circ (b_2, a_2) = (b_1 b_2, a_1 (b_2 \phi) a_2),$$

the result is a group isomorphic to G.

The semidirect product of A by B using the action homomorphism $\phi : B \to \mathrm{Aut}(A)$ is written as $A \rtimes B$, or (if we want to stress the action) $A \rtimes_\phi B$.

However, not every extension is split. For example, C_4 and $C_2 \times C_2$ are both extensions of C_2 by C_2 with trivial action (since both are abelian); the second, but not the first, is a semidirect product. So we must look further. The complement H is a special kind of set of coset representatives for A in G. So we choose an arbitrary set of coset representatives, and describe the extension.

Let S be a set of coset representatives. We will always assume that we have chosen the identity as the representative of the coset A. We also use the notation $\overline{g}$ for the representative of the coset containing g. Note that the set of cosets (and hence S) is in one-to-one correspondence with B. We write the representative of the coset corresponding to $b \in B$ as $s(b)$.

Now any element of G has a unique representation of the form $s(b)a$, for $b \in B$, $a \in A$. Also, $s(b)$, acting by conjugation on A, induces the automorphism $b\phi$. Now

$$(b_1) a_1 s(b_2) a_2 = (s(b_1) s(b_2))(a_1 (b_2 \phi) a_2),$$

and the only difference is that now we do not have $s(b_1) s(b_2) = s(b_1 b_2)$. However, it is true that $s(b_1) s(b_2)$ lies in the coset corresponding to $b_1 b_2$; so we can write

$$s(b_1) s(b_2) = s(b_1 b_2) f(b_1, b_2),$$

where f is a function from $B \times B$ to A (that is, a function of two variables in B taking values in A).

The function f is called a **factor set**.

Now it is clear that, if we know the action ϕ and the factor set f, then the group is uniquely determined; taking its elements to be $B \times A$ as before, the group operation $\circ$ is given by

$$(b_1, a_1) \circ (b_2, a_2) = (b_1 b_2, f(b_1, b_2) a_1 (b_2 \phi) a_2).$$

Note that, if the factor set is trivial (that is, it always takes the value 1), we have the semidirect product.

At this point, for clarity, we change notation. We write the abelian group A additively (so that 0 is the identity, and $-a$ the inverse of a), and we write a^b instead of $a(b\phi)$.

Theorem 7.16 *The function $f : B \times B \to A$ is a factor set if and only if it satisfies*

(a) $f(1,b) = f(b,1) = 0$;
(b) $f(b_1, b_2b_3) + f(b_2, b_3) = f(b_1b_2, b_3) + f(b_1, b_2)^{b_3}$.

Proof (a) follows from our choice of the identity as coset representative for the coset A. (b) is obtained by (carefully) evaluating $((b_1, 1) \circ (b_2, 1)) \circ (b_3, 1)$ and $(b_1, 1) \circ ((b_2, 1) \circ (b_3, 1))$ and equating the results.

Conversely, if (a) and (b) hold, then the operation $\circ$ defined above makes the set $A \times B$ into a group which is an extension of A by B. $\qquad \square$

Our representation of the extension by a factor set depends on the choice of the coset representatives. How does the factor set change if we use different representatives? Suppose that the coset representatives $s(b)$ and $s'(b)$ give factor sets f and f' respectively. We have $s'(b) = s(b)d(b)$, where d is a function from B to A satisfying $d(1) = 0$. Then we have

$$s(b_1)d(b_1)s(b_2)(b_2) = s(b_1b_2)d(b_1b_2)f'(b_1, b_2).$$

After some calculation, and writing the result in additive notation, we obtain

$$f'(b_1, b_2) - f(b_1, b_2) = d(b_1)^{b_2} + d(b_2) - d(b_1b_2).$$

We call the factor sets f and f' **equivalent** if this condition holds. Thus, equivalent factor sets arise from the same group with (possibly) different choices of coset representatives.

It turns out that there is a convenient algebraic description of factor sets. Since they are functions, we can add them pointwise:

$$(f_1 \oplus f_2)(b_1, b_2) = f_1(b_1, b_2) + f_2(b_1, b_2).$$

The sum of factor sets is again a factor set (which can be checked using Theorem 7.16), and indeed the factor sets form a group $\mathcal{F}$ with this operation.

The factor sets equivalent to the zero element (those of the form $f(b_1, b_2) = d(b_1)^{b_2} + d(b_2) - d(b_1b_2)$ for some function d) are called **inner factor sets**. They form a subgroup $\mathcal{I}$ of $\mathcal{F}$. Now we define the **extension group** $E(B, A)$ of A by B (with the given action) to be the group $\mathcal{F}/\mathcal{I}$. Thus, elements of $E(B, A)$ describe extensions, and the zero element describes the split extension. In particular,

Every extension of A by B splits if and only if $E(B, A) = \{0\}$.

From this fact, we can obtain an important theorem on extensions:

Theorem 7.17 (Schur's Theorem) *Suppose that the abelian group A and the group B have coprime orders. Then any extension of A by B splits.*

Proof Let $|A| = m$ and $|B| = n$, with $(m, n) = 1$; suppose that $pm + qn = 1$ for some integers p and q. Let $f \in \mathcal{F}$ be a factor set. Since the values of f lie in a group of order m, Lagrange's Theorem shows that $mf = 0 \in \mathcal{I}$. We show that also $nf \in \mathcal{I}$. It follows that $f = (pm + qn)f \in \mathcal{I}$, so that $\mathcal{I} = \mathcal{F}$ and $E(B, A) = \mathcal{F}/\mathcal{I} = \{0\}$.

Define

$$d(b) = \sum_{x \in B} f(x, b).$$

Now sum the equation

$$f(b_1, b_2 b_3) + f(b_2, b_3) = f(b_1 b_2, b_3) + f(b_1, b_2)^{b_3}$$

(equation (b) of Theorem 7.16) over $b_1 \in B$, using the fact that $b_1 b_2$ runs over B as b_1 does: we obtain

$$d(b_2 b_3) + n f(b_2, b_3) = d(b_3) + d(b_2)^{b_3},$$

so that $n f(b_2, b_3) = d(b_2)^{b_3} + d(b_3) - d(b_2 b_3)$ is an inner factor set, as required. □

Remark The theorem is true without the restriction that A is abelian. If either A or B is soluble, then an inductive argument can be used to reduce the problem to the case handled by Schur. This was done by Zassenhaus, and the result is referred to as the **Schur–Zassenhaus Theorem**. In general, if the orders of A and B are coprime, then at least one of them must be odd, and the celebrated **Feit–Thompson Theorem** asserts that a group of odd order is soluble. But the proof of this theorem is several hundred pages long!

7.8 A glimpse at homological algebra. The arguments used above give a glimpse of an important area of algebra on which we have not yet touched, known as **homological algebra**. The calculations that we made with factor sets are not as *ad hoc* as they appear, but are part of a much larger scheme.

Let R be a ring, and M a right R-module. As part of the programme of studying R via its modules, one can define a sequence of abelian groups $H^n(R, M)$ for $n \geq 0$, called the **cohomology groups** of R with coefficients in M. This is in part inspired by algebraic topology, where (abelian) groups are used as invariants of topological spaces.

We cannot here even give the general definition, much less study the important properties of the cohomology groups. It will suffice to say how they generalise the extension group.

Let G be a group, and R a ring. We define the **group ring** to be the set of all finite sums $\sum r_i g_i$, where $r_i \in R$ and $g_i \in G$. Addition is defined coordinatewise: $\sum r_i g_i + \sum s_i g_i = \sum (r_i + s_i) g_i$. Multiplication is defined by extending the group operation linearly. This multiplication is often called **convolution**: it is given by the rule

$$\left(\sum r_i g_i \right) \cdot \left(\sum s_i g_i \right) = \left(\sum t_i g_i \right),$$

where

$$t_i = \sum_{g_j g_k = g_i} r_j s_k.$$

Now, if A is an abelian group on which G acts, then A becomes a right RG-module, by extending the action of G on A linearly.

Now the following is true:

The extension group $E(B, A)$ is equal to the second cohomology group $H^2(\mathbb{Z}B, A)$.

So the second cohomology group tells us about the splitting of extensions.

The first cohomology group also has a group-theoretic significance. In the split extension G of A by B (with the given action), there are usually several **complements** of A (subgroups H which satisfy $AH = G, A \cap H = \{1\}$). The number of conjugacy classes of such complements is equal to the order of the first cohomology group $H^1(\mathbb{Z}G, A)$. Indeed, there is a natural regular action of $H^1(\mathbb{Z}G, A)$ on the set of conjugacy classes.

For higher cohomology groups, the interpretations in terms of extensions are less transparent.

Exercise 7.1 Show that (GA3) is a consequence of (GA1) and (GA2).

Exercise 7.2 Show that the proof of Cayley's Theorem involves considering the action of G by right multiplication on the set of right cosets of the trivial subgroup $\{1\}$.

Exercise 7.3 Let H and K be subgroups of the group G.

(a) Show that $|HK| = |H| \cdot |K| / |H \cap K|$.
(b) Show that HK is a subgroup if and only if $HK = KH$.
(c) Suppose that $h^{-1}Kh = K$ for all $h \in H$. Show that HK is a subgroup.

Exercise 7.4 For each of the five regular solids, find a formula for the number of colourings of the faces with r colours, two colourings differing by a rotation being regarded as identical.

Exercise 7.5 What happens if we apply the Orbit-Counting Lemma to the action of G on itself by conjugation?

Exercise 7.6 Let P be a p-group, and N a non-trivial normal subgroup of P. By considering the action of P on N by conjugation, prove that $N \cap Z(P) \neq \{1\}$.

Exercise 7.7 (*) Let P be a Sylow p-subgroup of a finite group G, and let $H = N_G(P)$. Prove that $N_G(H) = H$.
   [*Hint*: if $g \in N_G(H)$, then $g^{-1}Pg$ is a Sylow subgroup of H; but H has only one Sylow p-subgroup.]

Exercise 7.8 (∗) Let G be a finite group. Show that the following properties of G are equivalent:

(a) G is the direct product of its Sylow subgroups;
(b) if H is a proper subgroup of G, then $N_G(H)$ properly contains H.

[*Hint*: Any p-group satisfies (b), and this property is preserved by direct products. For the converse, use the preceding question.]

Remark   Such a group is said to be **nilpotent**.

Exercise 7.9 Deduce from Corollary 7.8 that, if p is prime, then any subgroup of p-power order of a finite group G is contained in a Sylow p-subgroup of G.

Exercise 7.10 Let N be a normal subgroup of G, and P a Sylow p-subgroup of N. Show that $G = N_G(P)N$.
   [*Hint*: Take $g \in G$. Then $g^{-1}Pg$ is a Sylow subgroup of N, and so $g^{-1}Pg = n^{-1}Pn$ for some $n \in N$. Then $gn^{-1} = h \in N_G(P)$, and $g = hn$.]

Remark   The argument in this question is called the **Frattini argument**.

Exercise 7.11 Prove the Jordan–Hölder Theorem for infinite groups having composition series of finite length.

Exercise 7.12 Show that a conjugacy class C in S_n splits into two classes in A_n if and only if the cycle lengths of its members consist of distinct odd numbers.

Exercise 7.13 Show that, if an element $c \in S_n$ has cycle type which contains the number i with multiplicity a_i, for $i \leq n$, then the order of the centraliser of c is

$$ f = \prod_{i=1}^{n} a_i!\, i^{a_i}, $$

and hence that the conjugacy class of c has size $n!/f$. Deduce that the size of the conjugacy class is greater than $2(n-1)$ for $n \geq 6$.

Exercise 7.14 Show that the group $\mathrm{PSL}(2, q)$ has order $q(q^2 - 1)$ if q is a power of 2, and order $q(q^2 - 1)/2$ if q is an odd prime power.

Exercise 7.15 The groups A_5, $\mathrm{PSL}(2,4)$, and $\mathrm{PSL}(2,5)$ are all simple groups of order 60. Prove that they are all isomorphic.

Exercise 7.16 Prove that there is a unique simple group of order 60 (up to isomorphism).

Exercise 7.17 Prove that a semidirect product $A \rtimes_\phi B$, where the action ϕ is trivial, is the direct product $A \times B$.

Exercise 7.18 Prove Theorem 7.16

Exercise 7.19 Prove that $E(C_p, C_p) \cong C_p$. [The split extension $C_p \times C_p$ corresponds to the zero element of this group. The $p - 1$ non-zero elements are all associated with the non-split extension C_{p^2}.]

Exercise 7.20 (a) Prove that the sum of factor sets is a factor set.
(b) Prove that an inner factor set is a factor set.
(c) Prove that the sum of inner factor sets is an inner factor set.

Further ring theory

We have some unfinished business from Chapter 2: the proof that any principal ideal domain (PID) is a unique factorisation domain (UFD), and the proof of Gauss' Lemma. The first of these introduces a connection between factorisation and chain conditions in a ring, and leads us to the Hilbert Basis Theorem.

7.9 PID implies UFD. In Chapter 2, we gave part of the proof that a principal ideal domain is a unique factorisation domain: we showed the uniqueness, but not the existence, of factorisations into irreducibles. Here is the remainder of the proof. It depends on showing that a PID satisfies a condition which is very important in more advanced ring theory, the **ascending chain condition (ACC)** on ideals.

Proposition 7.18 *Let R be a PID and let $I_1, I_2, \ldots$ be ideals in R forming an ascending chain:*

$$I_1 \subset I_2 \subset \cdots .$$

Then the number of ideals in the chain is finite.

Proof Suppose that we have an infinite ascending chain of ideals; say, $I_1 \subset I_2 \subset \ldots$. Let $I = \bigcup I_n$ be the union of all the ideals in the chain. We apply the Ideal Test to I:

(a) Take $x \in I, r \in R$. Now $I = \bigcup I_n$, so $x \in I_n$ for some n. Since I_n is an ideal, we have $rx, xr \in I_n$, so $rx, xr \in I$.

(b) Take $x, y \in I$. Then $x \in I_n, y \in I_m$ for some m. Without loss of generality, $n \geq m$; then $I_m \subseteq I_n$, and so $y \in I_n$. Now $x - y \in I_n$, since I_n is an ideal; so $x - y \in I$.

Thus I is indeed an ideal.

Now R is a PID, so $I = \langle a \rangle$ for some element $a \in I$. Now $I = \bigcup_n I_n$, and so $a \in I_n$ for some n; so all multiples of a lie in I_n, whence $I = I_n$. But $I_n \subset I_{n+1} \subseteq I$, and we have a contradiction.

So an ascending chain of ideals must be finite. $\qquad\qquad\square$

Now we prove the theorem. Let R be a PID, and suppose a_0 is an element of R which is not zero or a unit and cannot be factorised into irreducibles. In particular, a_0 itself is not irreducible (or we would have a factorisation with only one term); say $a_0 = a_1 b_1$, where neither a_1 nor b_1 is zero or a unit. It cannot be that both a_1 and b_1 have factorisations, or else we would obtain a factorisation of a_0 by combining them. Suppose, without loss, that a_1 has no factorisation. Then $a_1 = a_2 b_2$, where we may suppose that a_2 has no factorisation; and so we may continue with $a_n = a_{n+1} b_{n+1}$ for all n.

Let $I_n = \langle a_n \rangle$. Since $a_n = a_{n+1} b_{n+1}$ and b_{n+1} is not a unit, we have $I_n \subset I_{n+1}$; so we have an infinite ascending chain of ideals, contrary to Proposition 7.18.

This establishes that every element in R (other than zero and units) has a factorisation into irreducibles. We have already proved that the factorisation is unique. So this completes the proof of Theorem 2.21.

7.10   Noetherian rings.   The crucial step in the proof in the last section is that, in a PID, there cannot be an infinite strictly ascending chain of ideals. This followed from the fact that each ideal is generated by a single element. With a little more effort, we can give a necessary and sufficient condition.

Theorem 7.19   *The following conditions on a ring R are equivalent:*

(a) there is no infinite strictly ascending chain of ideals of R;
(b) every ideal of R is generated by a finite number of elements.

Proof   The proof that (b) implies (a) follows closely the argument in the last section. Suppose that we have an ascending chain of ideals, say

$$I_1 \subseteq I_2 \subseteq \dots .$$

Then the union of this chain is an ideal I. Since we are assuming (b), the ideal I is finitely generated; say, $I = \langle r_1, r_2, \dots, r_m \rangle$. Now each of $r_1, \dots, r_m$ belongs to some ideal in the chain; say, $r_j \in I_{n_j}$ for $j = 1, \dots, m$. If n denotes the greatest of $n_1, \dots, n_m$, then all of $I_{n_1}, \dots, I_{n_m}$ are contained in I_n, by the fact that the ideals form an ascending chain. Hence $I = \langle r_1, \dots, r_m \rangle \subseteq I_n$. But obviously $I_n \subseteq I$, since I_n is a member of a chain whose union is I. We conclude that $I = I_n$, and the chain has at most n distinct terms in it.

Now we prove that (a) implies (b). Suppose that (a) holds for the ring R. Suppose, for a contradiction, that I is an ideal which is not finitely generated. Choose $r_1 \in I$. Then $\langle r_1 \rangle \subseteq I$; and the inclusion is strict, since otherwise I would be generated by just one element. Hence we can choose $r_2 \in I \setminus \langle r_1 \rangle$. Then $\langle r_1, r_2 \rangle \subseteq I$, and again the inclusion is strict.

Continuing in this way, we choose elements $r_1, r_2, \dots \in I$ such that each one is outside the ideal generated by its predecessors. Thus, if $I_n = \langle r_1, \dots, r_n \rangle$, the chain

$$I_1 \subset I_2 \subset \dots$$

is a strictly increasing chain of ideals, contrary to assumption.   □

In this proof, we made use of an innocent-looking principle: we showed that, for each n, the set $I \setminus I_n$ is non-empty, and proceeded to choose r_{n+1} from this set. This is justified by the **Axiom of Choice**, which we discussed at the end of the last chapter.

A ring R is called **Noetherian**, or is said to satisfy the **ascending chain condition on ideals** (or the **ACC**), if condition (a) of the above theorem holds. Thus, in particular, any principal ideal domain is Noetherian.

The last part of the proof in the preceding section shows the following:

Proposition 7.20 *Let R be a Noetherian integral domain. Then every element of R which is not zero or a unit can be factorised into irreducibles.*

Here is an example of a non-Noetherian integral domain, in which factorisations into irreducibles do not necessarily exist. Let F be a field, and let R be the ring of all expressions which are finite sums of terms of the form ax^q, where $a \in F$, q is a non-negative rational number, and x is an indeterminate. Addition and multiplication are defined in the 'obvious' way. (If we had said 'non-negative integer' rather than 'non-negative rational number', this would be just the polynomial ring $F[x]$.) Now x cannot be factorised into irreducibles: for

$$x = x^{1/2} \cdot x^{1/2} = x^{1/2} \cdot x^{1/4} \cdot x^{1/4} = \ldots$$

and none of the factors is a unit. From this, it is possible to extract an ideal which is not finitely generated, or an infinite ascending chain of ideals.

On the other hand, the following theorem ensures a supply of Noetherian rings.

Theorem 7.21 (Hilbert Basis Theorem) *Let R be a commutative Noetherian ring with identity. Then $R[x]$ is Noetherian.*

Proof Let J be an ideal in $R[x]$. Then it is not hard to show that the leading coefficients of polynomials of degree n in J form an ideal I_n of R. Moreover, $I_n \subseteq I_{n+1}$: for, if $f \in J$ has degree n, then $xf \in J$ has degree $n + 1$ and has the same leading coefficient as f.

Since R is Noetherian, the ascending chain $I_0 \subseteq I_1 \subseteq \ldots$ is finite. Say that $I_n = I_m$ for all $m > n$. Moreover, I_n is finitely generated. Let $f_1, \ldots, f_k$ be polynomials of degree n whose leading coefficients generate I_n.

Let $g \in J$ be a polynomial of degree $m \geq n$. Then, since $I_m = I_n$, there are elements $r_1, \ldots, r_k \in R$ such that $r_1 f_1 + \cdots + r_k f_k$ has the same leading coefficient as g. In other words, $g - x^{m-n}(r_1 f_1 + \cdots + r_k f_k)$ belongs to J and has degree less than m. By induction, every element of J is the sum of a polynomial with degree less than n and one of the form $f_1 h_1 + \cdots + f_k h_k$, for some $h_1, \ldots, h_k \in R[x]$.

Similarly, for each $p < n$ there is a finite set S_p of polynomials in J which have degree p and whose leading coefficients generate I_p. Arguing as above, any polynomial in J with degree less than n is a linear combination of $S_0 \cup \cdots \cup S_{n-1}$ (with coefficients in R).

So J is finitely generated.   □

The dual condition to ACC is also important. A commutative ring R is said to satisfy the **descending chain condition** or **DCC**, or to be **Artinian**, if every strictly descending chain $I_1 \supseteq I_2 \supseteq \ldots$ of ideals is finite. This condition turns out to be very strong:

Theorem 7.22 (Hopkins' Theorem) *If a commutative ring with identity is Artinian, then it is Noetherian, and has the property that there is a finite upper bound on the length of any chain of ideals.*

The converse is false: the ring $\mathbb{Z}$ is Noetherian but not Artinian.

A similar result holds for non-commutative rings. We do not discuss this here: see the books by McCoy or Cohn listed under Further Reading.

7.11 Gauss' Lemma. Let R be a UFD. We want to show that the polynomial ring $R[x]$ is a UFD.

Take a polynomial $f(x) = a_n x^n + \cdots + a_1 x + a_0 \in R[x]$. We define the **content** of f, written $C(f)$, to be the greatest common divisor (g.c.d.) of the coefficients $a_n, \ldots, a_1, a_0$. (Remember that greatest common divisors exist in a UFD, and are determined uniquely up to associates.) We say that f is **primitive** if its content is associate to 1 (that is, $C(f)$ is a unit). Then any polynomial f can be written $f = C(f) \cdot f_1$, where f_1 is primitive.

Proposition 7.23 *If f and g are primitive, then fg is primitive.*

(This is the crucial step in the proof; it is sometimes called Gauss' Lemma, rather than the theorem that $R[x]$ is a UFD.)

Proof Suppose that fg is not primitive; let p be an irreducible which divides its content $C(fg)$. Let

$$f = a_n x^n + \cdots + a_1 x + a_0,$$
$$g = b_m x^m + \cdots + b_1 x + b_0.$$

Now f and g are primitive, so p cannot divide all the coefficients of either polynomial. Choose r and s such that:

$$p \mid a_i \quad \text{for} \quad i < r \quad \text{but} \quad p \nmid a_r;$$
$$p \mid b_j \quad \text{for} \quad j < s \quad \text{but} \quad p \nmid b_s.$$

Consider the coefficient of x^{r+s} in fg. This is given by

$$c_{r+s} = \cdots + a_{r-1} b_{s+1} + a_r b_s + a_{r+1} b_{s-1} + \cdots.$$

Now p divides a_i for $i < r$, so p divides all the terms before $a_r b_s$ in the sum. Similarly, p divides b_j for $j < s$, so p divides all terms after $a_r b_s$. But p does not divide $a_r b_s$, by assumption. (In a UFD, if an irreducible p divides ab, then p divides a or p divides b.) So p does not divide the coefficient c_{r+s}. But this contradicts the assumption that p divides the content of fg. The contradiction shows that no such p can exist, so fg is primitive.   □

It follows that, for any two polynomials f and g, we have $C(fg) = C(f)C(g)$. For, if we write $f = C(f)f_1$ and $g = C(g)g_1$, where f_1 and g_1 are primitive, then $fg = C(f)C(g)f_1g_1$, and f_1g_1 is primitive, so the content of fg is $C(f)C(g)$.

Now the work of factorising a polynomial in $R[x]$ can be divided into two parts: factorise its content; and factorise a primitive polynomial. The content can be factorised uniquely, since R is a UFD. For primitive polynomials, we need to consider the field of fractions F of R.

Proposition 7.24 *Let R be a UFD with field of fractions F.*

(a) *A primitive polynomial in $R[x]$ is irreducible in $R[x]$ if and only if it is irreducible in $F[x]$.*

(b) *If f and g are primitive polynomials in $R[x]$ and f divides g in $F[x]$, then f divides g in $R[x]$.*

Proof (a) If f factorises in $R[x]$, then it factorises in $F[x]$. Conversely, suppose that $f = gh$ in $F[x]$. The coefficients in g and h are fractions of elements of R; let a and b be the least common multiples of their denominators. Then $ag, bh \in R[x]$, and $abf = (ag)(bh)$. Hence

$$ab = C(abf) = C(ag)C(bh).$$

Now we can write $ag = C(ag)g_1$, $bh = C(bh)h_1$, where g_1 and h_1 are primitive. Then

$$abf = C(ag)C(bh)g_1h_1.$$

From the two displayed equations and the fact that g_1h_1 is primitive, we see that $f = g_1h_1$, a factorisation of f in $R[x]$.

(b) Suppose that $f = gh$ with $h \in F[x]$. As above, let b be the least common multiple of the denominators of the coefficients in h. Then $bf = g.bh$ in $R[x]$, so

$$b = C(bh),$$
$$bf = gC(bh)h_1,$$

where h_1 is primitive. So $f = gh_1$ as required.   □

Thus, factorisation of primitive polynomials in $R[x]$ exactly mirrors their factorisation in $F[x]$, so is unique up to order and associates.

7.12   Eisenstein's criterion. Gauss' Lemma is very useful for the practical business of factorising polynomials. As an application, here is a simple proof, using Gauss' Lemma, of the theorem of Pythagoras.

Theorem 7.25 $\sqrt{2}$ *is irrational.*

Proof It is enough to show that the polynomial $x^2 - 2$ is irreducible over $\mathbb{Q}$, since if $q = \sqrt{2}$ were rational, then $x - q$ would be a factor of $x^2 - 2$. Now $x^2 - 2$ is a primitive polynomial over $\mathbb{Z}$, so, if it factorised in $\mathbb{Q}[x]$, then it would factorise in $\mathbb{Z}[x]$, by Gauss' Lemma. But, if $ax + b$ divides $x^2 - 2$ in $\mathbb{Z}[x]$, then a divides 1, and b divides -2; so the only possible factors are $x + 1$, $x - 1$, $x + 2$, or $x - 2$. None of these is a factor, since none of $\pm 1, \pm 2$ is equal to $\sqrt{2}$. $\qquad\square$

What Gauss' Lemma is telling us here is that, if the monic integer polynomial $x^2 - 2$ has a rational root, then this root is an integer.

In view of the importance of Gauss' Lemma, it is interesting that it has a very much simpler proof for integer polynomials, or indeed for polynomials over any PID.

Proposition 7.26 *Let R be a principal ideal domain, and let f and g be primitive polynomials in $R[x]$. Then fg is primitive.*

Proof Suppose not, so that there is an irreducible element $p \in R$ which divides every coefficient of fg. Now $F = R/(p)$ is a field. By considering the coefficients of a polynomial $f \in R[x]$ mod p, we obtain a polynomial $\overline{f} \in F[x]$.

Now $\overline{f}$ and $\overline{g}$ are non-zero, since p does not divide all the coefficients of f or g (these polynomials are primitive); but $\overline{f} \cdot \overline{g} = \overline{fg} = 0$, by assumption. This contradicts the fact that $F[x]$ is an integral domain. $\qquad\square$

The trick in the proof could be stated like this. The natural homomorphism from R to $R/(p) = F$ extends to a homomorphism from $R[x]$ to $F[x]$.

This trick has other uses too.

Theorem 7.27 (Eisenstein's criterion) *Let R be a principal ideal domain, and p an irreducible in R. Let*

$$f(x) = a_n x^n + \cdots + a_1 x + a_0$$

be a primitive polynomial in $R[x]$ with the following properties:

(a) p does not divide the leading coefficient a_n;
(b) p divides the other coefficients $a_{n-1}, \ldots, a_1, a_0$;
(c) p^2 does not divide the constant term a_0.

Then f is irreducible.

Proof Suppose that $f = gh$. By reducing mod p, we have $\overline{f} = \overline{g}\overline{h}$. By assumptions (a) and (b), we have $\overline{f} = \overline{a_n}x^n$ (all the other terms are equal to zero mod p). Now the ways in which a power of x can factorise are very limited: we must have $\overline{g} = \overline{b_m}x^m$ and $\overline{h} = \overline{c_{n-m}}x^{n-m}$, say, with $0 < m < n$. Thus we have

$$g(x) = b_m x^m + \cdots + b_1 x + b_0,$$

$$h(x) = c_{n-m} x^{n-m} + \cdots + c_1 x + c_0,$$

where all the coefficients except the leading ones are divisible by p.

Now consider the constant term. We have $a_0 = b_0 c_0$, and p divides both b_0 and c_0; so p^2 divides a_0, contradicting assumption (c). The theorem is proved. □

Example 1   Taking $p = 2$, we see that $x^2 - 2$ is irreducible over $\mathbb{Z}$ (and hence over $\mathbb{Q}$, by Gauss' Lemma).

More generally, if p is prime and $n > 1$, then $x^n - p$ is irreducible, so the nth root of p is irrational.

Example 2   Sometimes Eisenstein's criterion does not apply to the given polynomial, but can be made to do so by transforming it.

Consider the polynomial

$$f(x) = x^{p-1} + x^{p-2} + \cdots + x + 1,$$

where p is prime, whose roots are the complex pth roots of unity. Eisenstein's criterion does not apply, since 1 has no prime factors. Instead, we consider $g(x) = f(x + 1)$. We have $f(x) = (x^p - 1)/(x - 1)$, so $g(x) = ((x+1)^p - 1)/x$.

(a) The coefficient of x^{p-1} in $g(x)$ is 1.
(b) For $1 \le i \le p - 2$, the coefficient of x^i in $g(x)$ is the coefficient of x^{i+1} in $(x+1)^p$, namely $\binom{p}{i+1}$, which is divisible by p.
(c) The constant term in $g(x)$ is equal to $g(0) = f(1) = p$, which is divisible by p but not by p^2.

By Eisenstein's criterion, g (and hence also f) is irreducible.

Example 3   The polynomial $x^2 + y^2 - 1$ in $\mathbb{Q}[x, y]$ is irreducible. For we regard this expression as a polynomial in $R[x]$, where $R = \mathbb{Q}[y]$: it has the form $x^2 + 0x + (y+1)(y-1)$. Now Eisenstein's criterion applies, where we take p to be the irreducible $y - 1$ in R.

This argument works over any field in which $y + 1 \ne y - 1$; that is, any field of characteristic different from 2. It fails for a field of characteristic 2, and in this case we have $x^2 + y^2 - 1 = (x + y - 1)^2$.

7.13   A glimpse at algebraic geometry.   Algebraic geometry represents the flowering of the seed planted by Descartes when he turned geometry into algebra. It is a central subject of modern mathematics, and we can get no more than a brief glimpse here. Fortunately, good introductory books are available.

An algebraic curve in the plane is a set of points whose coordinates satisfy some polynomial equation. Examples such as a circle $x^2 + y^2 = 1$ and a parabola $y = x^2$ are familiar, but many more complicated examples have been studied.

When we look at higher dimensions, however, we see that the definition must be broadened. For example, in three dimensions, a polynomial equation defines a surface, and a curve may be defined as the intersection of two surfaces. Such a curve cannot be defined by a single equation. For example, the cylinders $x^2 + y^2 = 1$ and $x^2 + z^2 = 4$ in 3-dimensional Euclidean space meet in a pair of non-plane curves.

The approach we take is to consider all polynomials which vanish on the set of points in question. Accordingly, given a set S of polynomials in n variables over a field F, the **algebraic set** $A(S)$ defined by S is the set of all points in F^n on which all the polynomials vanish. (So the curves referred to above form $A(x^2 + y^2 - 1, x^2 + z^2 - 4)$.) Conversely, for any subset X of F^n, we let $I(X)$ be the set of all polynomials vanishing on X.

We see immediately that $I(X)$ is an ideal. By the Hilbert Basis Theorem (Theorem 7.21), we know that $I(X)$ is generated by a finite number of elements. Moreover, $A(\langle f, g \rangle) = A(\langle f \rangle) \cap A(\langle g \rangle)$. So *any algebraic set is the intersection of a finite number of algebraic sets defined by single polynomial equations.*

The ideal $I(X)$ has another important property: *if $f^n \in I(X)$ for some positive integer n, then $f \in I(X)$.* An ideal with this property is called a **radical ideal**.

We might hope that there is an exact correspondence between algebraic sets and radical ideals. In general, this is not so. For example, over the real numbers, the equations $x^2 + y^2 = 0$ and $x^2 + y^4 = 0$ both define the algebraic set consisting only of the origin. Of course, over the complex numbers, they define larger and quite different algebraic sets. So we should work over an algebraically closed field in order to obtain the nicest properties. Now Hilbert's **Nullstellensatz** ('Theorem on zeros') states the following:

Theorem 7.28 (Nullstellensatz) *Let F be an algebraically closed field. Let $f, g_1, \ldots, g_m$ be polynomials in n variables over F, and suppose that, for any $x \in F^n$,*

$$g_1(x) = \ldots = g_m(x) = 0 \qquad \Rightarrow \qquad f(x) = 0.$$

Then, for some positive integer k, we have $f^k \in \langle g_1, \ldots g_m \rangle$.

Corollary 7.29 *Let F be algebraically closed. Then the maps I and A defined above are mutually inverse bijections between the algebraic sets in F^n and the radical ideals in $F[x_1, \ldots, x_n]$.*

Thus any problem about algebraic sets over an algebraically closed field F can be translated into a problem about ideals in the polynomial ring. We cannot pursue this correspondence much further; we end with a few observations.

It is clear that the correspondence above interchanges inclusion: a larger algebraic set satisfies fewer equations. Adding ideals corresponds to intersecting algebraic sets, while multiplying ideals corresponds to taking the union of algebriac sets. Also, since $F[x_1, \ldots, x_n]$ is a unique factorisation domain, an algebraic set defined by a single polynomial is the union of a finite number of algebraic sets defined by irreducible polynomials.

The **coordinate ring** of an algebraic set A is the ring $F[x_1, \ldots, x_n]/I(A)$. It consists of all the functions on A which are **algebraic** (induced by polynomial functions on F^n). It is a commutative Noetherian ring, generated over F by n

elements (the images of the coordinate functions $x_1, \ldots, x_n$ under the canonical homomorphism). The next stage of development in algebraic geometry is to regard a set as a space in which to work, which is just as good as the original space F^n: its coordinate ring replaces the polynomial ring, and a similar correspondence between radical ideals and algebraic sets can be described. Now the same algebraic set (as specified by its coordinate ring) may be represented in many different ways in F^n (perhaps for different n); but now we can ignore these differences in representation and correspond on the structure of the algebraic set.

To end on a specific note, here is the algebraic proof of a fact which is intuitively clear geometrically. We define a **plane curve** to be an algebraic set $A(f)$, where $f \in F[x, y]$; it is **irreducible** if f is irreducible.

Theorem 7.30 (Bezout's Theorem) *Let $A(f)$ and $A(g)$ be plane curves over a field F such that $A(f)$ is irreducible and not contained in $A(g)$. Then $A(f)$ and $A(g)$ intersect in only finitely many points.*

Proof At least one of the variables, say x, must occur in f. Let K be the field of fractions of $F[y]$. Then we can view f and g as elements in $K[x]$. Gauss' Lemma shows that f is irreducible in $K[x]$. The hypothesis that $A(f)$ is not contained in $A(g)$ shows that f does not divide g. Hence f and g are coprime. Since $K[x]$ is a PID, there exist polynomials p and q in $K[x]$ such that $pf + qg = 1$.

Now the coefficients of p and q are rational functions in y. So we can multiply up by the least common multiple of their denominators (say, $h(y)$) to obtain

$$r(x, y)f(x, y) + s(x, y)g(x, y) = h(y),$$

where $r = hp$ and $s = hq$.

Let (a, b) be a point lying on both curves. Then $f(a, b) = g(a, b) = 0$, and so $h(b) = 0$. This equation has only a finite number of solutions; and for each solution b, the equation $f(x, b)$ has only a finite number of solutions in x. The theorem is proved. □

7.14 Local rings. In this section we consider only commutative rings with identity.

Definition A **local ring** is a commutative ring with identity which has a unique maximal ideal.

We shall give several constructions of local rings which are important in many parts of algebra: formal power series rings, p-adic integers, and localisations. We begin with a simple property of local rings.

Proposition 7.31 *Let I be a proper ideal of a commutative ring R with identity. Then R is a local ring with maximal ideal I if and only if every element outside I is a unit.*

Proof Suppose that R is a local ring and I its maximal ideal. For $a \notin I$, the ideal aR generated by a is not contained in I. By Krull's Theorem, if it were a proper ideal, it would be contained in a maximal ideal. So $aR = R$, whence there exists $b \in R$ with $ab = 1$.

Conversely, since no unit can lie in a proper ideal, it follows that if every element outside I is a unit, then every proper ideal is contained in I, so that I is the unique maximal ideal. $\square$

Definition Let F be a field. The **formal power series ring** over F is the set $F[[x]]$ of all infinite sequences $(a_n) = (a_0, a_1, a_2, \ldots)$ of elements of F; addition and multiplication are defined by

- $(a_n) + (b_n) = (c_n)$, where $c_n = a_n + b_n$;
- $(a_n) \cdot (b_n) = (d_n)$, where $d_n = \sum_{i=0}^{n} a_i b_{n-i}$.

We usually write the sequence $(a_0, a_1, \ldots)$ as $\sum_{n \geq 0} a_n x^n$; then the addition and multiplication appear natural. The set of elements $\sum a_n x^n$ of $F[[x]]$ with $a_0 = 0$ is easily seen to be an ideal. So the fact that $F[[x]]$ is a local ring follows from the next result.

Proposition 7.32 *A formal power series $\sum a_n x^n$ in $F[[x]]$ is invertible if $a_0 \neq 0$.*

Proof An inverse $\sum b_n x^n$ for the given sequence should satisfy $a_0 b_0 = 1$ and

$$a_0 b_n + a_1 b_{n-1} + \cdots + a_{n-1} b_1 + a_n b_0 = 0$$

for $n > 0$. The first equation gives $b_0 = a_0^{-1}$. The second allows us to find the other coefficients by induction: if we know $b_0, \ldots, b_{n-1}$, then

$$b_n = a_0^{-1} \sum_{i=1}^{n} a_i b_{n-i}.$$

The element $\sum b_n x^n$ given by this induction is clearly the inverse of the given element. $\square$

Remark If you have studied combinatorics, you may recognise the connection between recurrence relations and inverses of formal power series.

Since the set I of formal power series with zero constant term is a maximal ideal, we see that $F[[x]]/I$ is a field; in fact, it is the field F.

Next, let p be a prime number. The p-adic integers are 'limits' of consistent sequences of congruences modulo higher and higher powers of p. More precisely:

Definition A p-adic integer is a sequence $(a_1, a_2, \ldots)$, where $a_n \in \mathbb{Z}_{p^n}$, satisfying the condition that a_n and a_{n+1} are congruent modulo p^n for $n = 1, 2, \ldots$. (In other words, a_{n+1} represents the same element of $\mathbb{Z}_{p^n}$ as a_n does.) Addition and multiplication are defined componentwise: $(a_n) + (b_n) = (a_n + b_n)$, $(a_n) \cdot (b_n) = (a_n b_n)$. The ring of p-adic integers is denoted by $\mathfrak{O}_p$.

Remark I have to make an apology here about conflicting notation. Many mathematicians use the symbol $\mathbb{Z}_p$ for the ring of p-adic integers, but (following another very common convention) I have used this symbol for the ring of integers mod p. The letter $\mathfrak{O}$ is Gothic (or Fraktur) capital O. It is difficult to write without practice; I recommend a capital O in handwriting in your notes.

Proposition 7.33 *For any prime number p, $\mathfrak{O}_p$ is a local ring.*

Proof First note that the identity element of $\mathfrak{O}_p$ is the all-1 sequence (the constant sequence with value 1).

We have to show that a sequence (a_n) satisfying the defining condition, with $a_0 \neq 0$, is invertible.

By Euclid's Algorithm, we can find b_0 such that $a_0 b_0 \equiv_p 1$.

Suppose that we have found $b_1, \ldots, b_n$ such that $a_i b_i \equiv_{p^i} 1$ and $b_{i+1} \equiv_{p^i} b_i$. Let $b_{n+1} = b_n + x p^i$. Now

$$a_{n+1} b_n \equiv_{p^n} a_n b_n \equiv_{p^n} 1,$$

so $a_{n+1} b_n \equiv_{p^{n+1}} 1 + y p^n$. Then we find that

$$a_{n+1} b_{n+1} \equiv_{p^{n+1}} 1 + y p^n + x a_0 p^n + \text{ higher powers of } p,$$

so that if we choose x to satisfy $y + x a_0 \equiv_p 0$, we have succeeded. Thus by induction we have constructed an inverse. □

We will generalise this construction in Exercises 7.29 and 7.30.

The third construction is very general.

Definition Let R be a commutative ring with identity. A non-empty subset S of R is called **multiplicative** if $0 \notin S$ and, if $a, b \in S$, then $ab \in S$. If S is multiplicative, we define RS^{-1} in a similar way to the construction of the field of fractions of an integral domain: the elements are equivalence classes $[r, s]$ of ordered pairs (r, s) with $r \in R$ and $s \in S$, where (r_1, s_1) is equivalent to (r_2, s_2) if $r_1 s_2 = r_2 s_1$. Now define addition and multiplication of equivalence classes by

- $[r_1, s_1] + [r_2, s_2] = [r_1 s_2 + r_2 s_1, s_1 s_2]$,
- $[r_1, s_1] \cdot [r_2, s_2] = [r_1 r_2, s_1 s_2]$.

These operations are well defined and make RS^{-1} into a ring. The elements $[r, 1]$, for $r \in R$, form a subring isomorphic to R, and the elements $[s, 1]$, for $s \in S$, are invertible.

For example, if R is an integral domain, then the set of all non-zero elements of R is multiplicative, and the ring $R(R \setminus \{0\})^{-1}$ is the field of fractions of R.

Definition An ideal I of a commutative ring R with identity is said to be a **prime ideal** if, for any two elements $a, b \in R$, $ab \in I$ implies $a \in I$ or $b \in I$. Note that I is prime if and only if R/I is an integral domain.

If I is a prime ideal, then $S = R \backslash I$ is multiplicative. We define the *localisation* of R at I to be the ring $R_I = R(R \setminus I)^{-1}$.

Now it turns out that R_I is a local ring. For let

$$ J = \{[r, s] : r \in I, s \notin I\}. $$

Then J is an ideal of R_I, and it is easily checked that every element not in J is invertible.

Exercise 7.21 Eisenstein's criterion is in fact valid for polynomials over a unique factorisation domain. Prove this.

Exercise 7.22 In this exercise, we show that the following three assertions about an integer prime p are equivalent:

  (a) either $p = 2$ or p is congruent to 1 mod 4;
  (b) -1 is congruent to a square mod p;
  (c) p is the sum of two squares of integers.

Proof *(a) implies (b):* For $p = 2$ this is trivial. For $p = 4m + 1$, use the fact that the multiplicative group of $\mathbb{Z}_p$ is cyclic of order $4m$ (Proposition 7.45), and hence contains a cyclic subgroup of order 4.

  (b) implies (c): Suppose that p divides $a^2 + 1$. The ring $\mathbb{Z}[i]$ of Gaussian integers is Euclidean, and hence a PID. Let J be the ideal $(a + i, p)$. Let $J = (x + yi)$. Show that $x^2 + y^2 = p$.

  (c) implies (a): Any integer square is congruent to 0 or 1 mod 4. □

Exercise 7.23 Let R be a commutative ring, and I an ideal of R. The **radical** of I is defined to be the set of all $r \in R$ for which $r^n \in I$ for some positive integer n.

  (a) Prove that the radical of an ideal is a radical ideal.
  (b) If $g_1, \ldots, g_m \in F[x_1, \ldots, x_n]$, where F is algebraically closed, prove that $I(A(g_1, \ldots, g_m))$ is the radical of $\langle g_1, \ldots, g_m \rangle$.

Exercise 7.24 Prove that the polynomials of degree at most n in $F[x, y]$ form a vector space of dimension $(n+1)(n+2)/2$. Can you find the analogous formula for polynomials in k variables?

Exercise 7.25 Prove that $F[[x]]$ (for a field F) or $\mathfrak{O}_p$ (for a prime number p) are integral domains.

Exercise 7.26 Show that the inverse of $1 - x - x^2$ in $\mathbb{Q}[[x]]$ is $\sum F_n x^n$, where F_n is the nth *Fibonacci number* (see Exercise 1.52).

Exercise 7.27 Show that $\mathbb{Z}$ is a subring of $\mathfrak{O}_p$.

Exercise 7.28 Let p be an odd prime, and a an integer. Suppose that a has a square root in $\mathbb{Z}_p$. Prove that a has a square root in $\mathfrak{O}_p$.
    (*) What happens if $p = 2$?

Exercise 7.29 (*) This exercise and the next generalise the construction of the p-adic integers.

Let $R_1, R_2, \ldots$ be commutative rings with identity. Suppose that, for all $m \geq n$, we have a homomorphism $\theta_{m,n} : R_m \to R_n$ which is *surjective*. Suppose further that

- $\theta_{n,n}$ is the identity map for all n;
- for all $p \geq m \geq n$, $\theta_{p,m}\theta_{m,n} = \theta_{p,n}$.

Now let $\varprojlim R_n$ denote the set of all sequences $(r_1, r_2, \ldots)$, where $r_n \in R_n$ for all n and $r_m\theta_{m,n} = r_n$ for all $m \geq n$. Define componentwise addition and multiplication on this set, and show that it is a ring. (It is called the **inverse limit** of the family of rings and homomorphisms.)

Exercise 7.30 (*) Let $I_1 \supset I_2 \supset \ldots$ be a descending chain of ideals in a commutative ring R with identity, and suppose that $\bigcap I_n = \{0\}$. (Such a sequence is called a **filtration** of R.)

Show that the rings $R_n = R/I_n$ and natural homomorphisms $\theta_{m,n} : R_m \to R_n$ satisfy the conditions of the preceding exercise. (The ring $\hat{R} = \varprojlim(R/I_n)$ is called the **completion** of R with respect to the filtration.)

Show that R is embedded in $\hat{R}$ by the map $r \mapsto (I_n + r : n \in \mathbb{N})$.

Show that, if $R = \mathbb{Z}$ and $I_n = p^n\mathbb{Z}$, where p is prime, then $\hat{R} = \mathfrak{O}_p$.

Further field theory

We saw in Section 2.16 that the standard way to construct a field is by adjoining a root of a polynomial to a smaller field. In this section, we examine the procedure more closely, and iterate it to adjoin all the roots of a polynomial. We apply this to prove a theorem of Galois on the existence and uniqueness of finite fields. Apart from its intrinsic interest, this material is crucial to the two applications of algebra we discuss in Chapter 8: Galois theory (on solving polynomial equations) and coding theory (on correcting errors in message transmission).

7.15 Derivatives and repeated roots. Every student of calculus learns to differentiate polynomials. Using, for brevity, the notation Df for df/dx, we have:

$$\text{If } f(x) = a_nx^n + a_{n-1}x^{n-1} + \cdots + a_1x + a_0,$$
$$\text{then } Df(x) = na_nx^{n-1} + (n-1)a_{n-1}x^{n-2} + \cdots + a_1.$$

This statement makes sense for polynomials over any field (or indeed any ring), as long as we do not try to use arguments involving limits. (As usual, we take na to mean $a + a + \cdots + a$ (n terms) for positive integers n.)

In the spirit of the subject, we give an axiomatic treatment.

Theorem 7.34 *For any field F, there is a unique F-linear map $D : F[x] \to F[x]$ satisfying the following two conditions:*

(Der1) $D(fg) = f \cdot (Dg) + (Df) \cdot g$;
(Der2) $Dx = 1$.

Proof We have

$$D1 = D(1 \cdot 1) = 1 \cdot (D1) + (D1) \cdot 1,$$

so $D1 = 0$. By linearity, $Dc = 0$ for all $c \in F$. Also, $D(x^2) = x \cdot (Dx) + (Dx) \cdot x = 2x$; and an easy induction argument then shows that $D(x^n) = nx^{n-1}$ for all positive integers n. Hence, by linearity, D is given by the formula quoted earlier.

It remains to show that the map defined this way is F-linear and satisfies (Der1) and (Der2). The linearity and (Der2) are obvious. (Der1) follows by linearity if we can prove it in the case where f and g are powers of x: and this is done as follows:

$$D(x^m \cdot x^n) = D(x^{m+n}) = (m + n)x^{m+n-1},$$

$$x^m \cdot (Dx^n) + (Dx^m) \cdot x^n = mx^{m-1} \cdot x^n + x^m \cdot nx^{n-1} = (m + n)x^{m+n-1}.$$

$\square$

The use that we make of the derivative is the following. Contrary to one's expectation, perhaps, it can happen that, if $f(x)$ is an irreducible polynomial over F, then f can have two equal roots in some larger field. We want to decide when this can happen. So first we give a test for repeated roots.

Theorem 7.35 *A polynomial $f(x) \in F[x]$ has repeated roots (possibly in an extension field of F) if and only if the greatest common divisor of f and Df is not 1.*

Remark The greatest common divisor is computed in $F[x]$ by Euclid's Algorithm, as usual. If K is a larger field than F, we could make believe that this calculation was taking place in $K[x]$; the answer is the same, and lies in $F[x]$. So extending the field does not change the g.c.d.

Proof Suppose that α is a repeated root of f, so that $f(x) = (x - \alpha)^2 g(x)$ in some extension field of F. Then $Df = 2(x - \alpha)g + (x - \alpha)^2 Dg$. Hence $(x - \alpha)$ divides both f and Df, and their g.c.d. is not 1. (By our remark above, we do not have to specify the field in the last statement.)

Conversely, suppose that f has no repeated roots: so, in some extension field,

$$f(x) = c(x - \alpha_1)(x - \alpha_2) \cdots (x - \alpha_n),$$

where $\alpha_1, \alpha_2, \ldots, \alpha_n$ are all distinct. Up to a constant factor, any divisor of f is a product of some of the factors $(x - \alpha_i)$. But $D(x - \alpha_i) = 1$. So the product rule for the derivative shows that Df is the sum of n terms, where the ith term is c times the product of all $(x - \alpha_j)$ for $j \neq i$. Then $(x - \alpha_i)$ divides all terms except the ith, but does not divide the ith; so $(x - \alpha_i)$ does not divide Df. Thus the g.c.d. of f and Df is 1.   □

The **characteristic** of a field F is defined as follows: For positive integers n, let $n \cdot 1 = 1 + \cdots + 1$ (n terms). If there is a positive integer n such that $n \cdot 1 = 0$, then the characteristic is the least such n; otherwise the characteristic is zero. More concisely, it is the unique non-negative integer m generating the ideal $\{n \in \mathbb{Z} : n \cdot 1 = 0\}$ of $\mathbb{Z}$.

Proposition 7.36 *The characteristic of a field is zero or a prime.*

Proof Suppose that the characteristic is n, and that $n = rs$, with $r, s > 1$. Then $0 = n \cdot 1 = (r \cdot 1)(s \cdot 1)$, so either $r \cdot 1 = 0$ or $s \cdot 1 = 0$, contradicting the minimality of n.   □

Theorem 7.37 *Let f be an irreducible polynomial over the field F, and suppose that f has repeated roots in an extension of F. Then F has non-zero characteristic p (a prime), and there is a polynomial $g \in F[x]$ such that $f(x) = g(x^p)$.*

Proof Let f have repeated roots. Then $(f, Df) \neq 1$. But f is irreducible, so f divides Df. This implies that $Df = 0$; for, if not, then $\deg(Df) < \deg(f)$, but the divisibility implies $\deg(Df) \geq \deg(f)$.

Each term $a_i x^i$ of f gives rise to a term $i a_i x^{i-1}$ of Df. Since $Df = 0$, all of these terms must be zero. So, for every i, either $i = 0$ or $a_i = 0$. (Here $i = 0$ means that the element $i \cdot 1$ of F is equal to 0.)

If F has characteristic zero, then $i \cdot 1 = 0$ only if $i = 0$; so f is a constant polynomial, which contradicts the hypothesis that it is irreducible. So we may assume that F has non-zero characteristic p. Now $i \cdot 1 = 0$ only if i is divisible by p. So the only terms appearing in f are those $a_i x^i$ for which i is a multiple of p. This means that $f(x)$ is a polynomial in x^p, as claimed. (Note that, if $f(x) = g(x^p)$, then $Df = 0$.)   □

Let F be a field of prime characteristic p. We define the **Frobenius map** ϕ on F to be the pth power map: $c\phi = c^p$ for all $c \in F$.

Proposition 7.38 *The Frobenius map is an endomorphism of F (a homomorphism from F to F).*

Proof We have to show that

$$(a + b)^p = a^p + b^p,$$
$$(ab)^p = a^p b^p.$$

The second equation is obvious. For the first, we use the **Binomial Theorem**:

$$(a+b)^p = a^p + \binom{p}{1} \cdot a^{p-1}b + \binom{p}{2} \cdot a^{p-2}b^2 + \cdots + b^p.$$

The first and last terms give $a^p + b^p$, which is what we require. All the intermediate terms include binomial coefficients $\binom{p}{i}$, for $1 \le i \le p-1$. Now $\binom{p}{i} = p!/i!(p-i)!$, and the numerator (but not the denominator) is divisible by p; so p divides $\binom{p}{i}$, whence $\binom{p}{i} \cdot c = 0$ for any $c \in F$. The result is proved.

Since the Frobenius map is a homomorphism, its kernel is an ideal of F. But the only ideals of the field F are F and $\{0\}$. Clearly $1\phi = 1$, so the kernel is not F. Thus, $\mathrm{Ker}(\phi) = \{0\}$, and ϕ is one-to-one.

In general, ϕ is not necessarily onto. However, things are much simpler if it is. Accordingly, we make a definition:

Definition   The field F is said to be **perfect** if either:

(a) F has characteristic zero; or
(b) F has non-zero characteristic p, and every element of F has a pth root in F.

Note that the condition in (b) says precisely that the Frobenius map is onto, and hence is an automorphism of F.

The connection with repeated roots is as follows:

Theorem 7.39   *Let F be a perfect field. Then an irreducible polynomial over F has no repeated roots in any extension field of F.*

Proof   Let F be perfect, and suppose (for a contradiction) that f is irreducible and has repeated roots. By Theorem 7.37, $f(x) = g(x^p)$ for some polynomial g. Let

$$g(x) = a_n x^n + \cdots + a_1 x + a_0.$$

Since F is perfect, we can choose $b_0, b_1, \ldots, b_n$ such that $b_i^p = a_i$ for $i = 0, \ldots, n$. Now set

$$h(x) = b_n x^n + \cdots + b_1 x + b_0.$$

Since the Frobenius map is a homomorphism, we have

$$h(x)^p = b_n^p x^{np} + \cdots + b_1^p x^p + b_0^p = g(x^p) = f(x).$$

Since f is the pth power of a polynomial in $F[x]$, it is not irreducible, contrary to assumption. $\square$

It is quite difficult to find an imperfect field (see Exercise 7.31). By definition, all fields of characteristic zero are perfect. Also, the following holds:

Proposition 7.40   *A finite field is perfect.*

Proof The Frobenius map is always one-to-one; and a one-to-one map from a finite set to itself is necessarily onto. □

7.16 Splitting fields. A splitting field of a polynomial f is a 'smallest' field containing all the roots of f. This is only defined over a 'base field' F containing the coefficients of f. For example, if we regard $x^2 + 1$ as a real polynomial, its splitting field is $\mathbb{C}$; but if we regard it as a rational polynomial, its splitting field is the much smaller field $\mathbb{Q}(i) = \{a + bi : a, b \in \mathbb{Q}\}$.

Our goal in this section is to show that splitting fields exist and are unique (over a specified base field), up to isomorphism. But, because the base field is part of the data, we want to redefine the concept of isomorphism slightly:

Definition Let K, L be fields containing a subfield F. An F-**isomorphism** $\theta : K \to L$ is a field isomorphism from K to L which satisfies $c\theta = c$ for all $c \in F$.

Remark If we regard K and L as F-vector spaces, then an F-isomorphism is an F-linear transformation between them.

Definition Let f be a polynomial of degree $n > 0$ over F. A **splitting field** of f over F is a field K containing F such that

(a) $f(x) = c(x - \alpha_1) \cdots (x - \alpha_n)$ in $K[x]$ (so F 'splits' into linear factors over K);
(b) no proper subfield of K containing F has this property (so K is generated by F and the roots $\alpha_1, \ldots, \alpha_n$ of f).

Theorem 7.41 *Let f be a non-constant polynomial over F. Then f has a splitting field over F; and any two such splitting fields are F-isomorphic.*

Proof It is easy to see that there is a splitting field. For we can adjoin a root of an irreducible polynomial to a field, as explained in Section 2.16. Now adjoin a root α of an irreducible factor of f. Over $F(\alpha)$, we have $f(x) = (x - \alpha)g(x)$, where $\deg(g) = \deg(f) - 1$. Inductively add roots of g until f splits into linear factors. Now take the smallest field containing F and all the roots of f; this will be a splitting field. □

To prove the uniqueness up to F-isomorphism, we actually prove something which looks much more complicated than this, but is designed to streamline the induction:

Proposition 7.42 *Let $\theta : F \to F_1$ be an isomorphism of fields. Let $f(x)$ be a polynomial over F, and $f_1(x)$ the corresponding polynomial over F_1 (that is, if $f(x) = \sum a_i x^i$, then $f_1(x) = \sum (a_i \theta) x^i$). Let K and K_1 be splitting fields of f and f_1 over F and F_1 respectively. Then there is an isomorphism $\phi : K \to K_1$ whose restriction to F is θ.*

Proof We need to know that each step (adjoining one root of an irreducible polynomial) produces a unique field up to F-isomorphism. For this, we use the fact that $F(\alpha)$ is F-isomorphic to the 'standard' extension $F[x]/(f)$, by the map taking the coset $(f) + g$ to the element $g(\alpha)$.

It follows that, if $\theta : F \to F_1$ is an isomorphism of fields, f an irreducible polynomial over F, f_1 the corresponding polynomial over F_1, and α and α_1 roots of f and f_1 respectively, then there is an isomorphism from $F(\alpha)$ to $F_1(\alpha_1)$ extending θ. For θ induces in an obvious way an isomorphism from $F[x]$ to $F_1[x]$, which we shall also call θ, and maps f to f_1, and hence takes the ideal (f) of $F[x]$ to the ideal (f_1) of $F_1[x]$. Thus, $\theta : F \to F_1$ extends to an isomorphism $\bar{\theta} : F[x]/(f) \to F_1[x]/(f_1)$. Now we obtain the required map by composing:

- the F-isomorphism from $F(\alpha)$ to $F[x]/(f)$;
- $\bar{\theta}$;
- the F_1-isomorphism from $F_1[x]/(f_1)$ to $F_1(\alpha_1)$.

□

With this technical detail out of the way, the proof of the proposition (by induction on the degree of f) is straightforward. Let $\theta : F \to F_1$ be an isomorphism. Let α be a root of f in K, and α_1 a root of *the corresponding irreducible factor* of f_1 in K_1. Then, as noted, θ extends to an isomorphism $\psi : F(\alpha) \to F_1(\alpha_1)$, which maps α to α_1.

Now let $f(x) = (x - \alpha)g(x)$ (in $F(\alpha)[x]$), and $f_1(x) = (x - \alpha_1)g_1(x)$ (in $F_1(\alpha_1)[x]$). Then g and g_1 are corresponding polynomials under ψ, and K and K_1 are splitting fields of g and g_1 over $F(\alpha)$ and $F_1(\alpha_1)$ respectively. By induction, ψ extends to an isomorphism $\phi : K \to K_1$; and ϕ extends θ, as required.

Taking $F = F_1$ and θ to be the identity map, we obtain the theorem.

7.17   Finite fields.   We have seen that finite groups have a rich and varied structure, so that we cannot say, even to a very good approximation, how many there are of any given order. Finite fields, however, are much more restricted. In this section we will give the complete classification of finite fields, due to Galois, and investigate some of their properties.

Theorem 7.43 (Galois' Theorem on finite fields)   *The order of a finite field is a prime power.*

Conversely, there is a unique finite field of any given prime power order (up to isomorphism).

Proof   Let F be a finite field.

The characteristic of F must be non-zero. For the elements $n \cdot 1$, for $n \geq 0$, cannot all be distinct; and, if $m \cdot 1 = n \cdot 1$, with $m \neq n$, then $(m - n) \cdot 1 = 0$.

Let the characteristic be p (noting that p is prime). Now the elements $n \cdot 1$, for $n = 0, 1, \ldots, p - 1$, form a subfield $\mathbb{F}_p$ of F isomorphic to $\mathbb{Z}_p$. (The map $n \mapsto n \cdot 1$ is a ring homomorphism from $\mathbb{Z}$ to F: its kernel is $p\mathbb{Z}$, by definition of the characteristic.)

Now F is an extension field of $\mathbb{F}_p$, and so is a vector space over $\mathbb{F}_p$. Clearly, it has finite dimension, say n. Then F is isomorphic (as $\mathbb{F}_p$-vector space) to the space $\mathbb{F}_p^n$ of all n-tuples of elements of $\mathbb{F}_p$. This isomorphism tells us about the addition in F, but not the multiplication; so we have more work to do. But at least we know that $|F| = p^n$ is a prime power; so the first part of the theorem is established.

We also see that, if a field F has order a power of the prime p, then it has characteristic p, and contains a subfield isomorphic to $\mathbb{F}_p$.

Now we show that a field of order p^n, if it exists, is unique. The $p^n - 1$ non-zero elements of F form the multiplicative group; by Lagrange's Theorem, $c^{p^n-1} = 1$ for any non-zero element $c \in F$. Hence $c^{p^n} = c$ for any such c. But this also holds for $c = 0$. We conclude that the polynomial $x^{p^n} - x$ has all the p^n elements of F as its roots. So F is a splitting field for this polynomial over $\mathbb{F}_p$ (any smaller field could not contain all the roots!) By the uniqueness of splitting field (Theorem 7.41), F is unique, up to isomorphism.

It remains to show that fields of all possible prime power orders exist. This can be done by showing that, for any n, there is an irreducible polynomial of degree n over $\mathbb{F}_p$. However, we now have the machinery in place for a simpler proof.

Let p be prime and n a positive integer. Let $\mathbb{F}_p = \mathbb{Z}_p$, and let F be the splitting field of the polynomial $x^{p^n} - x$ over $\mathbb{F}_p$. We will show that $|F| = p^n$.

We have $D(x^{p^n} - x) = p^n x^{p^n-1} - 1 = -1$, since the characteristic is p. Hence $x^{p^n} - x$ is coprime with its derivative, and so it has p^n distinct roots in its splitting field F. Let S be the set of these roots. We show that S is a field. By minimality of the splitting field, it follows that $S = F$, and so that $|F| = p^n$, as required.

Let a and b be roots of $x^{p^n} - x$; that is, $a^{p^n} = a$ and $b^{p^n} = b$. We have to show that $a + b$, ab, and (if $a \neq 0$) $1/a$, are also roots. For this purpose, we use the Frobenius map, which is a homomorphism ϕ of F defined by $c\phi = c^p$. Applying ϕ n times, we have $c\phi^n = c^{p^n}$. This is also a homomorphism; so

$$(a + b)^{p^n} = (a + b)\phi^n = a\phi^n + b\phi^n = a^{p^n} + b^{p^n} = a + b,$$

$$(ab)^{p^n} = (ab)\phi^n = a\phi^n b\phi^n = a^{p^n} b^{p^n} = ab,$$

$$(1/a)^{p^n} = (1/a)\phi^n = 1/(a\phi^n) = 1/a^{p^n} = 1/a,$$

the last equation holding if $a \neq 0$. So $a + b$, ab, and $1/a$ (if $a \neq 0$) are roots of $x^{p^n} - x$, as required. $\qquad\square$

The unique field of order p^n is called the **Galois field** $\mathrm{GF}(p^n)$, after its discoverer. (Sometimes the notation $\mathbb{F}_{p^n}$ is used instead.)

We prove some structural properties of Galois fields.

Theorem 7.44 *Let p and q be primes, m and n positive integers.*

(a) The additive group of $\mathrm{GF}(p^n)$ is isomorphic to the direct sum of n cyclic groups of order p.

(b) The multiplicative group of $\mathrm{GF}(p^n)$ is cyclic of order $p^n - 1$.

(c) The automorphism group of $\text{GF}(p^n)$ *is cyclic of order* n, *generated by the Frobenius map.*

(d) $\text{GF}(q^m)$ *is a subfield of* $\text{GF}(p^n)$ *if and only if* $p = q$ *and* m *divides* n.

Proof (a) In the proof of Galois' Theorem, we worked out that $\text{GF}(p^n)$ is additively isomorphic to $(\mathbb{Z}_p)^n$. This is exactly what is required for (a).

(b) We will prove a more general result after this theorem.

(c) The Frobenius map ϕ is an automorphism. (This is a translation of the fact that $\text{GF}(p^n)$, being a finite field, is perfect.) Now ϕ^n is the identity map on $F = \text{GF}(p^n)$: for ϕ^n maps each element c to c^{p^n}, and we showed that $c^{p^n} = c$ for all $c \in F$. No smaller power of ϕ is the identity: for ϕ^m maps c to c^{p^m}, and the equation $x^{p^m} = x$ has at most p^m roots for $m < n$. So ϕ generates a cyclic group of order n of automorphisms of F.

It is harder to show that this is the full automorphism group. This follows from a theorem that we will meet when we consider Galois Theory in Chapter 8. Here is a more direct proof. By (b), the multiplicative group of $F = \text{GF}(p^n)$ is cyclic, generated (say) by a. Now $F = \mathbb{F}_p(a)$, since no proper subfield can contain a. Since $[F : \mathbb{F}_p] = n$, the element a satisfies a polynomial of degree n over $\mathbb{F}_p$. Any automorphism of F must map a to one of the n roots of this polynomial. Only the identity can fix a, since an automorphism fixing a must fix every power of a. So different automorphisms map a to different roots, and there are at most n automorphisms. Since we have a group of order n already (generated by the Frobenius map), it must be the full automorphism group.

(d) Suppose that $\text{GF}(q^m)$ is a subfield of $\text{GF}(p^n)$. Applying Lagrange's Theorem to the additive groups shows that q^m divides p^n; so $p = q$ (since p and q are prime). Applying Lagrange's Theorem to the multiplicative group shows that $p^m - 1$ divides $p^n - 1$. We claim that this implies that m divides n.

Let $n = mt + r$, where $0 \leq r \leq m - 1$. Since $x - 1$ divides $x^t - 1$ for any integer x, we see that $p^m - 1$ divides $p^{mt} - 1$, and hence divides $p^n - p^r$. It also divides $p^n - 1$ by assumption; so it divides $p^r - 1$. But $0 \leq p^r - 1 < p^m - 1$; so we must have $p^r - 1 = 0$, whence $r = 0$ and m divides n.

Part (b) follows from a more general result:

Proposition 7.45 *A finite subgroup of the multiplicative group of a field is cyclic.*

Proof We give two proofs. Both depend on the fact that a field contains at most n different nth roots of unity. For an nth root of unity satisfies the equation $x^n - 1 = 0$, and this polynomial of degree n has at most n roots.

First proof This proof uses Theorem 5.14, the structure theorem for finitely generated abelian groups. Let G be a subgroup of order n of the multiplicative group of a field F. Then

$$G \cong C_{d_1} \times C_{d_2} \times \cdots \times C_{d_k},$$

where $d_1 d_2 \cdots d_k = n$ and d_i divides d_{i+1} for $i = 1, \ldots, k-1$. (We use multiplicative, rather than additive, notation, since the group operation is multiplication.) Let p be a prime dividing d_1. Then p divides d_i for all i. So each factor C_{d_i} contains a cyclic subgroup of order p. Each subgroup contains $p-1$ elements of order p. Together with the identity, we obtain (at least) $1 + k(p-1)$ elements of order dividing p. Hence $1 + k(p-1) \le p$, whence $k = 1$ and G is cyclic. $\qquad\square$

Second proof This proof is more elementary. Let $\psi(m)$ be the number of elements of order precisely m in G. Also, let $\phi(m)$ be Euler's function, the number of non-negative integers less than m which are coprime to m. We show that:

(a) $\sum_{m \mid n} \phi(m) = n$;
(b) $\sum_{m \mid n} \psi(m) = n$;
(c) $\psi(m) \le \phi(m)$ for all $m \mid n$.

It follows that $\psi(m) = \phi(m)$ for all $m \mid n$. In particular, $\psi(n) = \phi(n) > 0$. So G contains an element of order n, and must be cyclic.

Proof of (a): We ask, how many non-negative integers $k < n$ have the property that the g.c.d. of k and n is n/m, for any divisor m of n? Putting $e = n/m$, we see that the g.c.d. of k/e and n/e is 1; so there are $\phi(n/e) = \phi(m)$ such integers. Summing over m must give n, since all the integers $0, 1, \ldots, n-1$ occur.

Proof of (b): Each element of G has some order which divides n.

Proof of (c): This is obvious if $\psi(m) = 0$, so suppose not. Then there is an element of order m in G. It generates a cyclic group H of order m, containing m solutions of $x^m = 1$. So all solutions of $x^m = 1$ lie in H. In particular, all elements of order precisely m lie in H. But a cyclic group of order m contains exactly $\phi(m)$ elements of order m. (If $H = \langle h \rangle$, then h^l has order m if and only if $(l, m) = 1$.) So $\psi(m) = \phi(m)$ in this case. $\qquad\square$

Proposition 7.46 *Let q be a prime power. Then the polynomial $x^{q^n} - x$ over $\mathrm{GF}(q)$ is the product of all the monic irreducible polynomials over $\mathrm{GF}(q)$ whose degrees divide n.*

Proof The roots of $x^{q^n} - x$ are all the elements of $\mathrm{GF}(q^n)$. Each of these generates $\mathrm{GF}(q^m)$ for some m dividing n, and hence satisfies an irreducible polynomial of degree m. Conversely, any root of an irreducible polynomial of degree m dividing n generates $\mathrm{GF}(q^m)$, and hence is contained in $\mathrm{GF}(q^n)$. $\qquad\square$

Example Let $q = 2$ and $n = 4$. There are two irreducible polynomials of degree 1 over $\mathrm{GF}(2) = \mathbb{Z}_2$, namely x and $x - 1$. There is one irreducible of degree 2, whose roots are the two elements of $\mathrm{GF}(4) \setminus \mathrm{GF}(2)$; namely, $x^2 + x + 1$; and

$$x^4 - x = x(x-1)(x^2 + x + 1).$$

There are three irreducibles of degree 4, namely $x^4 + x + 1$, $x^4 + x^3 + 1$, and $x^4 + x^3 + x^2 + x + 1$; the product of these three polynomials with $x^4 - x$ is $x^{16} - x$. Roots of the third irreducible have order 5, since

$$(x - 1)(x^4 + x^3 + x^2 + x + 1) = x^5 - 1.$$

Roots of the other two irreducibles are *primitive*; that is, they have order 15 and generate the multiplicative group of GF(16).

7.18 Wedderburn's Theorem. Wedderburn's Theorem allows us to extend our classification of finite fields to a classification of finite division rings very cheaply.

Theorem 7.47 (Wedderburn's Theorem) *A finite division ring is a field (that is, it is commutative).*

Proof We need two preliminaries. The first concerns cyclotomic polynomials. The nth **cyclotomic polynomial** $\Phi_n(x)$ is the unique monic polynomial whose roots are the primitive nth roots of unity in $\mathbb{C}$. Since every nth root of unity is a primitive mth root for some divisor m of n, we have

$$x^n - 1 = \prod_{m \mid n} \Phi_m(x).$$

By induction, we see that $\Phi_m(x)$ is a polynomial over $\mathbb{Z}$. Its degree is Euler's function $\phi(n)$, the number of congruence classes mod n which are coprime to n.

The second is a revision of some group theory. Recall from Section 7.1 that any group G is a union of **conjugacy classes**, where g and h are conjugate if and only if $h = x^{-1}gx$ for some $x \in G$. The number of elements in the conjugacy class of G is $|G : C_G(g)|$, where $C_G(g)$ is the subgroup

$$\{x \in G : xg = gx\}.$$

Now let F be a finite division ring. It is easy to check that the **centre**

$$Z(F) = \{x \in F : xa = ax \text{ for all } a \in F\}$$

is a subfield of F. Moreover, for any $a \in F$, its **centraliser**

$$C_F(a) = \{x \in F : xa = ax\}$$

is a sub-division ring of F. Moreover, F itself, and any centraliser $C_F(a)$, is a vector space over $Z(F)$ (with the given addition and scalar multiplication by elements of $Z(F)$).

Let $|Z(F)| = q$, a prime power (since $Z(F)$ is a finite field). Then $|F| = q^n$, where n is the dimension of F as a $Z(F)$-vector space. Choose representatives $a_1, \ldots, a_r$ for the conjugacy classes (in the multiplicative group) of elements not in $Z(F)$. Suppose that $C_F(a_i)$ has dimension m_i over $Z(F)$, so that $|C_F(a_i)| = q^{m_i}$. Then the centraliser of a_i in the multiplicative group of F has order $q^{m_i} - 1$, so the size of the conjugacy class of a_i is $(q^n - 1)/(q^{m_i} - 1)$. Since every element of $F \setminus Z(F)$ lies in just one such class, we have the **class equation**

$$q^n - 1 = q - 1 + \sum_{i=1}^{r} \frac{q^n - 1}{q^{m_i} - 1}.$$

Now $\Phi_n(x)$ divides $x^n - 1$ (in $\mathbb{Z}[x]$), and also $\Phi_n(x)$ divides $(x^n - 1)/(x^{m_i} - 1)$ as $m_i < n$. So, in $\mathbb{Z}$, the non-zero integer $\Phi_n(q)$ divides $q^n - 1$ and $(q^n - 1)/(q^{m_i} - 1)$ for each i. It follows from the class equation that $\Phi_n(q)$ divides $q - 1$.

But this is impossible for $n > 1$. (If $n = 2$, then $\Phi_2(q) = q + 1$. If $n > 2$, then $\phi(n) > 1$, and $\Phi_n(q)$ is the product of $\phi(n)$ terms of the form $(q - \omega)$, where ω is a primitive nth root of unity; each such factor is larger than $q - 1$.)

So we must have $n = 1$, whence $F = Z(F)$ is commutative. $\square$

Exercise 7.31 Define the **field of rational functions** over a field F to be the field of fractions of the polynomial ring $F[x]$. (Its elements are of the form $f(x)/g(x)$, where f and g are polynomials and $g \neq 0$.) Denote it by $F(x)$.

Prove that the element x has no pth root in $F(x)$, for $p > 1$. Deduce that, if F has characteristic p, then $F(x)$ is imperfect.

Now let F have characteristic $p = 2$, and let $K = F(x)$. Show that the polynomial $y^2 - x$ in $K[y]$ is irreducible and has repeated roots.

Exercise 7.32 (a) Prove by induction that, if $f(x) \in F[x]$ has degree n and splitting field K, then $[K : F] \leq n!$.

(b) (∗) Prove by induction that, with the same hypotheses, $[K : F]$ divides $n!$.

Exercise 7.33 Let p be prime. Show that the g.c.d. of $p^m - 1$ and $p^n - 1$ is $p^k - 1$, where k is the g.c.d. of m and n.

Exercise 7.34 (a) Let a_n be the number of monic irreducible polynomials of degree n over $\mathrm{GF}(q)$. Prove that

$$\sum_{m \mid n} m a_m = q^n.$$

Hence, in the case $q = 2$, calculate a_n for $n \leq 6$.

(b) Let b_n be the number of monic **primitive** irreducible polynomials of degree n over $\mathrm{GF}(q)$ (that is, polynomials any one of whose roots generates the multiplicative group). Prove that

$$b_n = \phi(q^n - 1)/n.$$

Calculate b_n for $q = 2$ and $n \leq 6$.

Other structures

The title of this section could mean one of two things. So far, we have concentrated on groups, rings, fields, vector spaces, and modules. There are a few important types of algebras (though less important than those just listed) which have been studied: Lie algebras are perhaps the most notable of these. One approach would be a Cook's tour through some of these.

Another approach is to look for unifying principles in algebra. The following sections do that. First, we examine the notion of an algebra, as a set on which various operations are defined, and make the most general definition possible. At first sight, it is surprising how much elementary group theory and ring theory can be developed at this level of generality.

The last two sections are more radical departures. In the second, we examine algebras from the viewpoint of their subalgebras and congruences (kernels of homomorphisms), independently of the actual operations. Finally, we come to the viewpoint that knowing the homomorphisms tells us all about a class of algebras; and we do not need to know them as functions, but merely the rule for composition. In this way, we find ourselves doing elementary algebra again, but having climbed further up the mountain to reach a higher viewpoint.

7.19 Universal algebra. An algebra is a set carrying various operations. Recall that an *n*-ary operation on a set A is a function $\mu : A^n \to A$. The integer n is called the **arity** of μ. Binary operations are often written with infix notation, as we have seen in the case of groups and rings. In general, this is not possible, and we write operations on the right, as $(x_1, \ldots, x_n)\mu$ if μ is an *n*-ary operation. Given a family of operations with prescribed arities, we consider a **type** of algebras with these operations. The type is described by the list of arities of the operations. In fact, it is possible to dispense with brackets and commas and write $x_1 \cdots x_n \mu$; no ambiguity arises (see Exercise 7.38). But we will not adopt this convention.

We allow the possibility of **nullary** operators (of arity zero); these are just distinguished elements, sometimes referred to as **constants**. The identity element of a group, and the zero of a ring, are examples.

The class of all algebras of given type is unlikely to be interesting. So we specialise as follows: A **law** is an expression $w_1 = w_2$, where w_1 and w_2 are expressions involving variables and operations, which properly define elements of an algebra A when elements of A are substituted for the variables. A law is **satisfied** in A if the equation is valid for all substitutions of elements of A for the variables. Now a **variety** of algebras is the class of all algebras of a given type which satisfy a given collection of laws.

Many classes of algebras we have met are varieties.

Example Consider the variety of algebras with a binary operator μ, unary operator ι, and nullary operator ϵ, satisfying the laws

$$((x,y)\mu, z)\mu = (x, (y,z)\mu)\mu,$$
$$(x,\epsilon)\mu = (\epsilon, x)\mu = x,$$
$$(x, x\iota)\mu = (x\iota, x)\mu = \epsilon.$$

This is just the class of groups: μ is the group operation, ι is inversion, and ϵ is the identity. Thus groups form a variety of algebras of type $(0, 1, 2)$.

Similarly, abelian groups, rings, commutative rings, rings with identity, and so forth, form varieties (Exercise 7.35).

Other varieties are less obvious. In Chapter 2, we considered **Boolean rings**, which satisfy the law $xx = x$.

A group G is said to be **metabelian** if it has a normal subgroup N such that both N and G/N are abelian. Now a group is metabelian if and only if it satisfies the law

$$[[x_1, x_2], [x_3, x_4]] = 1,$$

where the **commutator** $[x, y]$ is defined to be $x^{-1}y^{-1}xy$ (Exercise 7.41). So metabelian groups form a variety.

Example   Fields do not form a variety: only non-zero elements have multiplicative inverses, and this fact cannot be expressed as a law. (This does not prove that fields cannot be made into a variety by some clever trickery. But we will see later that this is so.)

The set of operators is not necessarily finite. We do not need specially contrived examples for this.

Example   For a given field F, the class of vector spaces over F is a variety. It has a binary operation (addition), a unary operation (additive inverse), a nullary operation (zero), and, for each $c \in F$, a unary operation (multiplication by c).

Example   Let G be a group. A **G-set** is an algebra with a unary operator μ_g for each $g \in G$, satisfying the laws

$$x\mu_1 = x,$$

$$(x\mu_g)\mu_h = x\mu_{gh}.$$

(The first equation is a law; the second represents one law for each pair (g, h) of elements of G.) The G-sets form a variety, which is a familiar one: a G-set is just a set on which there is an action of G by permutations (see Section 7.1).

You may have one of two common reactions at this point. One is a feeling of freedom, or licence: anything goes. Indeed, mathematicians have studied a very wide variety of varieties, or closely related structures: for example, semigroups, quasigroups, partial groups, loops, sloops, squags, semirings, alternative rings, Lie rings, near-rings, planar ternary rings, Lie algebras, Jordan algebras, Boolean rings, quasifields, near-fields, near-domains, . . .

The other reaction is vertigo at the wide range of subject matter opened up by this definition. But, in all the above cases, there is a good mathematical reason for considering the class of algebras. I know of no instance where someone wrote down a set of axioms out of the blue and invented a lively and important theory. Axioms for a class of algebras always follow the introduction of the class for other reasons. Each of these classes played some role in mathematics before its axiomatic definition: Lie algebras in differential geometry, planar ternary rings in the theory of projective planes, Boolean algebras in logic, Jordan algebras in physics, and so on.

A strength of the universal viewpoint in algebra is that many arguments recur in similar form in different topics, and it is more efficient to do them once in the most general context. We remarked in Chapter 3 that beginning group theory (subgroups, homomorphisms, and so on) duplicates similar parts of ring theory. In fact, the arguments work much more generally. We learn something both from the generality of the arguments and from the modifications needed.

Definition   Let A be an algebra of a given type. A **subalgebra** of A is a subset which is closed under all the operators of A.

In the case of nullary operators, this asserts that the subalgebra must contain all the constants of the algebra. It is clear that a subalgebra of A is an algebra of the same type. Moreover, any law that is valid in A also holds in a subalgebra. Hence:

Proposition 7.48   *If an algebra A belongs to a variety $\mathcal{V}$, then so does any subalgebra of A.*

A **homomorphism** $\theta : A \to B$, where A and B are algebras of the same type, is a map from A to B satisfying

$$(a_1, \ldots, a_n)\mu_A\theta = (a_1\theta, \ldots, a_n\theta)\mu_B$$

for all $a_1, \ldots, a_n \in A$, for all n-ary operators μ, and for all n. (In this equation μ_A and μ_B are the operators on A and B which correspond. In future, we will adopt the practice we have used for groups, rings, and every other kind of structure, and suppress the subscripts.) Now the image of θ is a subalgebra of B. Moreover, any law which holds in A also holds in the image of θ.

For rings, groups, vector spaces, and modules, we defined the kernel of a homomorphism to be the inverse image of the identity, and showed that it is a subalgebra of A. In general, we cannot do this, since there may be no 'identity'— our algebras may have no constants, or several, and even if they exist they may not have appropriate properties.

The clue is the general definition of kernel in Chapter 1, as a partition (two elements in the same part if they have the same image). In the above special cases, this is the partition into cosets of the simpler 'kernel' (the inverse image of the identity). In general, we just work with the partition.

Definition   The **kernel** of a homomorphism θ is defined as the equivalence relation $\mathrm{KER}(\theta)$ which is given by the rule that $(x, y) \in \mathrm{KER}(\theta)$ if and only if $x\theta = y\theta$.

Although none of the parts of this partition may be a subalgebra, it still has a property generalised from the coset partition of a normal subgroup.

Definition   A **congruence** on an algebra A is an equivalence relation E on A with the property that, for any n-ary operation μ, if $(a_i, b_i) \in E$ for $i = 1, \ldots, n$, then

$$((a_1, \ldots, a_n)\mu, (b_1, \ldots, b_n)\mu) \in E.$$

Now it is clear that the partition $\mathrm{KER}(\theta)$ defined by a homomorphism is a congruence on A.

Definition   Given a congruence E on A, the **factor algebra** A/E is defined as follows: the elements of A/E are the classes of E; and, if $[a]$ denotes the congruence class of a, then

$$([a_1], \ldots, [a_n])\mu = [(a_1, \ldots, a_n)\mu].$$

That this is independent of the choice of representatives follows immediately from the definition of a congruence. The map $a \mapsto [a]$ from A to A/E is a homomorphism whose kernel is E and whose image is A/E. Now we have all the ingredients for the **First Isomorphism Theorem**:

Theorem 7.49 (First Isomorphism Theorem)   *Let* $\theta : A \to B$ *be a homomorphism. Then:*

(a) $\mathrm{Im}(\theta)$ *is a subalgebra of* B;
(b) $\mathrm{KER}(\theta)$ *is a congruence on* A;
(c) $A/\mathrm{KER}(\theta) \cong \mathrm{Im}(\theta)$.

We note that this theorem works in the class of all algebras of given type, or in any variety of algebras. The other isomorphism theorems also generalise, but we do not pursue this here.

We have seen that varieties are closed under taking subalgebras and factor algebras. They have another closure property as well.

Definition   Let I be a set, and suppose that, for each $i \in I$, we are given an algebra A_i, all of these algebras having the same type. The **Cartesian product** $\prod_{i \in I} A_i$ is defined to be the set of all functions $f : I \to \bigcup_{i \in I} A_i$ satisfying $f(i) \in A_i$ for all $i \in I$. (These functions are choice functions for the family $(A_i : i \in I)$ of sets. The Axiom of Choice guarantees that the Cartesian product of a family of non-empty sets is non-empty. For this reason, it is sometimes called the 'multiplicative axiom'.) For each n-ary operation μ in the type, we define $(f_1, \ldots, f_n)\mu$ to be the function given by

$$((f_1, \ldots, f_n)\mu)(i) = (f_1(i), \ldots, f_n(i))\mu.$$

Note that $f_1(i), \ldots, f_n(i)$ are elements of the algebra A_i, so we can apply to them the operation μ on this algebra. It is again easy to see that, if all the algebras A_i belong to a variety $\mathcal{V}$, then so does their Cartesian product.

This enables us to show that fields, integral domains, and so on do not form varieties: the Cartesian product of two fields is not even an integral domain. (Representing a function on a 2-element set as an ordered pair of values as usual, we see that

$$(a,0)(0,b) = (0,0),$$

so $(a,0)$ is a zero-divisor if $a \neq 0$.)

Remarkably, it turns out that these three closure properties characterise varieties:

Theorem 7.50 *A class of algebras (with a fixed set of operators) is a variety if and only if it is closed under isomorphism and under taking subalgebras, factor algebras, and Cartesian products.*

7.20   Lattices.   A **lattice** is an algebra with two binary operations, $\vee$ ('join') and $\wedge$ ('meet'), and two constants, 0 and 1, satisfying the following axioms:

Idempotent laws: $x \vee x = x = x \wedge x$.
Commutative laws: $x \vee y = y \vee x$ and $x \wedge y = y \wedge x$.
Associative laws: $(x \vee y) \vee z = x \vee (y \vee z)$ and $(x \wedge y) \wedge z = x \wedge (y \wedge z)$.
Identity laws: $x \vee 0 = x = x \wedge 1$.
Absorptive laws: $x \wedge (x \vee y) = x = x \vee (x \wedge y)$.

Note that these axioms are unchanged under the exchange of $\vee$ and $\wedge$, and 0 and 1. In this way, from any lattice L we obtain another lattice L^*, the **dual** of L.

Where do lattices come from? We will use them to describe the subalgebras, or the congruences, of an arbitrary algebra. But there is a more basic fact underlying this: a lattice is a special kind of partially ordered set. Recall that a **partial order** on a set X is a relation $\leq$ which is reflexive ($x \leq x$), antisymmetric ($x \leq y$ and $y \leq x$ imply $x = y$), and transitive ($x \leq y$ and $y \leq z$ imply $x \leq z$). If $(X, \leq)$ is a partially ordered set, and $x, y \in X$, we say that u is a **least upper bound**, or **supremum**, of x and y if:

(a) $x \leq u$ and $y \leq u$;
(b) if $x \leq v$ and $y \leq v$ then $u \leq v$.

Note that a least upper bound, if it exists, is unique; for if u and u' are both least upper bounds, then $u \leq u'$ and $u' \leq u$, whence $u = u'$. Dually, a **greatest lower bound**, or **infimum**, is an element w such that:

(a) $w \leq x$ and $w \leq y$;
(b) if $z \leq x$ and $z \leq y$ then $z \leq w$.

Again, if it exists, it is unique. A **least element** 0 satisfies $0 \leq x$ for all x; if it exists, it is unique. Similarly for a **greatest element**.

Theorem 7.51 *Let $(X, \leq)$ be a partially ordered set. Suppose that*

(a) X has a least element 0 and a greatest element 1;
(b) any two elements x, y have a least upper bound $x \vee y$ and a greatest lower bound $x \wedge y$.

Then $(X, \vee, \wedge, 0, 1)$ is a lattice.

Conversely, let $(X, \vee, \wedge, 0, 1)$ be a lattice. Set $x \leq y$ if $x \vee y = y$. Then $(X, \leq)$ is a partially ordered set satisfying conditions (a) and (b) above.

Moreover, the constructions above are mutually inverse.

Proof Starting with a partially ordered set with properties (a) and (b), we verify the lattice axioms. This is mostly straightforward. In the commutative laws, both sides represent the least upper bound (or greatest lower bound) of x and y. In the absorptive laws, $x \leq (x \vee y)$, so the greatest lower bound of x and $x \vee y$ is x.

Conversely, suppose that we are given a lattice. Note first that $x \vee y = y$ if and only $x \wedge y = x$, so either can be chosen as the definition of $x \leq y$. This follows from the absorptive laws: if $x \vee y = y$, then

$$x \wedge y = x \wedge (x \vee y) = x.$$

We show that the relation $\leq$ is a partial order. Reflexivity $x \leq x$ follows from the idempotent law $x \vee x = x$. Antisymmetry follows from the commutative law: if $x \vee y = y$ and $y \vee x = x$, then $x = y$. Transitivity follows from the associative law: if $x \vee y = y$ and $y \vee z = z$, then

$$x \vee z = x \vee (y \vee z) = (x \vee y) \vee z = y \vee z = z.$$

Finally, we have to show that the constructions are inverse. Suppose that we are given a lattice which arises from a partially ordered set. Then, if $x \leq y$, the least upper bound of x and y is y, so $x \vee y = y$, and conversely. So the partial order is uniquely determined.

In the other direction, we are given a partially ordered set arising from a lattice. We have to show that 0 and 1 are least and greatest elements, and that $x \vee y$ and $x \wedge y$ are the least upper bound and greatest lower bound of x and y. The first assertions are trivial: the identity laws $0 \vee x = x$ and $1 \wedge x = x$ imply that $0 \leq x$ and $x \leq 1$ for all x. For the second, we have

$$x \vee (x \vee y) = (x \vee x) \vee y = x \vee y,$$

so $x \leq x \vee y$. Similarly, $y \leq x \vee y$. If $x \leq v$ and $y \leq v$, then $x \vee v = v$ and $y \vee v = v$; so

$$(x \vee y) \vee v = x \vee (y \vee v) = x \vee v = v,$$

whence $x \vee y \leq v$. Thus $x \vee y$ is the least upper bound. The argument for the greatest lower bound is dual. $\square$

Remark If a lattice L arises from a partially ordered set $(X, \leq)$, then its dual L^* is obtained from the partially ordered set $(X, \geq)$ obtained by reversing the order.

Remark A finite lattice (or partially ordered set) can be represented by a **Hasse diagram** in the plane. The points of the lattice are represented by points in the plane, so that, if $a < b$, then the point b is higher (that is, larger Y-coordinate) than a. We join a to b by a line segment if b **covers** a; that is, $a < b$ but no element c satisfies $a < c < b$. An example is shown in Figure 7.1. Check that it is a lattice. Note that the order relation can be read off from the covering relation: if $a < z$, then there is a chain $a < b < \ldots < z$, each term covering the one before.

There are two very important examples of lattices, which we now describe.

Example The **subset lattice** of a set S: the elements are all subsets of S, and the partial order is inclusion. Thus, $x \vee y = x \cup y$, $x \wedge y = x \cap y$, $0 = \emptyset$, and $1 = S$. The lattice shown in Figure 7.1 is the lattice of subsets of a 3-element set.

These lattices have some additional properties, notably the following:

Distributive laws: $x \wedge (y \vee z) = (x \wedge y) \vee (x \wedge z)$ and $x \vee (y \wedge z) = (x \vee y) \wedge (x \vee z)$.

Any lattice satisfying these laws is called a **distributive lattice**. Clearly, any sublattice of the subset lattice is distributive. The converse also holds, at least for finite lattices:

Theorem 7.52 *Any finite distributive lattice is isomorphic to a sublattice of the subset lattice of a finite set.*

In the infinite case, some additional properties are required in order to obtain a similar characterisation.

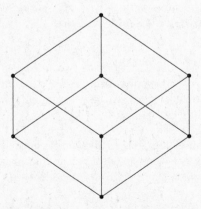

Fig. 7.1 A Hasse diagram

Note also that the dual of a distributive lattice is distributive.

Example Let A be an algebra (of some given type). Then the subalgebras of A form a lattice, the **subalgebra lattice** of A. The meet of two subalgebras is their intersection (which is always a subalgebra). The join of subalgebras B_1 and B_2 is usually not the union, which is not a subalgebra, but is, rather, the subalgebra **generated by** B_1 and B_2, which can be described as the intersection of all subalgebras containing them, or as the smallest set containing B_1 and B_2 and closed under the operations of A. The subalgebra 1 is A, while 0 is the unique minimal subalgebra (the subalgebra generated by the constants).

Example The **partition lattice** of a set S is defined as follows. The elements are the partitions of S (which we can regard as the equivalence relations on S.) For partitions π_1 and π_2, we take $\pi_1 \leq \pi_2$ if π_1 refines π_2, in the sense that any part of π_1 is contained in a part of π_2. (If we regard a partition as being an equivalence relation—that is a certain set of ordered pairs—this is just the inclusion order on $\mathcal{P}(S \times S)$.) So $\pi_1 \wedge \pi_1$ is the partition whose parts are all non-empty intersections of parts of π_1 with parts of π_2. The partition $\pi_1 \vee \pi_2$ is more difficult to describe: it is not just the union of the set of pairs, or the partition whose parts are all unions of parts of π_1 and π_2. Instead, join two points of S by an edge if they lie in the same part of either π_1 or π_2; then the parts of $\pi_1 \vee \pi_2$ are the connected components of this graph. The partition 0 is the one with singleton parts (the relation of equality), while the partition 1 has just one part, namely S.

Example More generally, if A is an algebra, then the congruences on A form a lattice, the **congruence lattice** of A. The order is as in the partition lattice, and the meet, 0, and 1 elements are the same; but the join $\pi_1 \vee \pi_2$ of two congruences π_1 and π_2 must be taken as the meet of all those congruences that are coarser than both.

Proposition 7.53 *The congruence lattice of a group, ring, or vector space is isomorphic to a sublattice of the subspace lattice.*

Proof In the case of a group, the classes of a congruence are the cosets of a normal subgroup; and the meet or join of two normal subgroups is a normal subgroup. So the normal subgroups form a sublattice of the subgroup lattice isomorphic to the congruence lattice. The arguments in the other cases are similar. $\square$

Remark In an abelian group, every subgroup is normal, and so the subgroup and congruence lattices are isomorphic. The same applies to a vector space, or a ring (such as $\mathbb{Z}$), in which every subring is an ideal.

We now consider two classes of lattices, the first more special than distributive lattices, and the second more general.

Boolean lattices form a special class of distributive lattices. In any lattice L, a **complement** of x is an element x' such that $x \vee x' = 1$ and $x \wedge x' = 0$. Complements may fail to exist, and they may not be unique. However, in a distributive lattice, an element has at most one complement. For, if x' and x^* are complements of x, then

$$x' = x' \wedge 1 = x' \wedge (x \vee x^*) = (x' \wedge x) \vee (x' \wedge x^*) = 0 \vee (x' \wedge x^*) = x' \wedge x^*,$$

so $x' \leq x^*$. Similarly, $x^* \leq x'$; so $x' = x^*$.

Note that 0 and 1 are complements of each other; and, if y is a complement of x, then x is a complement of y.

A **Boolean lattice** is a distributive lattice in which each element has a complement (necessarily unique).

It is usual to take complementation as an operation. This gives us the following definition. A **Boolean lattice** is a set X with operations $\vee, \wedge, 0, 1,'$ (of arities $2, 2, 0, 0, 1$) such that:

(a) $(X, \vee, \wedge, 0, 1)$ is a distributive lattice;
(b) $x \vee x' = 1$ and $x \wedge x' = 0$.

The subset lattice of any set S is a Boolean lattice: complementation is given by $x' = S \setminus x$. These examples are typical:

Theorem 7.54 *A finite Boolean lattice is isomorphic to the lattice of subsets of a set.*

This can be deduced from the representation theorem for distributive lattices, or can be proved directly (Exercise 7.44).

Now we turn to a larger class. A lattice is **modular** if it satisfies the following condition:

Modular law: If $x \leq z$, then $x \vee (y \wedge z) = (x \vee y) \wedge z$.

This property is weaker than the distributive law. It is self-dual, in that a lattice satisfies the modular law if and only if its dual does (interchange x and z in the statement to obtain its dual). As written, it is not a law, but it can be converted into one by noting that $x \leq z$ if and only if $x \vee u = z$ for some u; so the modular law can be written

$$x \vee (y \wedge (x \vee u)) = (x \vee y) \wedge (x \vee u).$$

So modular lattices form a variety. In particular, a sublattice of a modular lattice is modular. The connection of the words 'module' and 'modular' is explained in the first part of the next result.

Theorem 7.55 *(a) For any ring R, the submodule lattice of an R-module is modular.*
(b) The congruence lattice of a group is modular.

Remark In particular, this applies to the subspace lattice of a vector space, or the subgroup lattice of an abelian group (a $\mathbb{Z}$-module).

Proof (a) Let X, Y, Z be submodules of the R-module M, with $X \leq Z$. The join of two submodules is their sum. We have $X + (Y \cap Z) \leq (X + Y) \cap Z$, since this holds in any lattice. Take $z \in (X + Y) \cap Z$. Then $z \in Z$, and $z = x + y$, with $x \in X$, $y \in Y$. Since $x \in X \leq Z$, we have $y = z - x \in Z$, so $y \in Y \cap Z$, and $z = x + y \in X + (Y \cap Z)$, as required.

(b) The argument is similar.   □

In Theorem 7.52, we saw that every finite distributive lattice is embeddable in the lattice of subsets of a set. There is no similar theorem for modular lattices, since there is no one modular lattice which is sufficiently general to embed the subspace lattices of all finite vector spaces, for example. Nevertheless, there is an important characterisation theorem for a subfamily of the modular lattices, which we now state, after a few definitions.

An **atom** in a lattice is a minimal non-zero element. Thus, x is minimal if $x \neq 0$ but $y \leq x$ implies $y = x$ or $y = 0$. A lattice is **atomic** if every point is the join of a finite number of atoms. The **rank** of an element x in an atomic lattice is the smallest number of elements whose join is x; and the **rank** of the lattice is the rank of 1.

A **line** is a lattice consisting of 0, 1, and a set A of atoms, with $|A| > 1$. It is **proper** if $|A| > 2$. See Figure 7.2. Note that a line with two atoms is isomorphic to the direct product of the lattice $\{0, 1\}$ with itself. (If we permitted a line to have just one atom, such a line would not be atomic.)

A **projective plane** is an atomic modular lattice of rank 3 with the property that the meet of any two elements of rank 2 is an atom. (Alternatively, it is an atomic modular lattice of rank 3 whose dual has the same properties.) It is **proper** if any element of rank 2 is above at least three atoms. Calling atoms **points** and elements of rank 2 **lines**, we can draw geometric diagrams rather than Hasse diagrams: see Figure 7.3. The points in this diagram are the atoms of the lattice, and the six straight lines and one circle define seven sets of three atoms corresponding to the lattice elements of rank 2; the 0 and 1 of the lattice do not appear in the diagram.

Theorem 7.56 *Let L be an atomic modular lattice. Then L is isomorphic to the direct product of a finite number of lattices of the following form:*

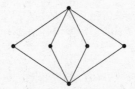

Fig. 7.2 A line

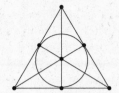

Fig. 7.3 A projective plane

(a) $L = \{0,1\}$;
(b) *a proper line;*
(c) *a proper projective plane;*
(d) *the submodule lattice of a finitely generated module over a division ring.*

Remark   By Wedderburn's Theorem, a finite division ring is a field. So for finite lattices, case (d) of the theorem becomes the subspace lattice of a finite-dimensional vector space over a finite field.

7.21   Category theory.   Category theory, sometimes dismissed as 'abstract nonsense', began in a very technical part of mathematics, algebraic topology. It has developed into an alternative foundation for the whole of mathematics (in the most extreme form), and certainly a unifying principle which is algebraic in nature but much wider in scope.

The underlying philosophy is that what is important about any class of mathematical structures is the structure-preserving maps between different objects in the class. For example, suppose that our structures are just sets. If $f : X \to Y$ and $g : Y \to Z$ are maps between sets, then there is a composite $fg : X \to Z$. Moreover, f is one-to-one if and only if there is a map $g : Y \to X$ with $fg = 1_X$ (where 1_X is the identity on X); and f is onto if and only if there exists $h : Y \to X$ with $hf = 1_Y$.

Similarly, other set-theoretic notions can be recognised. Here are some more examples. The associative law for groups is usually stated as a law (in the sense of universal algebra), asserting the equality of two expressions $(ab)c$ and $a(bc)$. Another version involves the maps λ_a and ρ_a defined by left and right multiplication by the element a: the associative law asserts that λ_a and ρ_c commute for any $a, c \in G$:

$$b\lambda_a\rho_c = (ab)c = a(bc) = b\rho_c\lambda_a.$$

This version uses a mixture of elements and maps. But the law can be stated using only maps. Let $\mu : G \times G \to G$ be the group operation. If $\alpha_i : G_i \to H_i$ are maps for $i = 1, 2$, then $\alpha_1 \times \alpha_2 : G_1 \times G_2 \to H_1 \times H_2$ is defined coordinatewise. Now the associative law asserts that

$$(1 \times \mu)\mu = (\mu \times 1)\mu,$$

where 1 is the identity map on G. (The left-hand side maps $(a, b, c) \mapsto (a, bc) \mapsto a(bc)$, and the right-hand side $(a, b, c) \mapsto (ab, c) \mapsto (ab)c$.)

In fact, we can define the Cartesian product of two sets using maps. Let X and Y be sets. The Cartesian product $X \times Y$ is a set P which has 'projections' π_1 and π_2 to X and Y respectively (by taking the first and second coordinate of each ordered pair). Moreover, if Z is any set and $\phi_1 : Z \to X$ and $\phi_2 : Z \to Y$ any maps, then there is a map $\psi : Z \to P$ such that $\psi\pi_i = \phi_i$ for $i = 1, 2$. (Set $z\psi = (z\phi_1, z\phi_2)$.) This property characterises the Cartesian product, up to isomorphism. Moreover, exactly the same properties and characterisation hold if we replace sets, maps, and Cartesian products by groups (or various other kinds of structures), homomorphisms, and direct products.

For a final example, a basis X of a vector space V over F is a linearly independent spanning set. However, bases are also characterised (and could be defined) by the following mapping property: any map from X into an F-vector space W can be uniquely extended to a linear map from V to W.

These examples give some insight into the viewpoint of category theory. The general definition is as follows:

A **category** consists of the following data:

- A set O of **objects**.
- A set M of **morphisms** or **arrows**.
- A pair of functions, dom (**domain**) and cod (**codomain**), from M to O.
- For each $x \in O$, an **identity morphism** 1_x.
- A partial operation of **composition** on M, the composition of f and g (if it exists) being written fg.

It satisfies the following axioms:

- The composition fg exists if and only if $\mathrm{cod}(f) = \mathrm{dom}(g)$. If this holds, then $\mathrm{dom}(fg) = \mathrm{dom}(f)$ and $\mathrm{cod}(fg) = \mathrm{cod}(g)$.
- If fg and gh are both defined, then $(fg)h = f(gh)$.
- $\mathrm{dom}(1_x) = \mathrm{cod}(1_x) = x$.
- If $\mathrm{dom}(f) = x$ and $\mathrm{cod}(f) = y$, then $1_x f = f = f 1_y$.

We abbreviate the information $\mathrm{dom}(f) = x$ and $\mathrm{cod}(f) = y$ by writing $f : x \to y$.

Part of the philosophy of category theory is that morphisms are more important than objects. In fact, a category can be defined using only the morphisms, the partial composition, and the identity morphisms; we identify the objects with their corresponding identity morphisms. See Exercise 7.56.

There are two, quite different, sources of examples of categories. Be careful to distinguish these, although the strength of category theory is that really no distinction needs to be made.

Classes of structures   Let O be a set of mathematical structures of some type. These may be universal algebras of a fixed type (such as groups, vector spaces over a given field, lattices). They may also, more generally, be topological spaces, differentiable manifolds, algebraic curves, and so on.

Let M be the class of all structure-preserving maps between members of O. (For algebras, we take M to consist of all homomorphisms; for topological spaces, all continuous maps; and so on.) For $f \in M$, we take $\operatorname{dom}(f)$ and $\operatorname{cod}(f)$ to be the usual domain and codomain of f, and take composition to be the usual composition of functions and 1_x to be the identity map on x.

It may also be possible to obtain a category by taking just some of the structure preserving maps. For example, we could take just the one-to-one maps, or the onto maps; for differentiable manifolds we could take the continuous functions, the differentiable functions, the smooth functions ...

A category of this sort, where the objects are sets (possibly with additional structure) and the morphisms are functions, is called a **concrete category**.

Remark There is a set-theoretic point which has to be mentioned here, although I will not elaborate on this. We want to consider the category of *all* groups, for example. But the class of all groups is not a set; if it were, we could not escape problems in the foundations of set theory, such as **Russell's paradox**. One way round this is to suppose that there is a very large 'universal' set U, in which all constructions which we want to perform can be made (some models of set theory provide such a set), and to consider the set of all groups which belong to U. If this remark means nothing to you, you may ignore it; if it intrigues you and you would like to know more, read a textbook on set theory and one on category theory and compare their approaches.

Individual structures It may surprise you to learn that a group G is an example of a category. Take a single object called $*$ (say), and take G to be the set of morphisms, with $\operatorname{dom}(g) = \operatorname{cod}(g) = *$ for all $g \in G$, and $1_* = 1$, the identity of G. Since there is only one object, any pair of morphisms can be composed.

With this example in mind, we could say that categories form just another type of algebraic object, more general than groups.

But now we have the option of turning the generality on itself. There is a category of all categories! (modulo the set-theoretic difficulties just discussed).

Categories are more general than groups in two respects: there can be more than one object, and morphisms need not be invertible. Two intermediate classes of structures are obtained by relaxing one or other of these conditions.

A **groupoid** is a category in which any morphism $f : x \to y$ has an inverse $g : y \to x$ (such that $fg = 1_x$ and $gf = 1_y$). An example is obtained by taking any class of structures as objects, and the isomorphisms as morphisms.

A **monoid** is a category with a single object. In other words, it is a set with a (total) operation of composition, satisfying the associative and identity axioms for a group, but not necessarily the inverse axiom. Thus, the endomorphisms of a single structure x (the homomorphisms from x to x) form a monoid.

For convenience, we will in future say 'Let $C = (O, M)$ be a category', meaning that O and M are the sets of objects and morphisms of C. Of course, the notation ignores part of the category structure (the composition and the identities), but old habits die hard.

Let $C = (O, M)$ be a category. A **subcategory** C' consists of a subset O' of O and a subset M' of M such that $C' = (O', M')$ is a category. An important special case occurs when M' consists of all the morphisms of C whose domain and codomain lie in O'. In this case, C' is called a **full subcategory**. For example, abelian groups form a full subcategory of the category of groups.

The philosophy of category theory is that morphisms carry the essential information about objects. Naturally enough, we next define 'morphisms between categories'.

Let $C = (O, M)$ and $C' = (O', M')$ be categories. A **functor** from C to C' consists of a pair of maps (denoted by the same symbol F) from O to O' and from M to M', satisfying the following conditions:

- $(fg)F = (fF)(gF)$ whenever fg is defined;
- $1_x F = 1_{xF}$ for all $x \in O$.

Note that the map on morphisms ('functions') is an important part of a functor, not just an appendage of the map on objects; the name 'functor' is intended to suggest this.

Functors are very common in mathematics. For example, let C be some concrete category of algebraic structures, and let C' be obtained by ignoring some of the structure. Then there is a **forgetful functor** from C to C'. For example, let C and C' be the categories of rings and abelian groups. Then the forgetful functor F maps a ring to its additive group, and a ring homomorphism to the same map (regarded merely as a group homomorphism). In particular, any concrete category has a forgetful functor to a category of sets.

Here are some further examples.

Derived group A different kind of functor maps groups to abelian groups. Let G be a group. The **derived group** G' is the subgroup generated by all commutators $g^{-1}h^{-1}gh$; it is the smallest normal subgroup with abelian factor group (see Section 6.1.4). Now there is a functor from groups to abelian groups which maps G to G/G'. Of course, we have to define the action of the functor on morphisms (see Exercise 7.60).

Unit group The functor U, from the category of rings with identity to the category of groups, maps a ring to its group of units. (Check that a ring homomorphism maps units to units and induces a group homomorphism on the group of units.)

General linear group More generally, for any n, the functor GL_n maps a ring R with identity to the group $\mathrm{GL}(n, R)$ of invertible $n \times n$ matrices over R.

Power set   The power set operation defines a functor from the category of sets to itself.

Homology   In algebraic topology, one defines, for each positive integer n and topological space X, an abelian group $H_n(X)$, the nth **homology group** of X. Homeomorphic spaces have isomorphic homology groups, and homology was originally a tool for telling topological spaces apart. It turns out that continuous maps between spaces induce homomorphisms between their homology groups. So H_n is a functor from topological spaces to abelian groups.

Group actions   Recall that any group G is a category with a single object $*$, in which the morphisms are the group elements. What is a functor F from G to the category of sets? $*F$ is a set Ω. For all $g \in G$, gF is a map from Ω to Ω, such that $(g_1 g_2)F = (g_1 F)(g_2 F)$ and $1F$ is the identity map on Ω. This is precisely the definition of a permutation action of G on Ω (see Axioms (GA1) and (GA2) in Section 7.1). So functors from G to sets are permutation representations (actions) of G.

More generally, functors from G to any category C of algebras are actions of G by automorphisms of an algebra in C. For example, if C consists of finite-dimensional vector spaces over F, then a functor from G to C is a representation of G by matrices over F.

We now move to the next level in this process. A natural transformation is a homomorphism between functors: with each object, it associates a morphism between the images of the object under the two functors. More precisely, let $F, G : C \to C'$ be functors, where $C = (O, M)$ and $C' = (O', M')$. A **natural transformation** $T : F \to G$ is a function from O to M' with the following properties:

- for any $x \in O$, $\mathrm{dom}(xT) = xF$ and $\mathrm{cod}(xT) = xG$;
- for any $f \in M$, with $\mathrm{dom}(f) = x$ and $\mathrm{cod}(f) = y$, we have

$$(fF)(yT) = (xT)(fG).$$

Note that $fF : xF \to yF$, $yT : yF \to yG$; and $xT : xF \to xG$, $fG : xG \to yG$, so the composite morphisms on both sides are defined. The condition can be represented by a **commutative diagram** as follows:

$$
\begin{array}{ccc}
xF & \xrightarrow{\;fF\;} & yF \\
{\scriptstyle xT}\downarrow & & \downarrow{\scriptstyle yT} \\
xG & \xrightarrow{\;fG\;} & yG
\end{array}
$$

This means that, if we start from an element of xF and map it to an element of yG by following the arrows along either possible route, the result will be the same (independent of the route taken).

It is probably still not clear what this definition means. In fact, mathematics abounds in important examples. Here are a few.

Determinant   Let C be the category of commutative rings with identity, C' the category of groups. As a small specialisation of an earlier example, both U (the group of units) and GL_n (the group of invertible $n \times n$ matrices) are functors from C to C'. We claim that det (determinant) is a natural transformation from GL_n to U. The first assertion of the definition is that, for any commutative ring R, det is a homomorphism from $\mathrm{GL}(n, R)$ to $U(R)$: this, as we have seen, is a fundamental property of the determinant, namely

$$\det(AB) = \det(A)\det(B).$$

The second property connects this with ring homomorphisms. If $f : R \to S$ is a homomorphism, we have

$$(\det(A))f = \det(Af),$$

where f denotes also the induced maps $\mathrm{GL}_n(R) \to \mathrm{GL}_n(S)$ and $U(R) \to U(S)$ (which we might, more consistently, call $\mathrm{GL}_n(f)$ and $U(f)$).

Group actions   Let G be a group. We saw that a functor from G to the category of sets is just a permutation action of G on a set Ω. A natural transformation between two such functors (actions on sets Ω_1 and Ω_2) is a G-homomorphism between two such actions: that is, a map $T : \Omega_1 \to \Omega_2$ such that $(\alpha g)T = (\alpha T)g$ for all $\alpha \in \Omega_1$.

Double duals   The **dual space** of an F-vector space V is the vector space V' of all linear maps from V to F. Duality is not a functor as we have defined it, since it 'reverses arrows'; that is, if $f : V \to W$ is linear, then $f' : W' \to V'$ is defined by

$$v(\phi f') = (vf)\phi$$

for $\phi \in W'$, $v \in V$. (Duality is what is known as a **contravariant functor**.) If D denotes duality, then D^2 is a functor from the category of vector spaces to itself. Also, there is a natural transformation T from the identity to D^2: VT is the map $V \to V''$ under which the image of $v \in V$ is the map $\phi \to v\phi$ from V' to F.

This makes precise the notion that there is a natural embedding of a space into its second dual, independent of any choice of basis. If we want to embed a space in its dual, we must make some choices: the embedding is not 'natural'.

Exercise 7.35   Show that the classes of rings, commutative rings, rings with identity, and abelian groups are varieties.

Exercise 7.36   Let $\mathcal{V}$ be a variety of algebras with two unary operations α_1 and α_2 and a binary operation β, satisfying the laws

$$(x,y)\beta\alpha_1 = x, \qquad (x,y)\beta\alpha_2 = y, \qquad (z\alpha_1, z\alpha_2)\beta = z.$$

Prove that any algebra in $\mathcal{V}$ with more than one element is infinite.

Exercise 7.37 A **quasigroup** is an algebra with three binary operations μ, λ, ρ satisfying the laws

$$(x, (x, y)\mu)\lambda = y, \qquad (x, (x, y)\lambda)\mu = y,$$
$$((x, y)\mu, y)\rho = x, \qquad ((x, y)\rho, y)\mu = x.$$

(a) Show that, in a quasigroup, the three equations $(x, y)\mu = z$, $(x, z)\lambda = y$, $(z, y)\rho = x$ are equivalent. (Thus, λ and ρ are 'left and right division' with respect to the 'multiplication' μ.)

(b) Show that the operation table of a binary operation μ on A has the property that each element occurs exactly once in each row or column if and only if there are operations λ and ρ such that the three operations μ, λ, ρ define a quasigroup on A.

[A table with the property described in this problem is a **Latin square**.]

Exercise 7.38 ($*$) This exercise shows that, in an algebra of given type, we can write elements unambiguously without needing brackets.

Consider an alphabet consisting of a set of variables and a set of operation symbols, each operation symbol having a given non-negative arity. Define the **variability** of a word in this alphabet to be the integer obtained by subtracting from its length the sum of the arities of the operation symbols that it contains. A **prefix** of a word is obtained by deleting any number of symbols from the end of the word. Show that a word w represents a (unique) element of an algebra (after substituting elements of the algebra for the variables) if and only if the following two conditions hold:

(a) w has variability 1;
(b) every non-empty prefix of w has positive variability.

Devise a **decoding algorithm** which tests whether a word satisfies this condition and, if so, parses the word.

Exercise 7.39 ($**$) Let A be the set of all words satisfying conditions (a) and (b) above. For each n-ary operator symbol μ, and any n elements $a_1, \ldots, a_n \in A$, show that $a_1 \cdots a_n \mu \in A$. Hence show that A is an algebra with the given collection of operators. Show that, if B is any algebra of the same type, and we choose an element $b_i \in B$ corresponding to each variable x_i, then there is a unique homomorphism from A to B which maps x_i to b_i for each i.

[A is called the **free algebra** of the given type with the given set of variables as generators.]

Exercise 7.40 Does the family of unique factorisation domains form a variety?

Exercise 7.41 In a group G, the **commutator** $[x, y]$ of elements x and y is the element $x^{-1}y^{-1}xy$.

(a) Show that $[x, y] = 1$ if and only if x and y commute.
(b) Hence show that G is abelian if and only if it satisfies the law $[x_1, x_2] = 1$.
(c) Let N be a normal subgroup of G such that G/N is abelian. Show that all commutators $[x, y]$ belong to N. Hence show that, if also N is abelian (so that G is metabelian), then any two commutators commute, and G satisfies the law $[[x_1, x_2], [x_3, x_4]] = 1$.
(d) Show that the subgroup H generated by all commutators in G (the **derived group** or **commutator subgroup** of G) is normal, and that G/H is abelian.

(e) Show that, if G satisfies the law $[[x_1, x_2], [x_3, x_4]] = 1$, then its derived group is abelian, and hence G is metabelian.

*(f) Generalise the above to show that soluble groups of derived length at most d form a variety.

Exercise 7.42 A congruence on an algebra A is an equivalence relation on A; that is, a subset E of $A \times A$. Prove that it is a **subalgebra** of $A \times A$. Is every subalgebra of $A \times A$ a congruence? If not, can you formulate necessary and sufficient conditions?

Exercise 7.43 Let X be a Boolean lattice. Define new operations $+$ and $\cdot$ on X by the rules $x + y = x \vee y \wedge (x \wedge y)'$, $x \cdot y = x \wedge y$. Prove that $(X, +, \cdot)$ is a Boolean ring (a ring satisfying the law $x^2 = x$). Show also that any Boolean ring gives rise to a Boolean lattice by taking $x \vee y = x + y + xy$, $x \wedge y = xy$.

Prove that the categories of Boolean lattices and Boolean rings are naturally isomorphic.

Exercise 7.44 (∗) In this exercise, we show that a finite Boolean lattice is isomorphic to the lattice of subsets of a set.

We must construct a set S, and a subset $s(x)$ of S for each $x \in S$, such that:

- Every subset of S has the form $s(x)$ for a unique $x \in X$.
- $s(0) = \emptyset$ and $s(1) = S$.
- $s(x \vee y) = s(x) \cup s(y)$ and $s(x \wedge y) = s(x) \cap s(y)$.
- $s(x') = S \setminus s(x)$.

An **ideal** in a Boolean lattice X is a subset I of X such that:

- if $a, b \in I$ then $a \vee b \in I$;
- if $a \in I$ and $x \in X$ then $a \wedge x \in I$.

Note the analogy with ideals in a ring (with $\vee$ and $\wedge$ taking the place of $+$ and $\cdot$). Let S be the set of all maximal ideals of X (those contained in no larger ideal), and for $x \in X$, let $s(x)$ denote the set of maximal ideals containing x.

Show that the properties listed above are indeed satisfied.

Exercise 7.45 (∗) Prove that a distributive lattice is modular.

Exercise 7.46 Prove that, in any lattice, if $x \leq z$, then

$$x \vee (y \wedge z) \leq (x \vee y) \wedge z.$$

[*Hint*: Show that, if $a_i \leq b_j$ for $i, j = 1, 2$, then $a_1 \vee a_2 \leq b_1 \wedge b_2$.]

Exercise 7.47 Complete the proof of Theorem 7.55(b).

Exercise 7.48 True or false?

(a) A partition lattice is modular.
(b) A subalgebra lattice is modular.

Exercise 7.49 Let S be a set, and F a field. Let V be the vector space of all functions from S to F, with pointwise operations. For each partition π of S, let V_π denote the

subspace

$$\{f \in V : f(x) = f(y) \text{ whenever } x, y \text{ are in the same part of } \pi\}.$$

Prove that

(a) The map $\pi \mapsto V_\pi$ is one-to-one;
(b) $V_{\pi_1 \vee \pi_2} = V_{\pi_1} \wedge V_{\pi_2}$.

Is it true that the dual of the partition lattice of S is embeddable in the subspace lattice of V?

Exercise 7.50 A lattice is **complete** if any subset has a least upper bound and a greatest lower bound.

(a) Show that a finite lattice is complete.
(b) Show that a lattice in which any subset has a greatest lower bound is complete.
(c) Show that the subalgebra lattice of an algebra is complete. What about the congruence lattice?
(d) Give an example of a lattice which is not complete.

Exercise 7.51 Prove that a finite lattice is Boolean if and only if it is the direct product of copies of $\{0, 1\}$.

Exercise 7.52 (a) Prove that the subspace lattice of a 2-dimensional vector space over F is a line with $|F| + 1$ points.
(b) Prove that the subspace lattice of a 3-dimensional vector space is a projective plane.
(c) Show that the lattice shown in Figure 7.3 is the lattice of subspaces of $\mathrm{GF}(2)^3$.

Exercise 7.53 Let L be an atomic lattice, with A the set of atoms. For any $x \in L$, let $S(x)$ denote the set of atoms $a \in A$ satisfying $a \leq x$.

(a) Prove that $S(x) \cap S(y) = S(x \wedge y)$.
(b) If L is Boolean, show that $S(x') = A \setminus S(x)$, and deduce that $S(x \vee y) = S(x) \cup S(y)$. Hence prove Theorem 7.54. [You have to show that a finite Boolean lattice is atomic.]

Exercise 7.54 Let M_3 be the three-point line and N_5 the pentagon (Figure 7.4). Prove that M_3 is not distributive and N_5 is not modular.

Exercise 7.55 ($*$) Show that, if a finite line is the subgroup lattice of a group, then the number of atoms is $p + 1$, where p is prime. Which groups have such a subgroup lattice?

Fig. 7.4 Two lattices

Exercise 7.56 Given a set M of **morphisms** with a partial composition and a subset I of identities, suppose that the following conditions are satisfied:

- For any $f, g, h \in M$, if fg and gh are defined, then $(fg)h$ and $f(gh)$ are defined and are equal.
- For any $f \in M$, there are unique identities i and j such that if and fj are defined; and $if = fj = f$.
- For $f, g \in M$, fg is defined if and only if there is an identity j such that fj and jg are defined.
- For any identity i, ii is defined.

Prove that M is the set of morphisms of a category.

Exercise 7.57 Let X_1 and X_2 be groups. Let P be a group and let $\pi_i : P \to X_i$ ($i = 1, 2$) be homomorphisms. Suppose that, if Z is any group and $\phi_i : Z \to X_i$ ($i = 1, 2$) are homomorphisms, then there is a homomorphism $\psi : Z \to P$ such that $\psi \pi_i = \phi_i$ for $i = 1, 2$.

Prove that P is isomorphic to $X_1 \times X_2$.

Exercise 7.58 ($*$) We can turn the preceding exercise into a definition. Let X_1 and X_2 be objects in a category. A **direct product** of X_1 and X_2 is an object P and a pair $\pi_i : P \to X_i$ ($i = 1, 2$) of morphisms such that, for any object Z and morphisms $\phi_i : Z \to X_i$ ($i = 1, 2$), there is a unique morphism $\psi : Z \to P$ such that $\psi \pi_i = \phi_i$ for $i = 1, 2$.

Show that any two direct products of X_1 and X_2 are isomorphic.

Give an example of two objects in a category which do not have a direct product.

Exercise 7.59 ($**$) Give a similar definition of **inverse limit** in a category (see Exercise 7.29).

Exercise 7.60 Let $\theta : G \to H$ be a homomorphism of groups. Prove that $G'\theta \le H'$. Hence show that θ induces a unique homomorphism $\theta^* : G/G' \to H/H'$.

Hence show how to define a functor from groups to abelian groups which maps G to G/G' for any group G.

Exercise 7.61 Why is there no natural way to define a functor from groups to abelian groups which maps G to $Z(G)$ (the centre of G)?

Exercise 7.62 A **preorder** is a reflexive and transitive relation on a set.

(a) Show that any preordered set (X, P) is a category, with object set X and morphism set P, with $\mathrm{dom}(x, y) = x$ and $\mathrm{cod}(x, y) = y$ for all $(x, y) \in P$, and $1_x = (x, x)$.

(b) Show that a category is a preorder if and only if there is at most one morphism with any given domain and codomain.

Exercise 7.63 Prove that GL_n is a functor.

8 Applications

From the surprisingly many applications of abstract algebra, I have chosen just two. One of these is the construction of Évariste Galois in the early nineteenth century; it explains why there is no formula for the solution of a polynomial equation of degree 5 or greater (comparable to the familiar formula $x = (-b \pm \sqrt{b^2 - 4ac})/2a$ for the solution of the quadratic equation $ax^2 + bx + c = 0$). The other is a much more recent development, the theory of error-correcting codes, for transmitting information through noisy channels.

Coding theory

8.1 Codes. One of the most famous applications of coding theory occurred in the exploration of the outer planets of the solar system by unmanned space probes. These probes carried scientific equipment and cameras. The information about temperatures, magnetic fields, and so on was very important to astronomers, but the pictures of Jupiter, Saturn, and their moons made us all aware of the existence of other worlds with their distinctive characters.

Typically, one of these probes had a generator capable of producing a few hundred watts of electric power, of which only a few tens of watts was available to the transmitter responsible for sending the information back to earth. This weak signal had to be separated from the radio 'noise' produced by the universe, and the useful information filtered from it. Naturally, sometimes the signal received was incorrect as a result of this interference. The job of error correction is to ensure that the correct information is received.

The procedure can be divided into a number of stages.

Stage 1: Generation of messages For simplicity, all information is sent as a sequence of 'words' or blocks of zeros and ones. A picture is divided into a large grid of small squares or 'pixels'. Each pixel is then scanned and the intensity of each of the three primary colours measured. This measurement is then digitised to be an integer in the range $[0, 255]$ say, and the resulting integer converted into 8-bit binary form. In this way, a picture becomes a very long string of zeros and ones, which can be chopped up into blocks of fixed length.

Stage 2: Encoding Each block is then translated into a longer block called a 'codeword'. This slows down the transmission time, since more bits have to be sent; the redundancy is used for error correction. The encoding is devised so that any two codewords look very different. Then even if a few bits are changed during transmission, what is received should look more like the transmitted codeword than any other, and decoding is possible.

Stage 3: Transmission The codewords are sent by the transmitter, and received by Earth-based equipment (possibly with some errors).

Stage 4: Decoding As explained, the receiver has the list of all possible codewords, and can compare the received word to them to find which is most similar.

Stage 5: Recovery of the information From the codeword, we can translate back into the binary string, and interpret it as a picture (or as scientific data, as appropriate).

We now develop the mathematical language to describe this.

Definition Let F be a set of symbols, called the **alphabet** (with $|F| = q > 1$), and let n be a positive integer. A **word** of length n over F is an n-tuple of symbols from F. It is common to write a word as $a_1 a_2 \cdots a_n$ instead of $(a_1, a_2, \ldots, a_n)$.

A **code** C of length n is a subset of the set F^n of all words of length n, subject to the condition that $|C| > 1$. Its elements are called **codewords**.

Remarks The reason for requiring that $|C| > 1$ is that, if only one message could be sent, no information could ever be conveyed (except the information that the transmitter is operating). In the space probe, the alphabet F is the **binary alphabet** $\{0, 1\}$.

Definition Let v, w be words of length n. The **Hamming distance** $d(v, w)$ from v to w is the number of coordinates where v and w differ:

$$d(v, w) = |\{i : 1 \le i \le n, v_i \ne w_i\}|.$$

Remark The motivation here is that we regard a single 'error' in transmission as changing one letter in a word. So the Hamming distance $d(v, w)$ is the number of errors which would be required to change the transmitted word v into the received word w. We suppose that our system has the property that it is unlikely that a large number of errors will occur; so, with high probability, the Hamming distance between transmitted and received words is not too great.

Proposition 8.1 *(a) For any words v, w, we have $d(v, w) \ge 0$, and $d(v, w) = 0$ if and only if $v = w$.*
(b) For any words v, w, we have $d(w, v) = d(v, w)$.
*(c) (The **triangle inequality**.) For any words u, v, w, we have $d(u, v) + d(v, w) \ge d(u, w)$.*

Proof (a) and (b) are trivial.

For (c), note that $d(u, v)$ errors will turn u into v, and a further $d(v, w)$ errors turn v into w. But some coordinates may have been changed twice, so the distance from u to w may be smaller than the sum of $d(u, v)$ and $d(v, w)$. □

Remark In topology, a **metric space** is defined to be a set M with a 'distance function' or **metric** d from $M \times M$ to the non-negative real numbers satisfying conditions (a), (b), and (c) of the proposition. So we have shown that the set of words of length n, equipped with Hamming distance, is a metric space.

Definition Let e be a positive integer. The code C of length n is said to be e-**error-correcting** if the following holds: for any word w of length n, there is *at most one* codeword $c \in C$ which satisfies $d(w, c) \le e$.

The reason for the name is as follows: Suppose that C is e-error-correcting. Suppose that we use C in a communication system in which we know, or can be fairly certain, that not more than e errors will occur during the transmission of a single word. Then these errors will be corrected. For suppose that c is transmitted and w received. Then by our assumption, $d(c, w) \le e$. Since C is e-error-correcting, any other codeword c' satisfies $d(c', w) > e$. So c is the nearest codeword to the received word, and the decoding is correct.

Definition The **minimum distance** of a code C is the least distance between two distinct codewords in C.

Theorem 8.2 *The code C is e-error-correcting if and only if its minimum distance is $2e + 1$ or greater.*

Proof Suppose that C is *not* e-error-correcting, so that there exist a word w and two different codewords c_1 and c_2 both at distance e or less from w: that is, $d(c_1, w) \le e$ and $d(c_2, w) \le e$. By the properties of Hamming distance, $d(w, c_2) \le e$, and so $d(c_1, c_2) \le e + e = 2e$. Hence it is not true that C has minimum distance $2e + 1$ or more.

Conversely, suppose that C has minimum distance $d \le 2e$. Choose f to be the integer part of $d/2$. Then $f \le e$ and $d - f \le e$. There are two codewords c_1 and c_2 with $d(c_1, c_2) = d$. Thus we can get from c_1 to c_2 by changing d coordinates. Let these be changed one at a time, and let w be the word obtained when f coordinates have been changed. Then $d(c_1, w) = f \le e$, and $d(c_2, w) = d - f \le e$. So C is not e-error-correcting. $\square$

Thus, we want a good code to have large minimum distance (so that it will correct as many errors as possible). We also want it to have as many codewords as possible: the more codewords, the faster information can be sent.

To measure this, we define the **rate** of a code C of length n over an alphabet of q symbols to be $\log_q |C|/n$. The motivation is that, if $|C| = q^k$, then q^k messages can be encoded. Without the encoding, each message could be sent as a k-tuple; after coding, it becomes an n-tuple, and so transmission is k/n times as fast as without the encoding.

The tension between error correction and rate is expressed in various inequalities connecting the minimum distance and size of a code. Here are two of the simplest:

Theorem 8.3 *Let C be a code of length n over an alphabet of q symbols, having minimum distance d.*

*(a) (**Hamming bound**): If $d \geq 2e+1$ (that is, if C is e-error-correcting), then*

$$|C| \leq q^n \Big/ \sum_{i=0}^{e} \binom{n}{i}(q-1)^i.$$

*(b) (**Singleton bound**): $|C| \leq q^{n-d+1}$.*

Proof (a) Let c be a codeword. We count the number of words w such that $d(c, w) \leq e$. How many satisfy $d(c, w) = i$? We have to make i errors, which we can do by choosing i coordinates to change (in $\binom{n}{i}$ ways), and then changing the entry in each of these coordinates to a different symbol in the alphabet ($q - 1$ choices for each coordinate, so $(q - 1)^i$ altogether). Multiplying these numbers and summing over i, the number of words w which satisfy $d(c, w) \leq e$ is

$$\sum_{i=0}^{e} \binom{n}{i}(q-1)^i.$$

We can regard these words as forming a 'ball' of radius e having the codeword c as its centre.

If we do this for all codewords, there is no overlap among the words we find. For, by assumption, C is e-error-correcting; so no word is at distance e or less from more than one codeword. Geometrically, the balls are packed into the space without overlap. So the number of words accounted for is

$$|C| \cdot \sum_{i=0}^{e} \binom{n}{i}(q-1)^i.$$

But this cannot exceed the total number q^n of words.

(b) Look at the codewords through a window which shows only the first $n - d + 1$ coordinates. The pieces we see are all different. For if the first $n - d + 1$ coordinates of c_1 and c_2 are the same, then these codewords cannot differ in more than the last $n - (n - d + 1) = d - 1$ coordinates; so $d(c_1, c_2) \leq d - 1$, contrary to assumption.

So the number of codewords does not exceed the total number q^{n-d+1} of things that can be seen through the window. $\qquad\square$

Remark Codes which attain these bounds are of particular importance. Such codes are called **perfect** for the Hamming bound, or **maximum-distance**

separable (MDS for short) for the Singleton bound. We will see interesting examples shortly, but here is a simple example. The **repetition code** of length n consists of all codewords $aa \cdots a$ (for $a \in F$). The idea is that, to get the message through, repeat it lots of times! This code has q codewords (one for each symbol), and the minimum distance is equal to the length n. The Singleton bound for $d = n$ asserts that $|C| \leq q^{n-n+1} = q$, so this bound is met. In the case $q = 2$, if n is odd, say $n = 2e + 1$, the Hamming bound is also attained.

There is more to coding theory than just resolving this tension between large minimum distance and many codewords. The codes should not be too difficult to implement. (Recall the space probe: the encoding must be done by a simple low-powered machine.) It turns out that concepts from algebra are crucial for this.

8.2 Linear codes. Let us formalise the encoding and decoding processes. Let S be a set of messages, and C a code of length n over an alphabet F with q symbols. Then encoding is a function $\epsilon : S \to C$, which is one-to-one (since if two messages were assigned to the same codeword, the receiver could never decide which had been intended). In fact, we normally assume that the encoding function is a bijection, since codewords which are never used could be removed from the code. In this case, there is an inverse map ϵ^{-1} which translates codewords back to messages.

Decoding is a function $\delta : F^n \to C$. No formal restrictions are made; but often we assume that it is **nearest-neighbour decoding**. This means that $\delta(w)$ is always a codeword which is as near as possible to w (one which minimises $d(c, w)$ over all $c \in C$). If there is more than one codeword at the smallest distance from w, then $\delta(w)$ should be one such, but we do not specify which one. We could if required follow δ with the inverse of ϵ to get a map from F^n to the set S of messages.

Usually, F is the so-called **binary alphabet** $\{0, 1\}$, which can be regarded as the field $\mathbb{Z}_2$. We will be more general, and assume only that F is a finite field. Then the set F^n is an n-dimensional vector space over F. We will see that there are good reasons for assuming that C is a subspace of F^n. If it is a k-dimensional subspace, then $|C| = q^k$. So there are also q^k messages in S, and we may assume that they are all the k-tuples of elements of F. Then it is natural to make the encoding map a linear transformation!

Definition Let the alphabet F be a finite field. The code C is a **linear code** if it is a subspace of the F-vector space F^n.

There are other advantages too.

Definition Suppose that the alphabet is a finite field. The **weight** $\mathrm{wt}(w)$ of a word w is the number of non-zero coordinates of w. The **minimum weight** of a code is the smallest non-zero weight of any codeword.

Proposition 8.4 *The minimum weight and minimum distance of a linear code are equal.*

Proof First we show that $d(v, w) = \mathrm{wt}(v - w)$. This holds because the ith coordinates of v and w are unequal if and only if the ith coordinate of $v - w$ is non-zero.

Now let C have minimum distance d and minimum weight f. Let $d(c_1, c_2) = d$. Then $\mathrm{wt}(c_1 - c_2) = d$, and $c_1 - c_2 \in C$ by linearity; so $d \geq f$ (as f is the minimum weight). Conversely, let $\mathrm{wt}(c) = f$. Then, by definition, $d(c, 0) = f$, where 0 is the all-zero word; and $0 \in C$ by linearity, so $f \geq d$ (as d is the minimum distance). We conclude that $d = f$, as required. $\square$

Thus, instead of comparing all pairs of codewords to find the minimum distance, in a linear code it is only necessary to look at all codewords to find the minimum weight.

Also, if c is transmitted and w received, then $w = c + x$, where the weight of x is equal to the number of errors which occurred.

How do we describe a linear code C?

Since C is nothing but a subspace of F^n, we choose a basis for it, a set of k words of length n. We can arrange these vectors as the rows of a $k \times n$ matrix G, called a **generator matrix** for C. Thus, formally, a generator matrix for C is a matrix whose rows form a basis for C; and C is the row space of G.

The reason for the term 'generator matrix' is as follows: Any codeword can be written uniquely as $x_1 g_1 + \cdots + x_k g_k$, where $g_1, \ldots, g_k$ are the rows of C. More briefly, this is xG, where $x = x_1 \cdots x_k \in F^k$. Thus, if the set S of messages to be sent is the set F^k of all words of length k, then the encoding map ϵ is just the linear map $x \mapsto xG$ from F^k to F^n: it is one-to-one, and its image is C.

In the binary case, the matrix multiplication involved in computing the encoding map can be performed by a very simple circuit, which can be built into a space probe.

If we apply elementary row operations to G, we do not change its row space, and hence the result is still a generator matrix for C. (We do, however, change the encoding map.) By the results of Section 4.10, we can assume that G is in reduced echelon form. Now, given a matrix in reduced echelon form, we can apply column permutations to bring the columns containing the leading 1s to the front, obtaining a matrix of the form $(I \ A)$ in block form. Of course, a column permutation does change the code: it has the effect of applying the same permutation to all codewords. However, this does not change weights or Hamming distances, which are the important things as far as coding theory goes; the new code is as good as the old one. So we may assume, if necessary, that the generator matrix is in the **standard form** $G = (I \ A)$.

If G is in standard form, then the encoding map is given by

$$x \mapsto xG = (x \ xA).$$

We see that the first k symbols of the codeword are precisely the message being sent. This makes the recovery of the message from the codeword after decoding particularly simple. These k symbols are called the **information digits**, and the remaining $n - k$ symbols the **check digits**. Encoding consists of taking the message and adding to it some new digits for error correction.

It is time for an example.

Example 1 Over the binary field $\mathbb{Z}_2$, let

$$G = \begin{pmatrix} 1 & 0 & 0 & 0 & 0 & 1 & 1 \\ 0 & 1 & 0 & 0 & 1 & 0 & 1 \\ 0 & 0 & 1 & 0 & 1 & 1 & 0 \\ 0 & 0 & 0 & 1 & 1 & 1 & 1 \end{pmatrix}.$$

Then G is the generator matrix, in standard form, of a binary linear code C with length 7 and dimension 4. By listing all $2^4 = 16$ codewords, it is easily checked that the minimum weight is 3, so that C is 1-error-correcting. Encoding takes a message $x_1 x_2 x_3 x_4$ to a codeword $x_1 \cdots x_7$, where

$$x_5 = x_2 + x_3 + x_4,$$
$$x_6 = x_1 + x_3 + x_4,$$
$$x_7 = x_1 + x_2 + x_4.$$

This code attains the Hamming bound of Theorem 8.3(a), since

$$|C| = 16 = 2^7/(1 + 7(2 - 1)).$$

This means that the balls of radius 1 with centres at the codewords cover the whole of F^7, so that any word whatever is at distance 0 or 1 from a unique codeword.

Hence, to decode, we could take the received word, look through the list of 16 codewords to find the one which is equal to it or differs in one place only, decode to this codeword, and take the first four digits as the message.

In the next section, we give a more efficient decoding method.

8.3 Syndrome decoding. We begin by giving another description of a code. The motivation comes from either linear algebra or coding theory.

In terms of linear algebra, we described a subspace of F^n as the row space of a matrix, in other words, the image of a linear transformation. It would be just as natural to use the kernel instead.

In coding theory terms, the motivation is even more convincing. The received word has the form 'codeword plus error'; we want to remove the error and reveal the codeword. However, the error is unknown, and we know that the codeword is chosen from a known subspace. So, instead, we remove the codeword to reveal the error, and then find the codeword by subtracting the error from the received word. Accordingly, we want a function f such that f maps every codeword to zero but f is one-to-one on possible error patterns. Of course, f should also be linear, so that

$$f(\text{codeword} + \text{error}) = f(\text{codeword}) + f(\text{error}) = f(\text{error}),$$

from which the error can be determined.

Definition Let C be a linear code with length n and dimension k. A **check matrix** for C is a $(n - k) \times n$ matrix H with the property that, for any word $w \in F^n$, we have $wH^\top = 0$ if and only if $w \in C$.

Note that we take H, like G, to have n columns, so that we have to transpose it in the equation.

The word $wH^\top$ is called the **syndrome** of w.

Proposition 8.5 *Let H be a check matrix for a linear e-error-correcting code. Then, if w_1 and w_2 are any two words with weight at most e which have the same syndrome, then $w_1 = w_2$.*

Proof If $w_1 H^\top = w_2 H^\top$, then $(w_1 - w_2)H^\top = 0$, and so $w_1 - w_2 \in C$. But $w_1 - w_2$ has weight at most $2e$, whereas C has minimum (non-zero) weight at least $2e + 1$. So, necessarily, $w_1 - w_2 = 0$.   □

Thus our condition that different error patterns have different images is satisfied, at least for the errors that we can correct.

The decoding now works as follows: Given the received word, calculate its syndrome, work out the error pattern which would produce that syndrome (for example, look it up in a table), and subtract that error pattern from the received word to give the codeword.

In our example, it can be checked that the matrix

$$H = \begin{pmatrix} 0 & 0 & 0 & 1 & 1 & 1 & 1 \\ 0 & 1 & 1 & 0 & 0 & 1 & 1 \\ 1 & 0 & 1 & 0 & 1 & 0 & 1 \end{pmatrix}$$

is a check matrix for the code of Example 1 earlier. Since the code is 1-error-correcting, the relevant error patterns are 0 and the word e_i with 1 in the ith position and 0 elsewere, for $i = 1, \ldots, 7$. The syndrome of 0 is 0; the syndrome of e_i is the ith row of $H^\top$, which by inspection happens to be the base 2 representation of the integer i. So decoding is particularly simple:

Given a received word w, calculate its syndrome $wH^\top$. If it is zero, then no error occurred; otherwise, if the syndrome is the base 2 representation of i, then the ith digit is incorrect.

Suppose, in our example, that we wish to send the message 1101. This is encoded as 1101001. Suppose that an error occurs in the second position, giving the received word 1001001. Multiplying by $H^\top$ gives the syndrome 010, which is the second row of $H^\top$, or the base 2 representation of 2. So we correct the word to 1101001, and extract the information 1101 correctly.

If, on the other hand, two errors occurred, say in positions 2 and 5, the received word would be 1001101, with syndrome 111; we would decode by changing the seventh digit, giving 1001100, and extract the wrong message 1001. Our use of this code depends on the assumption that it is very unlikely that two or more errors will occur during the transmission of a single word.

It is possible to look at syndrome decoding in another way. If an unknown codeword is sent, and an error x occurs, then the received word belongs to the coset $C + x$. Now C is the kernel of the linear transformation from F^n to F^{n-k} represented by $H^\top$, so each such coset maps to a single word in F^{n-k}. We can correct any assumed set of error patterns which are mapped one-to-one by $H^\top$; that is, at most one from each coset. It is natural to choose the word of smallest weight in a coset, as the most likely error pattern to occur. Such a word is called a **coset leader**. (If two or more words have the minimum weight in a coset, choose among them arbitrarily.) Then any syndrome specfies a unique coset, and we can decode using a table of syndromes and corresponding coset leaders.

Syndrome decoding can be used for any linear code, although in particular cases it may not necessarily be the most efficient way.

The minimum weight of a linear code can be found from its check matrix as follows.

Proposition 8.6 *Let C be a linear code with check matrix H. Then C has minimum weight δ or greater if and only if any $\delta - 1$ columns of H are linearly independent.*

Proof Let $h_1, \ldots, h_n$ be the columns of H. Then $c_1 \cdots c_n$ is a codeword if and only if $c_1 h_1 + \cdots + c_n h_n = 0$. So codewords of weight f correspond to dependence relations among sets of f columns; and the minimum weight is equal to the minimum size of a set of linearly dependent columns. □

Using the check matrix, it is easy to construct an important family of codes, the Hamming codes. Let r be given, let $F = \mathrm{GF}(q)$, and let $(F^r)^\top$ be the r-dimensional vector space of all columns of length r (that is, the $r \times 1$ matrices). Then $|X| = (q^r - 1)$. Let $X = (F^r)^\top \setminus \{0\}$. Call two elements of X **equivalent** if one is a scalar multiple of the other. Each equivalence class contains $q - 1$ elements (since there are $q - 1$ non-zero scalars); so there are $n = (q^r - 1)/(q - 1)$ equivalence classes. Let Y be a set of representatives of the equivalence classes. Thus Y consists of one non-zero vector from each 1-dimensional subspace of $(F^*)^\top$. Let H be the $r \times n$ matrix whose columns are the elements of Y. Then we define the q-ary **Hamming code** of length n to be the linear code with check matrix H.

Proposition 8.7 *Hamming codes are perfect 1-error-correcting (that is, they attain the Hamming bound).*

Proof By construction, no column is zero, and no column is a multiple of another. So any two columns are linearly independent. By Proposition 8.6, the code has minimum weight at least 3, and so it is 1-error-correcting. Since

$$|C| = q^{n-r} = q^n/(1 + n(q - 1)),$$

the Hamming bound of Theorem 8.3(a) is attained.   □

We can make the choice of columns definite by taking from each equivalence class the unique vector whose first non-zero element is 1. For example, the ternary (over GF(3)) Hamming code of length $(3^3 - 1)/(3 - 1) = 13$ has check matrix

$$\begin{pmatrix} 0 & 0 & 0 & 0 & 1 & 1 & 1 & 1 & 1 & 1 & 1 & 1 & 1 \\ 0 & 1 & 1 & 1 & 0 & 0 & 0 & 1 & 1 & 1 & 2 & 2 & 2 \\ 1 & 0 & 1 & 2 & 0 & 1 & 2 & 0 & 1 & 2 & 0 & 1 & 2 \end{pmatrix}.$$

One problem remains: How do we find a check matrix for a code?

Theorem 8.8 *(a) Let G and H be matrices of size $k \times n$ and $(n - k) \times n$ respectively over a finite field F, both having linearly independent rows. Then G and H are the generator and check matrices for the same code if and only if $GH^\top = 0$.*

(b) Let $G = (I\ A)$ be a generator matrix of a code C, in standard form. Then a check matrix of the same code is $H = (-A^\top\ I)$.

Proof   (a) By assumption, the row space C of G and the null space C' of H both have dimension k. Now $GH^\top = 0$ is equivalent to the assertion that every row of G lies in C'; that is, that $C \subseteq C'$. So the result follows.

(b) The identity blocks in G and H ensure that their rows are linearly independent; and $GH^\top = -IA + AI = 0$.   □

This gives a simple way to compute H, if G is in standard form.

In general, apply elementary row operations to G to bring it to reduced echelon form. If it is in standard form, proceed as before. Otherwise, apply a permutation π to its columns to bring it to standard form; then construct the H for this standard form of G, and apply the inverse of π to its columns.

8.4   Cyclic codes.   Cyclic codes form a subclass of the class of linear codes. For these codes, an even more precise algebraic description is possible, leading to improved decoding algorithms.

Definition   Let C be a linear code of length over a field F. Then C is a **cyclic code** if, for every word $w = a_1 a_2 \cdots a_n \in C$, the cyclic shift $a_n a_1 a_2 \cdots a_{n-1}$ is also in C.

We translate into algebra in the following way. It is convenient to change notation, and number the coordinates from 0 to $n - 1$, instead of from 1 to n. Now, with each word $w = a_0 a_1 \cdots a_{n-1} \in F^n$, we associate the polynomial $w(x) = a_0 + a_1 x + \cdots + a_{n-1} x^{n-1} \in F[x]$.

Let I be the ideal of $F[x]$ generated by $x^n - 1$, and let R be the factor ring $F[x]/I$. Now each coset of I in R has a unique representative which is a polynomial of degree at most $n - 1$ (or zero). For, given any polynomial $f(x)$, we can use the division algorithm to write

$$f(x) = (x^n - 1)q(x) + r(x),$$

where $r = 0$ or $\deg(r) < n$; and $f(x)$ and $r(x)$ lie in the same coset of I.

So there is a natural bijection between $R = F[x]/I$ and the set of polynomials of degree at most $n - 1$ (together with 0), and hence with the set F^n of words of length n. We will switch freely between these sets.

Proposition 8.9 *A code C of length n is cyclic if and only if the corresponding elements of R form an ideal.*

Proof Multiplication by x corresponds to the cyclic shift. For consider a word $w = a_0 a_1 \cdots a_{n-1}$. The corresponding polynomial is $a_0 + a_1 x + \cdots + a_{n-1} x^{n-1}$. Multiplying by x gives $a_0 x + a_1 x^2 + \cdots + a_{n-1} x^n$. Now x^n and 1 lie in the same coset of $I = (x^n - 1)$, so are equal in the factor ring. Thus, in R, the above 'polynomial' is equal to $a_{n-1} + a_0 x + a_1 x^2 + \cdots$, which corresponds to the word $a_{n-1} a_0 a_1 \cdots$, the cyclic shift of w.

Thus, if C is an ideal, it is closed under addition, and under multiplication by any scalar (hence it is a linear code), and under multiplication by x, in other words cyclic shift (and hence it is a cyclic code). Conversely, suppose that C is a cyclic code. Then it is closed under addition, and under multiplication by any scalar or by x. Combining these two operations, we can build any polynomial, so C is closed under multiplying by any polynomial, and so is an ideal. $\square$

The problem now is to describe the ideals in the ring R. First, we observe that they are all principal.

Proposition 8.10 *Let R be a commutative ring with identity in which every ideal is generated by a single element. Then the same properties hold for any factor ring of R.*

Proof Consider a factor ring R/I. By the Second Isomorphism Theorem, its ideals are all of the form J/I, where J is an ideal of R containing I. Now, if $J = (r)$, then $J/I = (I + r)$. $\square$

We cannot assume that the factor ring R/I is a principal ideal domain, however, even if R is; for it may not be an integral domain. For example, a factor ring of $\mathbb{Z}$ has the form $\mathbb{Z}_m$ for some m; all of its ideals are principal (by the proposition), but it is an integral domain only if m is prime (in which case it is a field).

Proposition 8.11 *Any ideal of $F[x]/(x^n - 1)$ is generated by (the coset containing) a monic polynomial $g(x)$ which divides $x^n - 1$. There is a unique such polynomial for any ideal.*

Remark The polynomial $g(x)$ is called the **generator polynomial** of the cyclic code corresponding to the ideal $(g(x))$.

Proof Suppose that the ideal I is generated by $f(x)$. Let $g(x)$ be the g.c.d. of $f(x)$ and $x^n - 1$ (in $F[x]$), chosen to be monic. Then g divides f, so (g) contains f. But also, by the Euclidean Algorithm, $g(x) = a(x)f(x) + b(x)(x^n - 1)$. In the factor ring R, this equation says $g(x) = a(x)f(x)$; so also f divides g, and (f) contains (g). Thus, g generates the ideal I. It is by definition a monic polynomial dividing $x^n - 1$.

The uniqueness follows from the Second Isomorphism Theorem. By assuming that our polynomial divides $x^n - 1$, we see that the ideal of $F[x]$ that it generates contains $(x^n - 1)$. So, if g_1 and g_2 were two such polynomials, they would generate the same ideal of $F[x]$. Hence they would be associates in $F[x]$; that is, they would differ only by a scalar factor. Since both are monic, they would be equal.   □

So to construct all cyclic codes of length n, we must factorise $x^n - 1$ into irreducibles in $F[x]$, then list all divisors of $x^n - 1$ (the products of some of these irreducibles), and for each divisor, form the corresponding ideal of R.

Theorem 8.12 *Suppose that $g(x)$ is the generator polynomial of the cyclic code C. Let*

$$g(x) = a_{n-k}x^{n-k} + a_{n-k-1}x^{n-k-1} + \cdots + a_0,$$

with $a_{n-k} = 1$, and let $g(x)h(x) = x^n - 1$, where

$$h(x) = b_k x^k + b_{k-1}x^{k-1} + \cdots + b_0,$$

with $b_k = 1$. Then a generator matrix G and a check matrix H for C are given by

$$G = \begin{pmatrix} a_0 & a_1 & \cdots & a_{n-k} & 0 & \cdots & 0 \\ 0 & a_0 & a_1 & \cdots & a_{n-k} & \cdots & 0 \\ \ddots & \ddots & \ddots & \ddots & \ddots & \ddots & \ddots \\ 0 & \cdots & 0 & a_0 & a_1 & \cdots & a_{n-k} \end{pmatrix},$$

$$H = \begin{pmatrix} b_k & b_{k-1} & \cdots & b_0 & 0 & \cdots & 0 \\ 0 & b_k & b_{k-1} & \cdots & b_0 & \cdots & 0 \\ \ddots & \ddots & \ddots & \ddots & \ddots & \ddots & \ddots \\ 0 & \cdots & 0 & b_k & b_{k-1} & \cdots & b_0 \end{pmatrix}.$$

Remark Both G and H are in echelon form. So the theorem implies that $\dim(C) = k = n - \deg(g(x))$.

Proof The rows of G correspond to the polynomials $g(x), xg(x), \ldots, x^{k-1}g(x)$, so they all belong to C.

Take any word $x \in C$, corresponding to a polynomial $f(x)g(x)$ (mod $x^n - 1$). Write $f(x) = h(x)q(x) + r(x)$, where $r = 0$ or $\deg(r) < k$. Then $f(x)g(x) = (x^n - 1)q(x) + r(x)g(x)$, and this is congruent mod $x^n - 1$ to $r(x)g(x)$, which is a linear combination of the polynomials $x^i g(x)$ for $i < k$. Hence w is a linear combination of the rows of G. So the row space of G is C, as claimed.

The (i, j) entry of G is a_{j-i}, with the convention that $a_l = 0$ if l is outside the range $[0, n - k]$. With a similar convention, the (i, j) entry of H is b_{k-j+i}. Hence the (i, j) entry of $GH^\top$ is

$$\sum_l a_{l-i} b_{k-l+j} = \sum_m a_m b_{k-i+j-m}.$$

This is the $(k - i + j)$th coefficient of the product gh. Now $1 \le i \le k$ and $1 \le j \le n - k$, so

$$k - k + 1 = 1 \le k - i + j \le k - 1 + (n - k) = n - 1.$$

But $g(x)h(x) = x^n - 1$, so all the relevant coefficients are zero. Thus $GH^\top = 0$, and it follows from Theorem 8.8 that H is a check matrix for C. $\square$

Example We consider binary cyclic codes of length 7. We have

$$x^7 - 1 = (x - 1)(x^3 + x + 1)(x^3 + x^2 + 1).$$

Thus there are $2^3 = 8$ cyclic codes, corresponding to the divisors of $x^7 - 1$ as follows:

$g(x) = 1$. This code is generated by 1000000 and its cyclic shifts, and so it is the whole of $\mathbb{F}_2^7$.

$g(x) = x - 1$. The code is generated by the word 1100000 and its cyclic shifts, and consists of all words of even weight. The dimension is 6 and the minimum weight is 2.

$g(x) = x^3 + x + 1$. This code consists of the zero word, 1101000 and its cyclic shifts, and every word obtained by interchanging zeros and ones. These 16 words form a code with dimension 4 and minimum weight 3 (and hence is 1-error-correcting). This is the code that we met in Example 1 earlier.

$g(x) = x^3 + x^2 + 1$. This code is obtained from the previous one by reversing all the codewords; so it is 'equivalent' (it has the same dimension and minimum weight).

$g(x) = (x - 1)(x^3 + x + 1)$. This code consists of the eight words of even weight in the earlier example (with $g(x) = x^3 + x + 1$). It has minimum weight 4.

$g(x) = (x - 1)(x^3 + x^2 + 1)$. This is the reverse of the preceding code.

$g(x) = x^6 + \cdots + x + 1$. This is the repetition code spanned by the all-1 vector.

$g(x) = x^7 + 1$. This is not a code at all, since it contains only the zero vector: we require a code to have at least two codewords!

8.5 BCH codes. In the last two sections, we have seen ways to construct codes for which the length and dimension are easy to calculate. Finding the minimum distance (or minimum weight), however, is much harder. In this section we examine a construction which enables one to specify a length n and a minimum distance d, and find a code of length n and minimum distance at least d. Moreover, we can also give a lower bound for the dimension of the constructed code. The construction was given independently by Bose and Ray-Chaudhuri and by Hocquenghem; so the codes should be called BRH codes, but the term 'BCH codes' has become standard.

The BCH codes depend on properties of finite fields. We now review these properties.

Definition Let n and q be coprime integers. The **order** of q mod n is the smallest positive integer e such that $q^e \equiv_n 1$. (Note that the condition $(n, q) = 1$ implies that q is a unit mod n; the order of q mod n is just the order of q as an element of the group of units of $\mathbb{Z}_n$.)

Definition A **primitive** nth root of unity in a field F is an element a whose order in the multiplicative group of F is precisely n (that is, $a^n = 1$ but $a^m \neq 1$ for $0 < m < n$).

Proposition 8.13 *Let q be a prime power, and let e be the order of q mod n. Then the smallest field which contains $\mathrm{GF}(q)$ and a primitive nth root of unity is $\mathrm{GF}(q^e)$.*

Proof Any field containing $\mathrm{GF}(q)$ has the form $\mathrm{GF}(q^m)$ for some m. If $\mathrm{GF}(q^m)$ contains a primitive nth root of unity, then its multiplicative group contains a subgroup of order n, and so n divides $q^m - 1$. Conversely, if n divides $q^m - 1$, then the multiplicative group of $\mathrm{GF}(q^m)$ (being cyclic) contains a cyclic subgroup of order n, so that $\mathrm{GF}(q^m)$ contains a primitive nth root of unity. $\square$

We also need the basic property of Vandermonde determinants (see Section 4.11): If $a_1, \ldots, a_n$ are distinct elements of a field F, and A is the $n \times n$ matrix with (i, j) entry a_i^{j-1} for $1 \leq i, j \leq n$, then $\det(A) \neq 0$.

The codes that we construct will be cyclic codes of length n over $\mathrm{GF}(q)$, where we assume that n and q are coprime. We are also given a positive integer δ. We define the **BCH code of length n and designed distance δ**. We will prove that the actual minimum distance is at least δ.

Let e be the order of q mod n, and let a be a primitive nth root of unity in $\mathrm{GF}(q^e)$. We take a representation of $\mathrm{GF}(q^e)$ by e-tuples of elements of $\mathrm{GF}(q)$, in the standard way: if α generates $\mathrm{GF}(q^e)$ over $\mathrm{GF}(q)$ and satisfies the polynomial $f(x) = 0$, where $\deg(f) = e$, then every element of $\mathrm{GF}(q^e)$ can be written uniquely as $c_0 + c_1\alpha + \cdots + c_{e-1}\alpha^{e-1}$, and can be represented by the e-tuple $c_0 c_1 \cdots c_{e-1}$. For technical reasons, we use here the transpose of this e-tuple: that is, we represent e by a $e \times 1$ matrix. (The actual element α used is unimportant, but we may choose $\alpha = a$.)

Now the **BCH code of length** n **and designed distance** δ **over** $\mathrm{GF}(q)$ is the code with check matrix

$$H = \begin{pmatrix} 1 & a & a^2 & \cdots & a^{n-1} \\ 1 & a^2 & a^4 & \cdots & a^{2(n-1)} \\ \cdots & \cdots & \cdots & \cdots & \\ 1 & a^{\delta-1} & a^{2(\delta-1)} & \cdots & a^{(\delta-1)(n-1)} \end{pmatrix}.$$

Each matrix element belongs to the field $\mathrm{GF}(q^e)$, and so is represented as an $e \times 1$ matrix over $\mathrm{GF}(q)$. Thus, H is a $e(\delta-1) \times n$ matrix over $\mathrm{GF}(q)$.

Theorem 8.14 *The BCH code of length n and designed distance δ over $\mathrm{GF}(q)$ has minimum distance at least δ, and has dimension at least $n - e(\delta - 1)$.*

Proof To show that the minimum distance of a code is at least δ, it is necessary and sufficient that any $\delta - 1$ columns of a check matrix for the code are linearly independent (see Proposition 8.6). Consider the determinant of the matrix (over $\mathrm{GF}(q^e)$) formed by columns $m_1, m_2, \ldots, m_{\delta-1}$. This is

$$\det \begin{pmatrix} a^{m_1} & \cdots & a^{m_{\delta-1}} \\ \cdots & \cdots & \cdots \\ a^{m_1(\delta-1)} & \cdots & a^{m_{\delta-1}(\delta-1)} \end{pmatrix}.$$

The ith column has a common factor $a^{m_i} \neq 0$. Taking out these factors, we obtain a Vandermonde determinant $V(a^{m_1}, \ldots, a^{m_{\delta-1}})$, which is also non-zero.

So the chosen columns are linearly independent over $\mathrm{GF}(q^e)$, and so certainly over the smaller field $\mathrm{GF}(q)$.

The dimension of the code is n minus the rank of the check matrix. This rank is not greater than the number $e(\delta - 1)$ of rows. (The rows may not be all independent.) $\square$

It is not obvious from the definition, but the following is true:

Proposition 8.15 *BCH codes are cyclic.*

Proof Any word $v = c_0 c_1 \cdots c_{n-1}$ corresponds, as in the last section, to a polynomial $f(x) = c_0 + c_1 x + \cdots + c_{n-1} x^{n-1}$. The conditions that the word lies in the BCH code can be written as

$$f(a^i) = c_0 + c_1 a^i + \cdots + c_{n-1} a^{i(n-1)} = 0,$$

for $i = 1, 2, \ldots, \delta - 1$. Let $g(x)$ be the least common multiple of the minimal polynomials of $a, a^2, \ldots, a^{\delta-1}$ over $\mathrm{GF}(q)$. Then the word corresponding to $f(x)$ lies in the BCH code if and only if $f(x)$ is divisible by $g(x)$. Moreover, the roots of $g(x)$ are nth roots of unity, so $g(x)$ divides $x^n - 1$. So the BCH code is the cyclic code with generator polynomial $g(x)$. $\square$

In many cases, we can use this observation to do better than the previous lower bound for the dimension. This depends on the following fact:

Proposition 8.16 *Let $f(x) \in \mathrm{GF}(q)[x]$, and suppose that α is a root of f in some extension field. Then α^q is also a root of f.*

Proof Let $q = p^n$, where p is prime. Then the map $x \mapsto x^q$ is the nth power of the Frobenius map, and hence is an automorphism of the field $K = \mathrm{GF}(q)(\alpha)$. Let $f(x) = a_n x^n + \cdots + a_0$, so that

$$f(\alpha) = a_n \alpha^n + \cdots + a_0 = 0.$$

We apply the automorphism $x \mapsto x^q$ to this equation, noting that $a_i^q = a_i$ for all i (since the coefficients a_i belong to $\mathrm{GF}(q)$, and so are roots of the polynomial $x^q - x$). So

$$f(\alpha^q) = a_n \alpha^{qn} + \cdots + a_0 = 0. \qquad \square$$

Now let $v = c_0 c_1 \cdots c_{n-1}$ be a word of length n. By the preceding result, if we can find two values i and j such that $j \equiv_n iq^m$ for some m, then the ith and jth conditions above are equivalent, and we can strike out the jth row of H. This does not affect the code, but means that the check matrix has fewer columns, so we obtain a larger lower bound for its dimension.

Example Consider the binary BCH code C with length 15 and designed distance 5. The codewords correspond to the polynomials having roots a, a^2, a^3, a^4, where a is a primitive 15th root of unity (in $\mathrm{GF}(2^4)$: note that the order of 2 mod 15 is 4.) This code has minimum weight 5 (and so is 2-error correcting). The lower bound we gave earlier for its dimension is $\dim(C) \geq 15 - 4 \cdot 4 = -1$: this is of course useless! But, by the previous result, a^2 and a^4 are unnecessary: it is enough to assume that a and a^3 are roots. This gives $\dim(C) \geq 15 - 4 \cdot 2 = 7$, so $|C| \geq 2^7 = 128$.

If we take a to be a root of the polynomial $x^4 + x + 1$ over $\mathrm{GF}(2)$, then a^3 (which is a 5th root of unity) is a root of $x^4 + x^3 + x^2 + x + 1$. So the generator matrix of the code is the product of these two polynomials. Since they are irreducible, we see that the dimension of C is exactly $15 - 8 = 7$.

The Hamming bound for a 2-error-correcting code of length 15 gives $|C| \leq 2^{15} / \left(1 + 15 + \binom{15}{2}\right) = 270.81\ldots$; since $|C|$ is a linear code, the number of codewords is a power of 2, so in fact $|C| \leq 256$. So the dimension of C is within one of best possible.

One important special case of BCH codes is that when $n = q - 1$. These codes were discovered earlier, and are called **Reed–Solomon codes**. In this case, the order of q mod n is clearly equal to 1. So the BCH bound for a code of designed distance δ gives $\dim(C) \geq n - \delta + 1$. On the other hand, if the true minimum distance is d, then $d \geq \delta$; and the Singleton bound (Theorem 8.3(b)) gives $|C| \leq q^{n-d+1}$, whence $\dim(C) \leq n - d + 1$. Summarising, we have

$$n - d + 1 \leq n - \delta + 1 \leq \dim(C) \leq n - d + 1,$$

so equality must hold throughout. Thus, we have

Proposition 8.17 *A Reed–Solomon code (that is, a BCH code of length $n = q - 1$ over $\mathrm{GF}(q)$) with designed distance δ has minimum distance δ and dimension $n - \delta + 1$. Hence it is maximum distance separable (that is, it attains the Singleton bound).*

Exercise 8.1 Consider the code over the alphabet $\{1, 2, 3\}$ whose words are 112233, 223311, 331122, 123123, 231231, 312312. Find the minimum distance of this code.

Exercise 8.2 A channel transmits binary digits in blocks of length 8. Because of synchronisation problems, errors are more likely in the first four bits than in the second four; the probability of incorrect transmission of a bit is $\frac{1}{10}$ for the first four bits and $\frac{1}{100}$ for the others. We use the following scheme to encode 4 bits of information for transmission through the channel. The input $a_1 a_2 a_3 a_4$ is encoded as $b_1 b_2 b_3 b_4 b_5 b_6 b_7 b_8$, where

(i) the first bit is repeated four times, that is,

$$b_1 = b_2 = b_3 = b_4 = a_1;$$

(ii) the next three bits are sent without change, that is,

$$b_5 = a_2, \; b_6 = a_3, \; b_7 = a_4;$$

(iii) the last bit is a 'parity check' for the three preceding, i.e. an even number of b_5, b_6, b_7, b_8 are equal to 1.

[For example, 1010 is encoded as 11110101.] Decoding is done as follows: Suppose that $c_1 c_2 c_3 c_4 c_5 c_6 c_7 c_8$ is received.

(a) If all or all but one of c_1, c_2, c_3, c_4 agree, we assume that their common value is a_1. If two of them are 0 and two are 1, we declare a *decoding failure* ('an error has occurred but we cannot correct it').

(b) If the last four bits have even parity, that is an even number of them are 1, we assume correct transmission, and set $a_2 = c_5, a_3 = c_6, a_4 = c_7$. Otherwise, we declare a decoding failure.

[For example, 11101111 is decoded as 1111, while 11101101 gives a decoding failure.]

Problems (a) Let C be the code, that is the set of all possible transmitted words. How many words are there in C? What is the rate of C?

(b) Prove that C is a linear code. What is its dimension? Write down a generator matrix for C.

(c) Calculate the probabilities of incorrect decoding and of decoding failure using this scheme. [*Hint*: The first four and the last four bits work quite independently, and so can be treated separately. In the first four,

0 or 1 error → correct, 2 errors → failure, 3 or 4 errors → incorrect;

while in the last four,

0 errors → correct, 1 or 3 errors → failure, 2 or 4 errors → incorrect.

Now make a 3 × 3 table; work out the probability of each of the 9 outcomes, and the overall result (correct, incorrect, or failure) in each case.]

Exercise 8.3 Let C be the linear code (over $\mathbb{Z}_3$) with generator matrix

$$G = \begin{pmatrix} 1 & 0 & 1 & 1 \\ 0 & 1 & 1 & 2 \end{pmatrix}.$$

(a) How many words are there in C?
(b) What is the minimum distance of C?
(c) What is the minimum weight of C?
(d) Find a check matrix for C (a matrix whose null space is C).
(e) Encode 12, and decode 1021, using C.

Exercise 8.4 A binary code C of length 8 has generator matrix

$$G = \begin{pmatrix} 1 & 1 & 1 & 1 & 1 & 1 & 1 & 1 \\ 0 & 0 & 0 & 0 & 1 & 1 & 1 & 1 \\ 0 & 0 & 1 & 1 & 0 & 0 & 1 & 1 \\ 0 & 1 & 0 & 1 & 0 & 1 & 0 & 1 \end{pmatrix}.$$

(a) Find a generator matrix in reduced echelon form.
(b) What is the minimum weight of C?
(c) Show that C can correct one error.
(d) Show that, if two errors occur during transmission, then C can detect this, but cannot locate the position of the errors.
(e) Calculate a check matrix for C.
(f) Which syndromes correspond to the occurrence of one error?
(g) Decode the received word 10101101.

Exercise 8.5 Show that a repetition code with $q = 2$ and n odd attains the Hamming bound.

Exercise 8.6 Verify the following table of values of the Hamming upper bound M for the maximum size of a binary code of length 10 which can correct up to e errors:

e	1	2	3	4
M	93	18	5	2

Prove that, in fact, a 3-error-correcting code of length 10 cannot contain more than two codewords. (Thus the Hamming bound is not always met!)

Construct a linear 1-error-correcting code of length 10 containing 32 codewords.

(*) Can you find one with 64 codewords?

Exercise 8.7 Show that, in any binary linear code, the set of codewords of even weight is a *linear* subcode (that is, a subspace of the vector space).

Exercise 8.8 (a) Let C be a linear code with check matrix H. Show that C is 1-error-correcting if and only if no column of H is zero and no column is a multiple of another.

(b) Suppose that the conditions of (a) hold. Verify the following rule for correcting one error:

Calculate the syndrome of the received word w. If it is zero, then w is correct. If it is a scalar multiple (say c) of the ith row of $H^\top$, then subtract c from the ith coordinate of w.

Exercise 8.9 (a) Let F be the finite field with four elements $\{0, 1, \omega, \overline{\omega}\}$. The arithmetic operations in F can be deduced from the rules

$$1 + 1 = 0, \qquad 1 + \omega = \omega^2 = \overline{\omega}.$$

Construct addition and multiplication tables for F.

(b) Let C be the linear code over F with generator matrix

$$\begin{pmatrix} 1 & 0 & 0 & 1 & 1 & 1 \\ 0 & 1 & 0 & 1 & \omega & \overline{\omega} \\ 0 & 0 & 1 & 1 & \overline{\omega} & \omega \end{pmatrix}.$$

Find the minimum weight of C, and a check matrix for C.

(c) Prove that no code over an alphabet of four symbols with the same length and minimum distance as C can contain more codewords than C.

Exercise 8.10 Prove that, if a perfect 2-error-correcting ternary code of length n exists, then $2n^2 + 1$ must be a power of 3.

Exercise 8.11 What is the dimension of the binary BCH code of designed distance 5 and length 31?

Exercise 8.12 A football pools competition requires contestants to predict the possible result (home win, away win, or draw) of each of n matches. Show that, in order to ensure that all or all but one of the predictions are correct, at least $3^n/(2n+1)$ entries are required. Explain why this bound can be met whenever $n = (3^d - 1)/2$ for some positive integer d.

Exercise 8.13 (*) Prove that a q-ary Hamming code of length n is cyclic if $(q - 1, n) = 1$.

Galois Theory

In Chapter 1 we saw the classical formula for the solution of a quadratic equation. Similar formulae for solving cubic and quartic equations were discovered by Tartaglia and Ferrari, and publicised by Cardano, in the Renaissance. Mathematicians searched unsuccessfully for such formulae for equations of higher degree. The fact that no such formulae can exist is just one detail in Galois Theory, which we now outline.

Galois himself, who has a good claim to be regarded as the founder of modern algebra, was killed in a duel in 1832 at the age of 21. Perhaps he would have performed better in the duel but for the fact that he had spent the night before in frantic activity, writing an account of all of his discoveries to his friend Chevalier. The letter was not published for 15 years, and Galois' work only found its rightful place in mathematics in the second half of the nineteenth century.

8.6   Normality and separability.   We are concerned here only with finite extensions of fields. Recall that the field K containing F is a **finite extension** of F if the dimension of K as F-vector space (forgetting mutiplication in K) is finite; this dimension is called the **degree** of the extension, written $[K : F]$.

Recall also that a **splitting field** of a polynomial $f(x)$ over F is a field generated over F by the roots of f. It is a finite extension with degree at most $n!$, where n is the degree of f. Galois theory is concerned with finding all the roots of a polynomial, so splitting fields are important. Our first job is to recognise them.

Definition   Let K be an extension of F. We say that K is a **normal** extension of F if, whenever f is an irreducible polynomial in $F[x]$ which has one root in K, then all the roots of f lie in K (that is, f splits into linear factors in $K[x]$).

'Normal' is a much overused word in mathematics. But at this point, there should be not too much risk of confusion between normal extensions of fields and normal subgroups of groups.

Normality is a property of **extensions** rather than individual fields. We often say 'The extension K/F is normal', although we have not defined an actual object K/F. We read K/F as 'K over F'.

For example, $\mathbb{C}$ is a normal extension of $\mathbb{R}$, or of $\mathbb{Q}$. However, if α is the real cube root of 2, then $K = \mathbb{Q}(\alpha)$ is not a normal extension of $\mathbb{Q}$. For the polynomial $x^3 - 2$ is irreducible over $\mathbb{Q}$, by Eisenstein's criterion; it has a root $\alpha \in K$; but it does not contain the other two roots of $x^3 - 2$, since they are non-real, whereas K is contained in the real numbers. In fact, we have the factorisation $x^3 - 2 = (x - \alpha)(x^2 + \alpha x + \alpha^2)$ in $K[x]$, where the second factor is irreducible in $K[x]$.

Theorem 8.18   *Let K be a finite extension of F. Then K/F is a normal extension if and only if K is the splitting field of a poynomial in $F[x]$.*

Proof   Suppose first that K/F is a normal extension. Since it is finite, K is generated over F by finitely many elements $a_1, \ldots, a_n$. Let $f_i(x)$ be the minimal polynomial of a_i over F. Then $f_i(x)$ is an irreducible polynomial in $F[x]$ with a root in K, so (by normality) it has all its roots in K.

Consider the polynomial $g(x) = f_1(x) \cdots f_n(x)$. We know that g splits in K. But g cannot split over any smaller field, since its splitting field contains all of $a_1, \ldots, a_n$, and these elements generate K. So K is the spliting field of g.

Conversely, suppose that K is the splitting field of a polynomial $g(x)$ over F. Arguing for a contradiction, suppose that $f(x)$ is an irreducible polynomial in $F[x]$ which has a root $\alpha \in K$ and another root β in an extension of K but not in K. Now K is obviously the splitting field of $g(x)$ over $F(\alpha)$. Also, the splitting field of $g(x)$ over $F(\beta)$ is $K(\beta)$ (since it is generated over $F(\beta)$ by the roots of g).

Now $F(\alpha) \cong F(\beta)$, since α and β are roots of the same irreducible polynomial over F. Moreover, there is an F-isomorphism from $F(\alpha)$ to $F(\beta)$ (see Section 7.16). By the uniqueness result for splitting fields, this F-isomorphism can be extended to an F-isomorphism between the splitting fields K and $K(\beta)$ of $g(x)$ over $F(\alpha)$ and $F(\beta)$ respectively. So $[K(\beta) : F] = [K : F]$. Since $K \subseteq K(\beta)$, this implies that $[K(\beta) : K] = 1$, so $\beta \in K$, contrary to assumption. The contradiction shows that our assumption is untenable, and K/F is normal. $\qquad\square$

There is another potential difficulty, concerned with repeated roots of irreducible polynomials. As we saw in Section 6.6, this cannot occur over a perfect field (this includes all fields of characteristic zero, and all finite fields). The fix is to define away the difficulty; it is somewhat technical, and you may want to skip the next few paragraphs.

Recall that an irreducible polynomial is called **separable** if its roots in a splitting field are all distinct. We now extend this definition. An arbitrary polynomial f over F is called **separable** if all its irreducible factors are separable. [*Note*: We do not require that all the roots of f are distinct. We do not mind if it has a repeated irreducible factor.] We say that an extension K/F is **separable** if, for every element $a \in K$, the minimal polynomial of a over F (which we know is irreducible) is separable.

How do these two definitions of separability relate to each other?

Theorem 8.19 *A finite normal extension K/F is separable if and only if K is the splitting field of a separable polynomial over K.*

Proof Examine the proof of the preceding theorem. If K/F is normal, then K is the splitting field over F of a polynomial g which is constructed as the product of the minimal polynomials of some elements of K. If K/F is separable, then all these minimal polynomials are separable, and hence so is g. $\qquad\square$

The converse is more difficult. Before tackling it, we develop some elementary properties of separability.

Proposition 8.20 *Let a be separable over F, and let K be a field containing F. Then a is separable over K.*

Proof The minimal polynomial of a over K divides its minimal polynomial over F. So, if the latter has no repeated roots, neither does the former. $\qquad\square$

Proposition 8.21 *Let F be a field of characteristic $p \neq 0$, and let a be algebraic over F. Then a is separable over F if and only if $F(a) = F(a^p)$.*

Proof Suppose that $F(a^p) = F(a)$. Then $a \in F(a^p)$, which is a finite extension of F; so $a = f(a^p)$ for some polynomial f. Now the polynomial $g(x) = f(x^p) - x$

has derivative -1, and hence has no repeated roots, and is satisfied by a; so a is separable over F.

Conversely, suppose that a is separable over F. Then it is separable over $F(b)$, where $b = a^p$. But a satisfies the polynomial $x^p - b = x^p - a^p = (x - a)^p$ over $F(b)$. Since it is separable, its minimal polynomial over $F(b)$ must be $(x - a)$; so $a \in F(b)$, and $F(a) = F(b)$. $\qquad\square$

Theorem 8.22 *Let K be an extension of F. Then the set of elements of K which are separable over F is a subfield of K containing F.*

Proof We must show that, if a and b are separable over F, then so are $a+b$, ab, and (if $a \neq 0$) a^{-1}. The argument is given for $a + b$ below. It is almost identical for ab, while a^{-1} is an easy exercise.

Suppose, for a contradiction, that a and b are separable but $a + b$ is not. Then F has non-zero characteristic, say p. By Proposition 8.20, we can enlarge F without changing the fact that a and b are separable. Since $F(a+b) \neq F((a+b)^p)$, the element $a + b$ is inseparable over $F((a+b)^p)$. So, replacing F by $F((a+b)^p)$ if necessary, we may assume that $(a + b)^p \in F$.

With this assumption, we have

$$F(a) = F(a^p) = F(b^p) = F(b),$$

the first and third equalities holding by Proposition 8.21, since a and b are separable over F. Hence $a + b \in F(a)$.

We show next that $c^p \in F$ for any $c \in F(a + b)$. Any such element is a polynomial in $a+b$, say $c = \sum d_i(a+b)^i$, with $d_i \in F$; then $c^p = \sum d_i^p(a+b)^{ip} \in F$ as claimed.

Now let $m = [F(a) : F(a+b)]$, and let $g(x)$ be the minimal polynomial of a over $F(a+b)$, an irreducible polynomial over $F(a+b)$. Let $g(x) = \sum c_j x^j$. Then $g(x)^p = \sum c_j^p x^{jp}$ is a polynomial of degree mp over F, and $g(a)^p = 0$. Since $[F(a) : F] = mp$, the minimal polynomial of a over F has degree mp, and so it must be $g(x)^p$. But this polynomial has at most m distinct roots, contradicting the fact that a is separable over F.

This contradiction shows that $a + b$ is separable. $\qquad\square$

Now we complete the proof of the theorem. Suppose that K is the splitting field of a separable polynomial $g(x)$ over F. Let L be the field consisting of all elements of K separable over F. Then L contains the roots of g by assumption; so $L = K$. $\qquad\square$

The theory that Galois developed applies to all finite normal separable extensions. So we make a definition:

Definition Let K/F be a field extension. We say that K/F is a **Galois extension** if it is finite, normal, and separable.

8.7 The main theorem. Galois Theory relates field extensions to groups. The groups arise in the following way:

Definition Let K/F be a Galois extension of fields. The **Galois group** of the extension, written $\mathrm{Gal}(K/F)$, is the group of all F-automorphisms of K; that is, all isomorphisms from K to itself which leave every element of F fixed.

Although the modern definition of a group was devised much later than the time of Galois, he understood what is meant by the statement 'the F-automorphisms of K form a group'. This group carries in its structure a lot of detailed information about the field extension. The key is the following piece of numerical information:

Theorem 8.23 *Let K/F be a Galois extension with Galois group G. Then $|G| = [K : F]$.*

Proof We use induction on $[K : F]$; the result is clearly true when $K = F$. So we assume the result for extensions of smaller degree than $n = [K : F]$.

Choose an element $\alpha \in K \setminus F$, and let its minimal polynomial $f(x)$ have degree m. Then $[F(\alpha) : F] = m$, and so $[K : F(\alpha)] = n/m$ Also, K is a Galois extension of $F(\alpha)$. So the group of $F(\alpha)$-automorphisms of K has order n/m.

Now G acts on the set Ω of roots of $f(x)$ in K, since the coefficients of this polynomial lie in F. The stabiliser of α fixes every element of $F(\alpha)$, and so is the Galois group of K over $F(\alpha)$. By induction, $|G_\alpha| = n/m$.

Also, the roots of f are all distinct (by separability), and lie in K (by normality); so $|\Omega| = m$. Moreover, Proposition 7.42 implies that G acts transitively on Ω. (We can regard K as the splitting field of a polynomial having f as one of its factors: now, if $\alpha, \beta \in \Omega$, then the F-isomorphism carrying α to β extends to an F-automorphism of K.)

By the Orbit–Stabiliser Theorem 7.2, we have

$$|G| = |\Omega| \cdot |G_\alpha| = m \cdot (n/m) = n,$$

and we are done. $\qquad\qquad\qquad\qquad\qquad\qquad\qquad\qquad\qquad\qquad\qquad\square$

The most important facts about the connection between a field extension K/F and its Galois group G is phrased in terms of subgroups H of G and subfields L intermediate between F and K. If L is such a subfield, then K is a Galois extension of L, and so the Galois group $\mathrm{Gal}(K/L)$ is a subgroup of G. (It consists of those automorphisms of K which fix, not only all of F, but all of L.) In the other direction, let H be a subgroup of G. Let $\mathrm{Fix}(H)$ be the set of elements of K which are fixed by all the automorphisms in H. Then $\mathrm{Fix}(H)$ is a subfield of K: for, if an automorphism fixes two elements a and b, then it also fixes their sum, difference, product, and quotient (if $b \neq 0$). Also, $\mathrm{Fix}(H)$ contains F, since all elements of G are by definition F-automorphisms.

Now we can state the Main Theorem:

Theorem 8.24 (Fundamental Theorem of Galois Theory) *Let K/F be a Galois extension with Galois group G. Then the maps*

$$L \mapsto \mathrm{Gal}(K/L),$$

$$H \mapsto \mathrm{Fix}(H),$$

are mutually inverse bijections between the set of subfields of K containing F and the set of subgroups of G. Moreover, we have

(a) $[K : L] = |\mathrm{Gal}(K/L)|$, $[L : F] = |G : \mathrm{Gal}(K/L)|$;
(b) *if L_1 and L_2 are intermediate fields, then $L_1 \subseteq L_2$ if and only if $\mathrm{Gal}(K/L_2) \subseteq \mathrm{Gal}(K/L_1)$;*
(c) *L/F is a normal extension if and only if $\mathrm{Gal}(K/L)$ is a normal subgroup of $\mathrm{Gal}(K/F)$;*
(d) *if the equivalent conditions of (c) hold, then $\mathrm{Gal}(L/F)$ is isomorphic to $\mathrm{Gal}(K/F)/\mathrm{Gal}(K/L)$.*

Remark Correspondences with property (b) (that is, correspondences between ordered sets which reverse the order) have become known as **Galois correspondences**. The most important example is the one described in this theorem. Note also the double occurrence of the word 'normal' in part (c). We see that the use of the same term to describe the two very different concepts in group theory and field theory is not accidental!

We need the following lemma. Its converse is also true, as we will see later.

Lemma 8.25 *Let K/F be a finite extension. Suppose that there are only finitely many fields L intermediate between F and K (that is, satisfying $F \subseteq L \subseteq K$). Then $K = F(a)$ for some $a \in K$.*

Proof First, if F is a finite field, then so is K. Thus the hypothesis is certainly true. We showed in Proposition 7.45 that the multiplicative group of K is cyclic. If a is a generator, then obviously $K = F(a)$, and so the conclusion is also true. So we may suppose that F is infinite.

Suppose that K/F has only finitely many intermediate fields. Then for any $a, b \in K$, we claim that there exists $t \in F$ such that $F(a, b) = F(a + tb)$. For the infinitely many intermediate fields $F(a + sb)$, for $s \in F$, cannot all be distinct; so there exist $s_1, s_2 \in F$ with $F(a + s_1 b) = F(a + s_2 b) = L$, say. Then L contains $((a+s_1 b)-(a+s_2 b))/(s_1 - s_2) = b$, and L contains $(s_2(a+s_1 b)-s_1(a+s_2 b))/(s_2 - s_1) = a$. So $L = F(a, b)$ as required.

Now K/F is finite, so $K = F(a_1, \ldots, a_t)$ for some $a_1, \ldots, a_t \in K$. Inductively use the above observation to replace these t generators by a single one. $\qquad\square$

Now we return to the proof of the Fundamental Theorem.

Proof For any intermediate field L, K is a Galois extension of L, and so $|\mathrm{Gal}(K/L)| = [K : L]$. But we have $L \subseteq \mathrm{Fix}(\mathrm{Gal}(K/L)) = L'$, say; and thus

$\text{Gal}(K/L) = \text{Gal}(K/L')$, and $|\text{Gal}(K/L)| = [K : L']$. It follows that $[L' : L] = 1$, so that $L = L'$. Thus, $\text{Fix}(\text{Gal}(K/L)) = L$.

This also shows that there are only finitely many fields L between F and K. By Lemma 8.25, $K = F(a)$ for some $a \in K$.

Now let H be any subgroup of G. Then K is a Galois extension of $\text{Fix}(H)$, with Galois group H' (say), where $H \leq H'$. If we can prove that equality holds, then $\text{Gal}(\text{Fix}(H)) = H$, and we have shown that Gal and Fix are mutually inverse bijections. This follows from the next fact, with $E = \text{Fix}(H)$:

Let K/E be a Galois extension with Galois group H. If $H' < H$, then $\text{Fix}(H') > E$.

Suppose, for a contradiction, that $\text{Fix}(H') = E$. Let $K = E(a)$ have degree n over E, and let $a = a_1, \ldots, a_n$ be the roots of the minimal polynomial of a over E. Now H permutes $a_1, \ldots, a_n$ transitively, and hence regularly. So the proper subgroup H' cannot act transitively on this set. Let $\{a_1, \ldots, a_r\}$ be an orbit. Then the coefficients of the monic polynomial with roots $a_1, \ldots, a_r$, being elementary symmetric functions of these quantities, are all fixed by H', and so lie in E. Then $a = a_1$ satisfies a polynomial of degree r over E, contradicting the fact that $[E(a) : E] = n$.

Parts (a) and (b) of the Fundamental Theorem are now clear.

We turn to part (c). For any $g \in G = \text{Gal}(K/F)$, the conjugate $\text{Gal}(K/L)^g$ fixes elementwise the field $Lg = \{ag : a \in L\}$. So $\text{Gal}(K/L)$ is a normal subgroup of $G = \text{Gal}(K/F)$ if and only if $Lg = L$ for all $g \in G$. But, if f is an irreducible polynomial over F with a root $\alpha \in L$, then every root of f is the image of α under an element of G; so the condition $Lg = L$ for all $g \in G$ is equivalent to the condition that L contains all the roots of any polynomial such as f; that is, that L is a normal extension of F. Thus (c) holds.

Finally, suppose that these conditions hold for L. Then any element of G fixes L as a set; so we have a homomorphism from $G = \text{Gal}(K/F)$ to $\text{Gal}(L/F)$ given by restricting elements of G to L. The kernel of this homomorphism is just the subgroup of G fixing L elementwise, that is, $\text{Gal}(K/L)$; and, since every F-automorphism of L extends to an F-automorphism of K by Proposition 7.42, the image of the homomorphism is $\text{Gal}(L/F)$. Thus (d) follows from the First Isomorphism Theorem. □

8.8 Solubility by radicals. We now come to the most famous application of the theory developed by Galois: a criterion for the solubility of equations by radicals. Throughout this section, we make the assumption that our fields have characteristic zero. (The theory needs only minor changes in non-zero characteristic: see Exercise 8.14.)

First, what is meant by solving an equation by radicals? Consider the familiar formula for the solutions of the quadratic equation $ax^2 + bx + c = 0$; namely,

$$x = \frac{-b \pm \sqrt{b^2 - 4ac}}{2a}.$$

The solutions are obtained from the coefficients of the polynomial by applying field operations and extracting the square root of an element $b^2 - 4ac$: this final step may require a field extension. For a more complicated example, consider the equation $x^4 - 4x^2 + 2 = 0$, with solutions $x = \pm\sqrt{2 \pm \sqrt{2}}$. The solutions lie in a field which is obtained from $\mathbb{Q}$ first by adjoining $\sqrt{2}$, and then adjoining $\sqrt{2 + \sqrt{2}}$. (Since

$$\sqrt{2 - \sqrt{2}} = \sqrt{2}/\sqrt{2 + \sqrt{2}},$$

the resulting field contains all the solutions.)

We make the following **definitions:**

(a) A field extension K/F is a **simple radical extension** if $K = F(a)$ for some $a \in K$ with $a^n \in F$, $[K : F] = n$, and F contains a primitive nth root of unity.

(b) A field extension K/F is a **radical extension** if there are fields $F = F_0 \subset F_1 \subset \ldots F_r = K$ such that F_i/F_{i-1} is a simple radical extension for $i = 1, \ldots, r$.

(c) The polynomial $f \in F[x]$ is **soluble by radicals over** F if its splitting field is contained in a radical extension of F.

Remark The condition about roots of unity in the definition of a simple radical extension is vacuous if $n = 2$, since the square roots of unity are ± 1, both in F. In general, it is possible to do without this condition, but we start with it (it makes the arguments much simpler) and then work around it at the end.

Recall the definition of a **soluble group** in Section 7.5: the finite group G is soluble if there is a series

$$G = G_0 > G_1 > \ldots > G_r = 1$$

of subgroups of G, where $G_i \lhd G_{i-1}$ and $G_{i-1}/G_i = \mathrm{Gal}(F_i/F_{i-1})$ is cyclic for $i = 1, \ldots, r$.

Theorem 8.26 *Let f be a polynomial over a field F of characteristic zero, with Galois group G. Suppose that F contains a primitive $|G|$th root of unity. Then f is soluble by radicals over F if and only if G is a soluble group.*

We begin the proof with a special case.

Lemma 8.27 *Let K/F be an extension of degree n, and assume that F contains a primitive nth root of unity. Then K/F is a simple radical extension if and only if it is normal with cyclic Galois group.*

Proof Suppose first that K/F is a simple radical extension, say $K = F(a)$, where $a^n = c \in F$. Let ω be a primitive nth root of unity. Then the roots of the polynomial $x^n - c$ are $a, a\omega, \ldots, a\omega^{n-1}$; so K is the splitting field of this polynomial, and is normal. Now we have $K = F(a) = F(a\omega^i)$ for any i; so there is a unique element σ_i of the Galois group $G = \mathrm{Gal}(K/F)$ mapping a to $a\omega^i$. Then $G = \{\sigma_0, \sigma_1, \ldots, \sigma_{n-1}\}$. Now σ_1^i maps a to $a\omega^i$; so $\sigma_1^i = \sigma_i$, and G is cyclic (generated by σ_1).

Conversely, suppose that K/F is normal, with cyclic Galois group $G = \langle \sigma \rangle$. We have to find an element $a \in K$ such that $a^n \in F$ and $K = F(a)$. It suffices to find $a \in K$ with $a \neq 0$ such that $a^\sigma = \omega a$, where ω is a primitive nth root of unity. For then $a^n \sigma = \omega^n a^n = a^n$, so $b = a^n \in \mathrm{Fix}(\sigma) = F$; and the images of a under powers of σ (the nth roots of b) are all the roots of the minimal polynomial of a, so $[F(a) : F] = n$.

We find such an a by a trick. According to Artin's Lemma 8.28 (see below), the n endomorphisms $1, \sigma, \sigma^2, \ldots, \sigma^{n-1}$ are linearly independent over K. So there exists $x \in K$ such that

$$a = x + x\sigma/\omega + x\sigma^2/\omega^2 + \cdots + x\sigma^{n-1}/\omega^{n-1} \neq 0.$$

(If this expression were zero for all x, then the linear combination $1 + \sigma/\omega + \cdots + \sigma^{n-1}/\omega^{n-1}$ would be the zero map.) Then

$$a\sigma = x\sigma + x\sigma^2/\omega + \cdots + x/\omega^{n-1} = a\omega,$$

as required. $\qquad\qquad\square$

Proposition 8.28 (Artin's Lemma) *Let $\sigma_1, \ldots, \sigma_n$ be distinct automorphisms of a field K. Then $\sigma_1, \ldots, \sigma_n$ are linearly independent over K (in the sense that, if*

$$\sum_{i=1}^{n} a_i(x\sigma_i) = 0$$

for all $x \in K$, where $a_1, \ldots, a_n \in K$, then $a_1 = \ldots = a_n = 0$.

Proof Suppose that we have such a relation, with not all the coefficients equal to zero. Suppose that the number m of non-zero coefficients is as small as possible. Clearly, $m > 1$. We derive a contradiction by producing another relation with fewer non-zero coefficients.

Assume that $a_1 \neq 0$ and $a_2 \neq 0$. Since the automorphisms are distinct, there exists an element $y \in K$ with $y\sigma_1 \neq y\sigma_2$. Now $(xy)\sigma_i = (x\sigma_i)(y\sigma_i)$ for each i. Applying the dependence relation to xy, we obtain

$$a_1(x\sigma_1)(y\sigma_1) + a_2(x\sigma_2)(y\sigma_2) + \cdots = 0$$

for all $x \in K$. Multiplying the original equation by $y\sigma_1$ and subtracting gives

$$a_2(y\sigma_2 - y\sigma_1)x\sigma_2 + \cdots = 0.$$

This relation has fewer non-zero terms than the original one, but is not identically zero (since $a_2 \neq 0$ and $y\sigma_1 \neq y\sigma_2$). So we have the required contradiction. $\square$

Now we prove the theorem.

Suppose first that f is solvable by radicals, and let

$$F = F_0 \subset F_1 \subset \ldots \subset F_r = K,$$

where F_i/F_{i-1} is a simple radical extension of degree n_i for $i = 1, \ldots, r$, and f splits in K. We may assume that K is a normal extension of F; let G be its Galois group. Let $G_i = \mathrm{Gal}(K/F_i)$ for $i = 0, \ldots, r$. Then, by the FTGT,

$$G = G_0 > G_1 > \ldots > G_r = 1,$$

where $G_i \triangleleft G_{i-1}$; and $G_{i-1}/G_i = \mathrm{Gal}(F_i/F_{i-1})$ is cyclic (of order n_i) for $i = 1, \ldots, r$, by Lemma 8.27. Thus, by definition, the group G is soluble. Now, if L is the splitting field of f, then $\mathrm{Gal}(L/F)$ is a homomorphic image of $\mathrm{Gal}(K/F) = G$ (again by the FTGT), and so is soluble.

Conversely, suppose that L is the splitting field of f, and that $G = \mathrm{Gal}(L/F)$ is soluble. Then, by definition, there are subgroups

$$G = G_0 > G_1 > \ldots > G_r = 1,$$

where $G_i \triangleleft G_{i-1}$ and G_{i-1}/G_i is cyclic (of order n_i, say) for $i = 1, \ldots, r$. Letting $F_i = \mathrm{Fix}(G_i)$, we see by the FTGT that $[F_i : F_{i-1}] = n_i$ and $\mathrm{Gal}(F_i/F_{i-1})$ is cyclic of order n_i. By Lemma 8.27, F_i/F_{i-1} is a simple radical extension; so K/F is a radical extension, and f is soluble by radicals over F. $\square$

Finally, we show that the assumption we used above, that the field F contains appropriate roots of unity, is unnecessary. We do this by showing that a root of unity lies in a radical extension.

Proposition 8.29 *Let ω be a primitive nth root of unity over F. Then:*

(a) $F(\omega)/F$ is a normal extension with abelian Galois group;
(b) ω is contained in a radical extension of F.

Proof (a) $F(\omega)$ is the splitting field of $x^n - 1$ over F, since the roots of this polynomial are the powers of ω. Let G be its Galois group. Any element σ of G maps ω to some power ω^r, and is uniquely determined by r; call this element σ_r. Now $\sigma_r\sigma_s$ and $\sigma_s\sigma_r$ both map ω to ω^{rs}; so they are equal. Thus G is abelian.

(b) Now the order of G is at most the number $\phi(n)$ of primitive nth roots of unity, and hence less than n. Arguing by induction, the $|G|$th roots of unity lie in a radical extension E of F. By Theorem 8.26, since an abelian group is soluble, $E(\omega)$ is a radical extension of E, and hence of F, as required. $\square$

Let us see how these theoretical results lead to a formula for the solution of a cubic or quartic equation. For this, we need a version of Newton's Theorem on symmetric functions. A **symmetric function** $s(x_1, \ldots, x_n)$ is a polynomial in $x_1, \ldots, x_n$ which is unchanged by any permutation of its arguments.

Theorem 8.30 (Newton's Theorem) *Let f be a polynomial of degree n. Then any symmetric function of the roots of f can be expressed as a polynomial in the coefficients of f.*

Remark If $f(x) = x^n + a_1 x^{n-1} + \cdots + a_n$, then $(-1)^i a_i$ is the ith **elementary symmetric function** of the roots of f; that is, the sum of all products of the roots i at a time. Newton's Theorem asserts that any symmetric function can be expressed as a polynomial in the elementary symmetric functions.

The cubic Let f be a polynomial of degree 3, with roots α, β, γ.

Since S_3 has a normal subgroup of index 2 which is cyclic of order 3, we should look for a function of the roots which has cyclic symmetry. Such a function is

$$\Delta = (\alpha - \beta)(\beta - \gamma)(\gamma - \alpha).$$

Now $D = \Delta^2$ is a symmetric function of the roots, and so can be calculated in terms of the coefficients; then Δ is found by extracting a square root.

Now the analysis of cyclic extensions tells us to look at the quantity $\alpha + \omega\beta + \omega^2\gamma = \rho$. Since ρ^3 has cyclic symmetry in the roots, it can be expressed in terms of Δ and the coefficients. Extracting the cube root gives ρ. We now have enough information to determine the roots.

The quartic Let f be a polynomial of degree 4, with roots $\alpha, \beta, \gamma, \delta$.

This time, the key observation is that S_4 has a normal Klein subgroup (consisting of the identity, $(\alpha\ \beta)(\gamma\ \delta)$, $(\alpha\ \gamma)(\beta\ \delta)$, and $(\alpha\ \delta)(\beta\ \gamma)$. We look for three functions of the roots reflecting this symmetry: we can take

$$\begin{aligned} \xi &= \alpha\beta + \gamma\delta, \\ \eta &= \alpha\gamma + \beta\delta, \\ \zeta &= \alpha\delta + \beta\gamma. \end{aligned}$$

Now the three elementary symmetric functions $\xi + \eta + \zeta$, $\xi\eta + \eta\zeta + \zeta\xi$, and $\xi\eta\zeta$, are symmetric functions of $\alpha, \beta, \gamma, \delta$, and so can be calculated. Then ξ, η, ζ are the roots of a cubic with known coefficients, and we already know how to solve this. Finally, knowing these three quantities, it is easy to find $\alpha, \ldots, \delta$ by solving quadratics.

There are various tricks which can be used to streamline these methods. For the cubic, see Exercise 8.16.

Finally, we can prove that quintics are not soluble by radicals. Of course, it suffices to write down a single quintic over $\mathbb{Q}$ whose splitting field is not a radical extension of $\mathbb{Q}$. We show that the quintic $f(x) = x^5 - 6x + 3$ has this property.

First, we observe that f is irreducible. This is immediate from Eisenstein's criterion (using the prime 3). So the Galois group G of the polynomial is a transitive subgroup of the symmetric group S_5, acting on the five roots of f. By the Orbit–Stabiliser Theorem, $|G|$ is divisible by 5. (This also follows by observing that, if α is a root of f, then the subfield $\mathbb{Q}(\alpha)$ of the splitting field has degree 5 over $\mathbb{Q}$.) By Sylow's (or Cauchy's) Theorem, G contains an element of order 5, which must be a 5-cycle.

Next, we use some elementary calculus to show that f has exactly three real roots. For the stationary points of f occur when $5x^4 - 6 = 0$; that is, $x = \pm \sqrt[4]{6/5}$. Since there are only two stationary points, f can have at most three real roots. But $f(-2) = -17$, $f(-1) = 8$, $f(1) = -2$, and $f(2) = 23$; by the Intermediate Value Theorem, there is a root in each of the intervals $(-2, -1)$, $(-1, 1)$, and $(1, 2)$.

Now complex conjugation is an automorphism of the splitting field of f, which fixes the three real roots and interchanges the two non-real ones; so it acts as a transposition.

It can be shown (see Exercise 8.17) that a subgroup of S_5 which contains both a 5-cycle and a transposition must be the whole of S_5. So the Galois group of f is S_5. Now S_5 is insoluble (since it contains the non-abelian simple group A_5 as a normal subgroup); so f is not soluble by radicals.

8.9   Ruler-and-compass revisited.

We now return to the subject of ruler-and-compass constructions. We proved in Section 6.8 a sufficient condition for a point p to be constructible with ruler and compass from a set S of points: its coordinates should lie in an extension of $\mathbb{Q}(S)$ with degree a power of 2. (Recall that $\mathbb{Q}(S)$ is the field generated over $\mathbb{Q}$ by the coordinates of the points of S; we assume that $(0,0), (1,0) \in S$.) This enabled us to prove the impossibility of certain classical construction problems. But the only tool we have so far to show that a construction is possible is to give an explicit algorithm for doing it. It is possible to improve this, by giving a necessary and sufficient condition.

Theorem 8.31 *Let S be a set of points in the Euclidean plane, containing $(0,0)$ and $(1,0)$. A point p can be constructed from S with ruler and compass if and only if its coordinates lie in a normal extension of $\mathbb{Q}(S)$ with degree a power of 2.*

The proof uses a property of 2-groups, observed in Section 7.3:

Lemma 8.32 *Let G be a group of order 2^r. Then G has a chain*

$$G = G_0 > G_1 > \ldots > G_r = 1$$

of subgroups with the property that $|G_{i-1} : G_i| = 2$ for $i = 1, \ldots, r$.

Note that it follows that $G_i \lhd G_{i-1}$ for all i, since a subgroup of index 2 is normal. In particular, G is soluble.

Proof By Lemma 8.32, there is a non-identity element g in $Z(G)$; we may assume that g has order 2. Then $\{1, g\}$ is a normal subgroup of G. We set $G_{r-1} =$

$\{1, g\}$. Now G/G_{r-1} has order 2^{r-1}. By induction, it has a chain of subgroups, each of index 2 in the preceding. By the Second Isomorphism Theorem, these subgroups have the form G_i/G_{r-1}, where the G_i are subgroups of G forming a chain with the required properties. □

Proof of the theorem Suppose first that p is constructible from S. Let $F_0 = \mathbb{Q}(S)$, and let F_i be the field generated by the coordinates of all points constructed at step i of the construction. We showed in Section 6.8 that $[F_i : F_{i-1}] \leq 2$. So the final extension has degree a power of 2, although it may not be normal. To rectify this, let $f_i(x)$ be the product of the minimal polynomials of all points constructed at the ith stage, over F_0, and K_i its splitting field. Then $K_0 = F_0$, and K_i is obtained from K_{i-1} by adjoining the solutions to a finite number of quadratics. Hence we may interpolate a finite number of intermediate fields between K_{i-1} and K_i, each of degree 2 over its predecessor. The final field K_r is normal over F_0 and contains the coordinates of p.

For example, suppose that we want to construct $\alpha = \sqrt[4]{2}$, Then $F_0 = \mathbb{Q}$, $F_1 = \mathbb{Q}(\sqrt{2})$, $F_2 = \mathbb{Q}(\alpha)$. Now F_2 is not a normal extension of F_0. But the minimal polynomial of α over $\mathbb{Q}$ is $x^4 - 2 = (x^2 - \sqrt{2})(x^2 + \sqrt{2})$. So $K_1 = F_1$, and K_2 is obtained by adjoining square roots of $\sqrt{2}$ and $-\sqrt{2}$. So we have the chain

$$F_0 = \mathbb{Q} \subset F_1 = \mathbb{Q}(\sqrt{2}) \subset L = \mathbb{Q}(\sqrt[4]{2}) \subset K_2 = \mathbb{Q}(\sqrt[4]{2}, \mathrm{i}).$$

To prove the converse, we note that any quadratic equation with positive discriminant can be solved with ruler and compass, by intersecting a line with a circle. (We may assume that the quadratic is $x^2 + bx + c = 0$, where $b^2 > 4c$. Now draw the circle with centre $(-b/2, 0)$ and radius $\sqrt{b^2 - 4c}/2$: it has equation $x^2 + bx + c + y^2 = 0$, so it intersects the x-axis in the required points.)

So, if the coordinates of p lie in an extension of $\mathbb{Q}(S)$ with a chain of intermediate fields as in the theorem, then they can be found by successively solving quadratics, and hence constructed with ruler and compass. □

Sometimes an alternative form of the theorem is convenient, where we represent points in the Euclidean plane by complex numbers: given a point p with coordinates (x, y), we let $c(p)$ be the complex number $x + \mathrm{i}y$. For a finite set S of points, let $c(S) = \{c(p) : p \in S\}$. Now the following theorem can be shown:

Theorem 8.33 *Let S be a set of points in the Euclidean plane, containing $(0, 0)$ and $(1, 0)$. A point p can be constructed from S with ruler and compass if and only if $c(p)$ lies in a normal extension of $\mathbb{Q}(c(S))$ with degree a power of 2.*

Hint We have $\mathbb{Q}(c(p)) \subseteq \mathbb{Q}(p)(\mathrm{i})$.

The Greeks knew how to construct regular polygons of various numbers of sides (for example, pentagons and hexagons), but were unable to construct various others (for example, heptagons). Using Theorem 8.33, it is possible to give an exact characterisation of the constructible regular polygons. We need the concept of a **Fermat prime**, a prime number of the form $F_n = 2^{2^n} + 1$ for some integer n. The first few Fermat primes are $F_0 = 3$, $F_1 = 5$, $F_2 = 17$, $F_3 = 257$, and $F_4 = 65537$. Surprisingly, no further Fermat primes are known, though several further numbers of this form are known to be composite. (For example, Euler observed that $641 = 5^4 + 2^4$ divides $5^4 2^{28} + 2^{32}$, and also $641 = 5 \cdot 2^7 + 1$ divides $5^4 \cdot 2^{28} - 1$, so 641 divides the difference F_5.)

Theorem 8.34 *A regular n-gon is constructible with ruler and compass if and only if n is the product of a power of 2 and a number of distinct Fermat primes.*

In particular, a regular 7-gon is not constructible, but a regular 17-gon is. This last fact was first observed by Gauss; he was so pleased with it that he made his career in mathematics, and asked for a regular 17-gon to be inscribed on his tombstone. This was the first significant advance in 'ruler-and-compass geometry' since the time of Euclid.

Proof I will not give the argument in detail, but outline the steps.
  (a) By Theorem 8.33, a regular n-gon is constructible by ruler and compass if and only if $[\mathbb{Q}(e^{2\pi i/n}) : \mathbb{Q}]$ is a power of 2.
  (b) $[\mathbb{Q}(e^{2\pi i/n}) : \mathbb{Q}] = \phi(n)$ (Euler's function). In fact, the minimal polynomial of $e^{2\pi i/n}$ over $\mathbb{Q}$ is the nth **cyclotomic polynomial** $\Phi_n(x)$, whose roots are all the primitive nth roots of 1, and whose degree is $\phi(n)$; and it can be shown that $\Phi_n(x)$ is irreducible over $\mathbb{Q}$.
  (c) If n_1 and n_2 are coprime, then $\phi(n_1 n_2) = \phi(n_1)\phi(n_2)$; and $\phi(p^a) = p^{a-1}(p-1)$ if p is prime.

Combining these steps, we see that the regular n-gon is constructible if and only if each prime power factor p^a of n has the property that $p^{a-1}(p-1)$ is a power of 2. This requires that either $p = 2$, or $a = 1$ and $p - 1$ is a power of 2. Now if a number $p = 2^n + 1$ is prime, then necessarily n is itself a power of 2, and p is a Fermat prime. This completes the proof. □

8.10   The Theorem of the primitive element.   One curious corollary of the Fundamental Theorem of Galois Theory is the following test for when a field extension can be generated by one element. We saw one half of this theorem earlier, in Lemma 8.25.

Theorem 8.35 (The Theorem of the Primitive Element)   *Let K/F be a finite extension. Then the following are equivalent:*

  (a) $K = F(a)$ for some $a \in K$;
  (b) there are only finitely many fields L intermediate between F and K (that is, satisfying $F \subseteq L \subseteq K$).

Proof   We showed in Lemma 8.25 that (b) implies (a). For the converse, let $K = F(a)$ and let f be the minimal polynomial of a over F. For any intermediate field L, let f_L be the minimal polynomial of f over L. Then f_L divides f. There are only a finite number of monic polynomials dividing f in $K[x]$. If we show that the polynomial f_L determines L, then it follows that the number of intermediate fields is finite.

Let L' be the field generated over F by the coefficients of f_L. Then $L' \subseteq L$. Now $\deg(f_L) = [L(a) : L] = [K : L]$. Also, f_L is irreducible over L, and hence over its subfield L'; so $[K : L'] = \deg(f_L)$. It follows that $[L : L'] = 1$, and so $L = L'$. $\qquad\square$

Proposition 8.36 *If K/F is a finite separable extension (in particular, if it is a Galois extension), then $K = F(a)$ for some $a \in K$.*

Proof If K/F is a Galois extension, then the number of intermediate fields is equal to the number of subgroups of the Galois group, by the FTGT, and hence finite; so the result follows from the Theorem of the Primitive Element.

In general, let $K = F(a_1, \ldots, a_t)$, and let $f_i(x)$ be the minimal polynomial of a_i over F. By assumption, each f_i is separable, and hence $f = f_1 \cdots f_t$ is separable. Thus, if L is a splitting field for f over F, then L/F is a Galois extension, so has only finitely many intermediate fields (as above). *A fortiori,* the same is true for K/F; now argue as before. $\qquad\square$

8.11 Appendix: The Fundamental Theorem of Algebra. The Fundamental Theorem of Algebra asserts that any non-constant polynomial over the complex numbers has a root in $\mathbb{C}$: in other words, $\mathbb{C}$ is algebraically closed.

As we saw in Section 6.5, this is not a theorem of algebra: its proof requires some analysis. A proof using Liouville's Theorem from complex analysis was given there. It is possible to reduce, but not eliminate, the required analysis, as is done here, in an application of Galois theory. First we list the two analytic facts that we require. Both are readily proved using the Intermediate Value Theorem.

Proposition 8.37 *(a) Any polynomial of odd degree over $\mathbb{R}$ has a root in $\mathbb{R}$.*
(b) Any positive real number has a real square root.

Sketch proof (a) Let $f = a_n x^n + \cdots$ with n odd and $a_n \neq 0$. If $a_n > 0$, then $f(x)$ is positive for large positive x, and negative for large negative x; by the Intermediate Value Theorem, $f(x) = 0$ for some value of x. The argument is similar if $a_n < 0$.

(b) If $a > 0$, the function $x^2 - a$ is positive for large positive x, and is negative for $x = 0$. $\qquad\square$

Corollary 8.38 *Any complex number has a complex square root.*

Proof Let $z = x + iy$ with $x, y \in \mathbb{R}$. Check that, if $a^2 = (\sqrt{x^2 + y^2} + x)/2$ and $b^2 = (\sqrt{x^2 + y^2} - x)/2$, and the signs of the square roots are appropriately chosen, then $(a + ib)^2 = z$. (The square roots exist by Proposition 8.37(b).) $\qquad\square$

We also list the group theory that we need:

(a) Sylow's Theorem (Theorem 7.5).
(b) If $|G|$ is a power of 2, greater than 1, then G has a subgroup of index 2. (This follows from the solubility of 2-groups.)

Now we prove the theorem. Indeed, we prove the stronger assertion that $\mathbb{R}$ has no finite extension of degree greater than 1 except for $\mathbb{C}$. So let F be a finite extension of $\mathbb{R}$, of degree n. By enlarging F if necessary, we may assume that $F/\mathbb{R}$ is a normal extension. Let G be its Galois group.

We claim first that $|G|$ is a power of 2. For let P be a Sylow 2-subgroup of G, and L the fixed field of P. Then $[L : \mathbb{R}] = |G : P|$ is odd. But, by Proposition 8.37(a), $\mathbb{R}$ has no non-trivial extension of odd degree. So $P = G$.

Let H be a subgroup of G of index 2, and M its fixed field. Then M is a quadratic extension of $\mathbb{R}$, so of the form $R(a)$, where $a^2 \in \mathbb{R}$. By Proposition 8.37(b), a is negative, and $-a$ has a real square root c. So $M = R(ci) = \mathbb{C}$. The Galois group of F over $\mathbb{C}$ is H.

If $|H| > 2$, let K be a subgroup of index 2 in H, and N its fixed field. But then N is a quadratic extension of $\mathbb{C}$, contradicting Corollary 8.38. So $H = 1$ and $F = \mathbb{C}$ as required.

Exercise 8.14 (a) Let F be a field of characteristic p. Show that any separable extension of F of degree p has the form $F(\alpha)$, where $\alpha^p = \alpha + a$ for some $a \in F$.

(b) Let F be a field of characteristic p, and K a Galois extension of F. Prove that there is a chain of subfields $F = F_0 \subset F_1 \subset \ldots \subset F_m = K$ where, for $i = 1, \ldots, m$, we have $F_i = F_{i-1}(\alpha_i)$, $[F_i : F_{i-1}] = n_i$, and either $\alpha_i^{n_i} \in F_{i-1}$ and $p \nmid n_i$, or $\alpha^p - \alpha \in F_{i-1}$ and $n_i = p$.

Exercise 8.15 (a) Calculate the Galois group of the polynomial $x^3 - 2$ over $\mathbb{Q}$. If K is its splitting field, find all the subfields of K.

(b) Show that the Galois group of the polynomial $x^3 + x^2 - 2x - 1$ over $\mathbb{Q}$ is cyclic of order 3.

Exercise 8.16 (a) Show that the cubic $x^3 + ax^2 + bx + c + 0$ (over a field of characteristic zero) can be reduced to one with $a = 0$ by a substitution of the form $y = x + k$ for some k ('completing the cube').

(b) Verify that, if ω is a primitive cube root of unity, then

$$x^3 + y^3 + z^3 - 3xyz = (x + y + z)(x + \omega y + \omega^2 z)(x + \omega^2 y + \omega z).$$

(c) Show that, if y and z satisfy $y^3 + z^3 = c$ and $yz = -b/3$, then the roots of the cubic $x^3 + bx + c = 0$ are $-y - z$, $-\omega y - \omega^2 z$, and $-\omega^2 y - \omega z$.

(d) Hence solve the general cubic by radicals.

Exercise 8.17 Show that a subgroup of S_5 which contains the 5-cycle $(1\ 2\ 3\ 4\ 5)$ and a transposition must contain all transpositions, and hence must be S_5.

Exercise 8.18 (a) Let G be a group and H a subgroup. Suppose that

- the intersection of the conjugates of H is 1;
- there is a chain

$$G = G_0 > G_1 > \ldots > G_r = H$$

with $|G_{i-1} : G_i| = 2$ for $i = 1, \ldots, r$.

Prove that $|G|$ is a power of 2.

(b) Call an extension K/F **constructible** if there is a chain of fields

$$F = F_0 \subset F_1 \subset \ldots \subset F_r = K$$

with $[F_i : F_{i-1}] = 2$ for $i = 1, \ldots, r$. Prove that, if the characteristic of F is not 2 and K/F is constructible, then there is a constructible normal extension L/F with $L \supseteq K$.

(c) Hence give another proof of Theorem 8.31.

Further reading

There is no shortage of good books to choose if you want to read further. What follows is only a very partial and personal choice from what is available.

The material in the first chapter is basic to all of mathematics, and is often covered in books on Discrete Mathematics, such as Biggs [2]. (Numbers refer to the bibliography below.)

A very important part of algebra, especially for applications, which has only had a brief treatment in this book, is linear algebra. There are very many textbooks on linear algebra; I suggest Kaye and Wilson [23] or Blyth and Robertson [3,4].

There are many general algebra books which go further than this one. Two examples are those of Cohn [10] and Lang [24].

Two books covering in greater detail some of the topics in this book are Stewart [39] on Galois Theory, and Hartley and Hawkes [18] on modules over Euclidean domains and their applications.

Books on general group theory include those of Macdonald [27] and Rose [33], or the more encyclopaedic two-volume work by Suzuki [40]. Beyond a certain point, the books become more specialised. The most detailed information about groups is obtained by studying their representations, either as groups of permutations (see Cameron [8], Dixon and Mortimer [15]) or as matrix groups (Curtis and Reiner [14], Serre [37]). Another way of looking at representations of a group G by matrices over a ring R is to study the finitely generated modules for the group ring RG: this fits in with the modern philosophy that rings are best studied via their module categories. Rotman's book [34] is an introduction to homological algebra.

There is a wide profusion of books on more specialised parts of group theory. To mention just a few, we have Gorenstein's account [17] of the classification of finite simple groups, Johnson [20] on presentations of groups, Wilson [42] on the finite simple groups, and Leedham-Green and McKay [25] on p-groups.

General accounts of ring theory are given by McCoy [29] and Cohn [10], and a more thorough coverage (in two volumes) by Rowen [35]. For algebraic geometry, long regarded as a fearsome subject for beginners, there are now some very good introductions, such as those of Shafarevich [38] and Reid [32].

MacLane's book [28] is a good introduction to categories; Enderton's book [16] on set theory gives you a firm foundation for Section 6.1 (and indeed for all of mathematics). For a brief introduction see Cameron [7]. Van Lint [26] will guide you further in coding theory, and Kaplansky [22] in Galois theory.

I have also given here details of books from which I have quoted in the text.

1. Abu Ja'far Muhammad ifn Musa al-Khwarizml, Hisab al-jabr w'al-muqabala, House of Wisdom, Baghdad, Ca-810.

2. Norman Biggs, *Discrete Mathematics* (2nd edition), Oxford University Press, New York, 1989.

3. T. S. Blyth and E. F. Robertson, *Basic Linear Algebra*, Springer, London, 1998.

4. T. S. Blyth and E. F. Robertson, *Further Linear Algebra*, Springer, London, 2002.

5. Brian Butterworth, *The Mathematical Brain*, Macmillan, London, 1999.

6. John Cage, *Silence: Lectures and Writings*, Calder and Boyars, London, 1968.

7. P. J. Cameron, *Sets, Logic and Categories*, Springer, London, 1999.

8. P. J. Cameron, *Permutation Groups*, Cambridge University Press, Cambridge, 1999.

9. Lewis Carroll, *The Complete Illustrated Lewis Carroll*, Wordsworth, Ware, 1996.

10. P. M. Cohn, *Algebra*, Wiley, London: Volume 1 (2nd edition 1982), Volume 2 (1977), Volume 3 (1991).

11. P. M. Cohn, *Introduction to ring theory*, Springer, London, 2000.

12. J. H. Conway and R. K. Guy, *The Book of Numbers*, Copernicus, New York, 1996.

13. Alfred W. Crosby, *The Measure of Reality: Quantification and Western Society 1250–1600*, Cambridge University Press, Cambridge, 1997.

14. C. W. Curtis and I. Reiner, *Representations of Groups and Associative Algebras*, Wiley, New York, 1962.

15. J. D. Dixon and B. Mortimer, *Permutation Groups*, Springer, New York, 1996.

16. H. B. Enderton, *Elements of Set Theory*, Academic Press, New York, 1977.

17. D. Gorenstein, *Finite Simple Groups: an Introduction to their Classification*, Plenum, New York, 1982.

18. B. Hartley and T. O. Hawkes, *Rings, Modules and Linear Algebra*, Chapman & Hall, London, 1970.

19. J. Jaynes, *The Origin of Consciousness in the Breakdown of the Bicameral Mind*, Houghton Mifflin, New York, 1976.

20. D. L. Johnson, *Presentations of Groups*, Cambridge University Press, Cambridge, 1990.

21. Robert Kanigel, *The Man Who Knew Infinity: A Life of the Genius Ramanujan*, Scribner, New York, 1991.

22. I. Kaplansky, *Fields and Rings*, University of Chicago Press, Chicago, 1969.

23. R. Kaye and R. A. Wilson, *Linear Algebra*, Oxford University Press, Oxford, 1998.

24. S. Lang, *Algebra* (3rd edition), Addison–Wesley, Reading, 1993.

25. C. R. Leedham-Green and S. McKay, *The Structure of Groups of Prime Power Order*, Oxford University Press, Oxford, 2002.

26. J. H. van Lint, *Introduction to Coding Theory* (2nd edition), Springer, Berlin, 1992.

27. I. D. Macdonald, *The Theory of Groups*, Oxford University Press, Oxford, 1968.

28. Saunders MacLane, *Categories for the Working Mathematician*, Springer, Berlin, 1971.

29. N. H. McCoy, *The Theory of Rings*, Chelsea, New York, 1973.

30. J. von Neumann and O. Morgenstern, *Theory of Games and Economic Behavior*, Princeton University Press, Princeton, 1944.

31. K. Petersen, *Ergodic Theory*, Cambridge University Press, Cambridge, 1983.

32. M. Reid, *Undergraduate Algebraic Geometry*, Cambridge University Press, Cambridge, 1988.

33. J. S. Rose, *A Course on Group Theory*, Cambridge University Press, Cambridge, 1978.

34. J. J. Rotman, *An Introduction to Homological Algebra*, Academic Press, New York, 1979.

35. L. Rowen, *Ring Theory* (two volumes), Academic Press, San Diego, 1988.

36. Bertrand Russell, *History of Western Philosophy*, George Allen and Unwin, London, 1961.

37. J.-P. Serre, *Linear Representations of Finite Groups*, Springer, New York, 1977.

38. I. R. Shafarevich, *Basic Algebraic Geometry I: Varieties in Projective Space*, Springer-Verlag, Berlin, 1994.

39. Ian Stewart, *Galois Theory* (2nd edition), Chapman & Hall, London, 1989.

40. M. Suzuki, *Group Theory* (two volumes), Springer, New York, 1982, 1986.

41. Hermann Weyl, *Symmetry*, Princeton University Press, Princeton, 1952 (reprinted 1989).

42. R. A. Wilson, *The Finite Simple Groups*, Cambridge University Press, Cambridge, 2008.

Further resources are available on the Web. I have listed a few below. Since Web addresses change, it is better to search for these by name.

GAP website:

http://www.gap-system.org

GAP is a programming system for doing computations with algebraic structures, especially groups. The name is an acronym for 'Groups, Algorithms and Programming'.

Online Atlas of Finite Groups:

http://brauer.maths.qmul.ac.uk/Atlas/v3/

Detailed information about many finite simple groups, especially simple groups or those 'close to' being simple.

Encyclopedia of Integer Sequences:

http://www.research.att.com/~njas/sequences/

Here you can look up any interesting numberical sequence, such as the number of groups of order n for $n = 1, 2, \ldots$.

MacTutor History of Mathematics:

http://www-groups.dcs.st-and.ac.uk/~history/

Here you can read the fascinating stories of the people who created modern algebra.

Index

Index